Thermodynamic Behavior of Electrolytes in Mixed Solvents—II

Thermodynamic Behavior of Electrolytes in Mixed Solvents—II

William F. Furter, EDITOR
Royal Military College of Canada

Based on a symposium sponsored by the Division of Industrial and Engineering Chemistry at the 175th Meeting of the American Chemical Society, Anaheim, California, March 13–16, 1978.

ADVANCES IN CHEMISTRY SERIES **177**

AMERICAN CHEMICAL SOCIETY
WASHINGTON, D. C. 1979

Library of Congress CIP Data

Thermodynamic behavior of electrolytes in Mixed Solvents—II.
(Advances in chemistry series; 177 ISSN 0065-2393)

Includes bibliographies and index.

1. Electrolytes—Congresses. 2. Solvents—Congresses.
I. Furter, William F. II. American Chemical Society. Division of Industrial and Engineering Chemistry. III. Series.

QD1.A355 no. 177 [QD565] 540'.8s [541'.372]
ISBN 0-8412-0428-4 79-10009 ADCSAJ 177 1-400
1979

PRINTED IN THE UNITED STATES OF AMERICA

FOREWORD

ADVANCES IN CHEMISTRY SERIES was founded in 1949 by the American Chemical Society as an outlet for symposia and collections of data in special areas of topical interest that could not be accommodated in the Society's journals. It provides a medium for symposia that would otherwise be fragmented, their papers distributed among several journals or not published at all. Papers are reviewed critically according to ACS editorial standards and receive the careful attention and processing characteristic of ACS publications. Volumes in the ADVANCES IN CHEMISTRY SERIES maintain the integrity of the symposia on which they are based; however, verbatim reproductions of previously published papers are not accepted. Papers may include reports of research as well as reviews since symposia may embrace both types of presentation.

CONTENTS

Preface ix

1. Extraction of Concentrated Solutions: Refining of Sugar and Recovery of Acetic Acid from Wood-Pulping Liquors 1
   Donald F. Othmer

2. Salt Effect in Vapor–Liquid Equilibrium at Fixed-Liquid Composition 11
   John A. Burns and William F. Furter

3. Prediction of Salt Effect on Vapor–Liquid Equilibrium: A Method Based on Solvation II 27
   Shuzo Ohe

4. Error Analysis of Isobaric Liquid–Vapor Equilibrium Data for Mixed Solvents Containing Salts at Saturation 39
   Derek Jaques

5. The Concept of Basicity in Mixtures of Water with Organic Solvents 53
   C. F. Wells

6. Conductance and Ionic Association of Several Electrolytes in Binary Mixtures Involving Sulfolane and Protic Solvents 77
   Giuseppe Petrella, Antonio Sacco, and Maurizio Castagnolo

7. Enthalpies of Solution of Some Electrolytes with a Large Cation in Mixtures of Water with Aprotic Solvents 99
   C. de Visser, W. J. M. Heuvelsland, and G. Somsen

8. The Apparent Charge of Ionic Micelles in Aqueous Binary Solvents 109
   Claude Treiner, Anne LeBesnerais, and Christian Micheletti

9. The Solvation of Chromium(III) Ion in Mixed Solvents 129
   Edward L. King

10. Preferential Solvation of Some Electrolytes by Water and Diethyl Ether in Sulfolane 145
    E. Milanova, S. Y. Lam, B. Desjardins, and R. L. Benoit

11. Viscosity of Dilute Solutions of Alkali Halides in Methanol–Water Mixtures 167
    Robert A. Stairs

12. Ternary System Phase Diagram Determinations Concerning Potassium Electrolyte Influence on Aqueous Solutions of Dioxane or Tetrahydrofuran 177
    John A. McNaney, Howard K. Zimmerman, and Paul H. Gross

13. Solubility of Carbon Dioxide in Aqueous Mixed-Salt Solutions 189
    Akira Yasunishi, Masakatsu Tsuji, and Eizo Sada

14. **Solubility of Sodium Chloride in an Inverse Micellar Solution of Pentanol** 205
Stig Friberg, Irena Buraszewska, and Eva Sjöblom

15. **Computational Techniques of Ionic Processes in Water–Organic Mixed Solvents** 215
Buddhadev Sen, Rabindra Nath Roy, James J. Gibbons, David A. Johnson, and Louis H. Adcock

16. **Thermodynamic Behavior of Hydrobromic Acid in Mixtures of Water and *N*-Methylacetamide from 5° to 45°C** 249
Richard A. Butler and Roger G. Bates

17. **Thermodynamic Behavior of Tetraalkylammonium Bromide and Hydrobromic Acid in Aqueous Media at 298.15 K** 263
R. N. Roy, J. J. Gibbons, J. Moeller, and R. Snelling

18. **Thermodynamic Study of Glycine in Different Tetrahydrofuran–Water Mixtures at Several Temperatures from 278.15 to 328.15 K** 277
R. N. Roy, J. J. Gibbons, J. L. Padron, K. Buechter, and S. Faszholz

19. **The Excess Gibbs Free Energy and the Activity Coefficients of the $Nd(NO_3)_3$–HNO_3–H_2O System at 25°C** 299
William G. O'Brien and Renato G. Bautista

20. **The Activity Behavior of the $NdNO_3$–HNO_3–H_2O–HDEHP–AMSCO System** 323
William G. O'Brien and Renato G. Bautista

21. **Thermodynamics of Transfer of Hydrogen Iodide in Binary Mixtures of Ethylene Glycol + Diethylene Glycol from emf Measurements** 345
C. Kalidas and P. Sivaprasad

22. **Electromotive Forces and Thermodynamic Functions of the Cell Pt, H_2 | HBr(m), X% Alcohol, Y% Water | AgBr–Ag in Pure and Mixed Solvents** 355
Allen F. Robinette and Edward S. Amis

Index 379

PREFACE

This is the second volume in a series dealing with the fundamental phenomena (and their applications) occurring when an electrolyte is dissolved in a mixed solvent, that is, in a solvent consisting of two (or more) liquid components. The first volume, ADVANCES IN CHEMISTRY SERIES 155, was published in 1976. Twenty-two chapters in the present volume represent contributions from nine different countries: the United States; Canada; Britain; Japan; Australia; France; India; Italy; and the Netherlands.

The various thermodynamic and physicochemical effects reported are often quite complex. The liquid phase continues to yield its mysteries with reluctance. A system composed of an electrolyte and a single-component solvent is complex enough; when the solvent consists of two or more components an additional range of complexity unfolds. This is largely the result of the greatly increased number of permutations and combinations possible in the interactions that may occur among the species present, both ionic and molecular, along with the usual tendency for these interactions to be composition dependent. Although much of the research represented here is basic, the potential for industrial application is immense; ranging, for instance, from applications in separating processes such as extractive distillation and liquid extraction, to electrochemica' ɔower sources such as fuel cells and batteries.

Some of the content deserves particular attention by industry at the present time. The first chapter by D. F. Othmer, 1978 winner of both the SCI Perkin Medal and the ACS Murphree Award, stresses the energy-saving potential of solvent or liquid–liquid extraction over more energy-consuming alternatives. Electrolytes often are used in such processing as complexing agents added to enhance interphase mass transfer of a solute species between two immiscible liquid phases. The three chapters dirctly following the one by Dr. Othmer examine the effects of nonvolatile electrolytes on the equilibrium vapor composition of mixed-solvent systems. Again there is a potential for major energy saving, this time in the replacement of generally high concentrations of liquid separating agent recirculated within conventional extractive distillation processes with much lower concentrations of a dissolved salt or other nonvolatile electrolyte as the separating agent.

The remaining eighteen chapters address a wide range of electrolyte effects on the liquid-phase properties of mixed-solvent systems, including such diverse but interrelated topics as solvation and liquid structure;

solubilities; activity coefficients; dissociation constants; distribution coefficients; free energies; viscosities; entropy; transfer functions; stability constants; acidity functions; enthalpies of solution; electrostatic theory; conductances; standard potentials; dielectric constants; various computational techniques; and related properties and behavior. A better understanding of such phenomena in this type of system must be obtained if efficient chemical engineering process design is to be achieved in systems where they are encountered.

Finally I would like to acknowledge the expert work of Joan Comstock and Candace Deren of the ACS Books Department in the production of this book and its predecessor.

Royal Military College of Canada
Kingston, Ontario K7L 2W3
March 22, 1979

WILLIAM F. FURTER

1

Extraction of Concentrated Solutions

Refining of Sugar and Recovery of Acetic Acid from Wood-Pulping Liquors

DONALD F. OTHMER

Distinguished Professor, Polytechnic Institute of New York,
Brooklyn, NY 11201

Separation of materials takes much of the energy used by chemical industry. Extraction often requires the least. An entirely water-miscible solvent may be immiscible with an aqueous solution that has a high concentration of a solid and may even extract another component therefrom. Thus, water-soluble acetone (alone or containing water-soluble ethanol) extracts impurities from raw sugar syrups or molasses (above 50% solids), refines sugar at less cost—especially in terms of energy, gives higher yields of sugar, and recovers valuable impurities that otherwise would be wasted. Acetone extracts acetic acid out of concentrated waste liquors obtained from semichemical or kraft-pulping processes after their acidification with sulfuric acid and is several times as efficient as conventional solvents. Operational profits may range from $15 to $20/ton of pulp produced.

Energy—as scarce and expensive as it is becoming—is used in the largest amounts in the process industries to separate the components of many diversified raw materials or feedstocks of most industries. It is used in the next largest amounts to separate the intermediates resulting from treatments with other materials for chemical or physical change or convenience. Finally, it is used in the least amounts to purify (i.e., separate) the products. Thus, the success of the chemical engineer in reducing these major requirements of energy—as its cost increases—will be of the greatest importance in maintaining the present selling prices of all resulting products—from pharmaceuticals to paper or pig iron and from acetic acid to zinc.

0-8412-0428-4/79/33-177-001$05.00/1

The development of separation techniques is also a fascinating exercise in problem solving for the chemical engineer, in which his skill and experience are very important on the bottom line—i.e., in the energy cost as well as the dollar cost of the finished product. Also, the units of energy required in the separation processes now, more than ever before, translate into very important percentages of the dollar sum.

To separate solutions of both liquids and of solids in a liquid (particularly water), two methods usually are considered first: (1) vaporization—i.e., evaporation or distillation—to utilize the different relative volatilities of the components, either normally or accentuated by another liquid in azeotropic or extractive distillation and (2) liquid–liquid extraction to take advantage of the relative preferential solubility of one component in an added liquid.

A priori it would seem that liquid–liquid extraction with the almost negligible energy costs associated with the transfer of one material in a liquid solution (preferentially from one solvent to another) would be always more economical than vaporization in terms of energy. However, since the added solvent usually has to be separated subsequently from both the extract layer and the raffinate layer by distillation, these thermal costs for the overall separation may be substantial.

Processes using a great deal much less energy than conventional processes have been developed through pilot plant stages for the separation from an aqueous solution of important industrial materials, one of which is present as a very concentrated or saturated solution of a solid, which remains in the raffinate, while the other is extracted therefrom by an added solvent. Advantage is taken of the greatly lessened mutual solubility of the extracting solvent and of the water in the concentrated solution. The dissolved solid in the concentrated aqueous solution increases the relative solubility of the component dissolved in, and to be extracted by, the solvent compared with that in the water and at the same time reduces greatly the solubility of the solvent in the aqueous phase.

Two examples illustrate the many places in industry where other, quite different methods of separation are used conventionally to separate such concentrated solutions. Large amounts of energy are used now; valuable materials are lost or destroyed when a smaller amount of one material or group of materials is separated from the larger amount of another material or group of materials.

Sugar and Molasses

About 100 million tons world wide of sugar per year (more than half from cane, the balance mainly from beets) must be separated from five to ten times as much water and refined by more or less complete separation from a dozen impurities that are present in varying amounts. Juice expressed from sugar cane in the tropics is concentrated; sugar is crystallized therefrom and is shipped as "raw" (about 97% sucrose) to be refined to above 99% in the country of its use. Most of the molasses remaining as mother liquor after crystallizing out the raw sugar also is shipped from the point of origin (*1*).

The refining of raw sugar by solvent extraction costs less in energy and materials and can be done in a plant that is much less expensive to operate than a conventional plant (*2*).

When refining by solvent extraction, the cane juice can be concentrated to 60% or higher total solids, impurities can be dissolved out to give a pure syrup for shipment, or a pure sugar may be crystallized therefrom. If, alternatively, raw syrup instead of raw sugar is shipped (at the lower cost for handling a liquid in tankers), it can be refined by solvent extraction after delivery. Either operation will give a refined sugar or syrup at a much lower cost, particularly in energy requirements.

Conventional Cane Sugar Refining. Raw sugar as it is unloaded from the ships, besides containing 85–97% sucrose, also contains a small amount of water, suspended materials, invert sugars (glucose), and other organic chemicals such as aconitic acid, sugar cane fats and waxes, chlorophyl, various vitamins, and other constituents of the original sugar cane plant that have come with the juice. Soluble impurities are present in a thin aqueous film on the crystal surfaces of saturated sugar–molasses mixture.

The raw sugar is melted—i.e., dissolved in water—then refined in a sequence of evaporations and crystallizations with intermediate operations for removing and destroying minor impurities. By treatment with materials to adsorb the impurities, the sugar is crystallized from the purified solution and removed from the final mother liquid (molasses), which contains all of the impurities that were not adsorbed. These impurities—including potentially valuable wax, vitamins, and aconitic acid—if not wasted substantially in the molasses are lost on the adsorbent solid, with considerable expense in destruction or removal.

Such refining of raw sugar cane is done on a tremendous scale; most of it in this country is done in a dozen huge plants, one of which may spend $200,000 to refine 2,000 tons per day, of which $20,000 per day may be for fuel. Thus, all standard refining methods are expensive because of:

(a) the large heat cost in evaporating the water used to dissolve or to melt the raw sugar, as well as the water that is added in subsequent washing steps;

(b) revivification cost of the solid adsorbing agent, such as bone black, carbon black, or various ion-exchange resins, and the loss by discard of these adsorbed materials and ultimately the loss or discard of the adsorbent itself;

(c) complete loss of the value of all impurities except those sold as part of the final unseparable dregs, the ultimate molasses;

(d) mechanical and other losses (inversion) of the sugar during the numerous steps;

(e) the large size of and investment in the plant owing to the numerous steps and the considerable equipment required;

(f) the large requirements of water, principally for cooling; and

(g) the removal of pollution possibilities in various aqueous discharges from the refinery.

Refining of Raw Sugar by Washing with Solvent. To identify the impurities, raw sugar crystals of a carefully selected size, hence with a readily calculable surface area per unit weight, had their thin surface layer dissolved sequentially by several washings with nearly saturated sugar syrups of very high purity. Almost all impurities dissolved in the first washings, and these impurities were almost entirely in a film on the surface of the crystals. After very slight dissolution of the crystal surfaces themselves, most of the grain was almost pure sucrose. Thus, it appeared that a nonaqueous solvent, herein called the first solvent, which would dissolve off and separate the impurities from the crystals, would save the high cost of the classic melting or complete dissolution and the subsequent evaporation of water, crystallization of sugar, and removal of impurities in a low value liquid residue—molasses.

Previously reported experiments showed that the most desirable first solvents have the same number of oxygen and carbon atoms in the molecule. Such compounds are completely water soluble and also are good solvents for the impurities. (Solvents with molecules that contain chlorine, nitrogen, or atoms other than carbon hydrogen, and oxygen are not useful.) Thus, methanol and acetic acid are good, and the low boiling point of methanol is an advantage (*3*).

From 0.1 to 8% water in the first solvent aids this extraction. More water improves the efficiency of the solvent but decreases its selectivity—i.e., the solubility of many impurities is increased considerably, but so is the solubility of sugar. Thus, the water reduces the yield of refined sugar but increases its purity.

A higher temperature of washing, up to the solvent's boiling point, increases the removal of impurities in the least time. The temperature must not be high enough to carmelize the sugar. Good washing is secured in 1–2 hr with an equal weight of methanol just below its boiling point of 65°C. Higher temperatures require a pressure operation, which greatly reduces the time required.

With glacial acetic acid, temperatures up to its boiling point, 118°C, may be used. This higher washing temperature—without pressure—removes impurities in less time than methanol, and at temperatures between 105° and 110°C, the raw sugar is refined in 45–90 min.

Almost complete recovery of the solvent is essential, and higher boiling solvents makes this easier, but all but about 1% of the methanol may be recovered in a suitable closed system. Methanol is distilled readily for its recovery from water and from aqueous sugar solutions. Distillation of acetic acid, if used as a solvent, removes water also from the dissolved impurities and leaves a first molasses. (Other methods of separation of the solvent may be used.)

A very small amount of surface-active agent improves the rate of washing—i.e., reduces the time. A sucrose ester of a fatty acid, such as sebacic, in amounts of 0.001–0.01% of raw sugar, reduces the time necessary for extraction by 25–30%. This stays in the extract layer.

Sugar refined by washing with methanol is a very light straw color. Acetic acid gives a refined sugar with a much lighter color and lower ash. This indicates that the acidic as well as the solvent nature of acetic acid is important. Hence, a very small amount of sulfuric acid was added to methanol when this was used to give a pH of about 1.25. This step eliminated most of the color in the refined sugar, increased its purity by 0.1–0.25%, and reduced its ash content from 0.12–0.15% to 0.05–0.1%. This acidification also reduced the total free and combined aconitic acid from 0.05–0.15% to 0.01–0.02%. Aconitic acid was related directly to remaining color and, to a lesser degree, to the remaining ash. Thus, addition of sulfuric acid to spring the aconitic acid present at a pH of 1.25–1.3 gave a purity of 99.8% sucrose, improved color, and a washing time of less than 30 min at a temperature of 30°–40°C (3).

This washing by a completely water-soluble solvent does not remove all of the water in the raw sugar: the solubility of water in methanol and vice-versa is depressed by this complete saturation with sugar and the salting out of impurities.

The centrifuged and washed crystals of highly refined sugar immediately gave above 99% of the sucrose of the raw sugar. However, 1 or 2% of this came back as recycle from subsequent solvent extractions; these extractions would be on a very small scale because almost all of the sucrose content was produced as refined sugar in the first step.

Solvent Extraction of Sugar Solution Coming from the Washing of Raw Sugar. The solvent or extract layer from the washing of impurities from the raw crystals was settled to remove dirt and other small solid particles. The solvent was evaporated for immediate recycle, leaving behind a first molasses.

Equally as economical as the washing of impurities off of the raw sugar crystals, in terms of thermal energy and other costs, is the liquid–liquid extraction of impurities from concentrated sugar solutions (including molasses) by a solvent that is entirely soluble in water. Again, the high concentration of the sugar salts-out the impurities and prevents the mutual miscibility of the water in the sugar solution with the solvent, which, by itself, is completely miscible in all proportions with pure water.

In this way, constituents of the first molasses produced by solvent washing of raw sugar crystals can be separated by a second solvent. Preferably it is one that removes the less water-soluble impurities—oils, fats, waxes, aconitic acid, chlorophyl, etc.—from the sucrose and invert sugar. Hydrocarbons and chlorinated hydrocarbons, ethers, higher alcohols, and ketones have disadvantages as a solvent—e.g., they often form relatively stable emulsions.

Acetone was first rejected because of its complete miscibility with water but was found to be the best on all counts when it was used with solutions above 50% total solids because of the salting-out effect. Thus, the first molasses, diluted to 75% total solids, can be extracted in a counter-current extractor with one-half as much acetone by volume. The partition coefficient for aconitic acid may be over 4 to 1 in favor of acetone. The acetone extract layer contains a little water, and when the acetone is distilled off, a semisolid residue is left.

Analysis of this residue showed only about 4% sucrose, 6% inverts, and 17.4% total acids. Most of this (15.7%) was obtained as aconitic acid by leaching with hot water and subsequent crystallization. The residue was oils, waxes, and chlorophyl and could be separated by other solvent extractions; the sucrose and inverts were left to be processed for their sugar content by combining them with the raffinate from the extraction of the first molasses. The first molasses also was evaporated to strip off acetone and to concentrate the syrup further to a grain formation. The raw sugar obtained was about 98% sucrose and was recycled to join the original raw sugar feed, of which it represented approximately 1.3%. The extract layer was a small amount of light-colored molasses, low in sucrose and high in invert sugars; analysis showed it to contain five to ten times as much of several vitamins as conventional blackstrap molasses.

Solvent Extraction of Conventional Sugar Solutions. Sugar syrup made by directly concentrating cane juice or the blackstrap molasses that results after the crystallization of sugar therefrom can be refined also (2) by acetone extraction to remove the nonsugar constituents of oils, fats, waxes, chlorophyll, acids, etc. When total solids in the syrups are about 50%, the acetone forms a second layer that contains a larger percentage of the impurities than the first layer. The mutual solubility of acetone with the syrup decreases at high temperatures and this improves the extraction. Since the viscosity of the syrup is reduced greatly at higher temperatures, the rate of extraction is greatly increased. Other entirely water-soluble liquids were tried—e.g., methanol, ethanol, and glacial acetic acid. By themselves they usually go into solution, then remove the water to precipitate crystals (very small) and to give a solution and a solid phase at equilibrium. However, when mixed with appropriate amounts of acetone, they also give two liquid phases and can be used to refine sugar syrups.

Thus, as first indicated, the salting-out effect on impurities of concentrated sugar solutions will allow the refining of a sugar syrup by simple evaporation of a sugar juice expressed from cane to a total solids content of more than 50% and subsequent extraction with acetone, either at the point of production or at a refinery to which the raw syrup is shipped in tankers. A refined syrup or sugar then can be made mode cheaply than by present methods.

As another example, after a reaction process, acetic acid remained in a 60% sucrose solution. Water-miscible acetone was used to extract the acetic acid, and the partition coefficient between the two layers (gram acetic per gram acetone/gram acetic per gram sugar solution) was over 5 compared with less than 1.0 for solvents conventionally used to extract acetic acid from industrial solutions. This example shows again the great ability of the high concentration of the dissolved solid to prevent the normal miscibility of water and acetone and to salt-out the acetic acid from the aqueous solution into the extracting solvent.

Acetic Acid

Acetic acid is produced industrially by the oxidation or other degradation of many organic materials, and it is involved as a solvent or reagent in the synthesis of many important compounds. Having the same ratio of oxygen to carbon atoms as does the carbon monoxide molecule, it is often an intermediate in many oxidation processes, both chemical and bacterial. Formic acid, which has the same ratio of oxygen to carbon as does carbon dioxide, was produced also but in lesser amounts because of its comparative instability. Numerous chemical reactions leave acetic acid or its salts in the spent aqueous solutions, together with a high concentration of other liquids or solids. The decomposition of lignocellulose by any process gives acetic acid and usually formic acid or their salts. For example, destructive distillation gives charcoal as the principal product, pulping gives cellulose for paper, alkaline fusion gives oxalates, and bacterial action gives numerous materials. Often large amounts of other solids are present also in the aqueous solution of acetic acid or its particular salt, depending on the raw material used. All of these may involve large amounts of other solids, which may salt-out the acetic acid (and formic acid if it is present) in an extraction for their removal.

The production of wood pulp from lignocellulosic materials by treatment with various chemical liquors, particularly the neutral sulfite semichemical process and the kraft or sulfate process, gives residual black liquors. These contain salts that carry acetic acid and formic acid equivalent to 5% or more of the dry weight of the wood.

Either the aliphatic acids or their salts is a most undesirable constituent of waste liquors that are to be discharged into streams since both acids and salts have a high biochemical oxygen demand (BOD), with consequent damage to aquatic life. Usually these liquors are evaporated to concentrate the solids present, including the salts. If they are then treated with sulfuric acid, the acetic acid and related acids will be freed or "sprung," giving the corresponding sulfates.

The acetic acid is then present in an aqueous solution with a high concentration of solids—i.e., sodium sulfate plus lignosulfonates and other materials derived from the degradation of lignin and other constituents of wood. Acetone is an excellent solvent here (*2*) even though it is completely soluble in water. The dissolved solids prevent the usual complete miscibility of acetone with water. Thus, the higher the concentration of solids, the less is the mutual solubility of acetone and water, and the greater is the selectivity of acetone for acetic and formic acids. The partition coefficient is very high—4 to 6—which is five to eight times that usually found for solvents in the conventional extraction of acetic acid solutions.

Acetone has the additional essential property that it does not emulsify during mixing with most concentrated aqueous solutions of solids in industrial extractors, and with these aqueous solutions it is better in this respect than any other solvent.

Furthermore, acetone, which remains dissolved in the raffinate solution after extraction, may be distilled readily therefrom because of its high volatility from aqueous solutions and particularly from the solution with its high concentration of solute. This high concentration greatly reduces the vapor pressure of water, and for the same reason, acetone can be evaporated or distilled readily from the extract layer. In fact, a substantial portion of that used as the solvent for the acetic acid can be evaporated from the extract layer in a simple pot still without distilling over an appreciable amount of acetic acid.

The liquids in one extract layer containing the acetic acid extracted from a solution that contains solids had a composition of about 70% acetone, 15% water, and 15% acetic acid (*2*). The first distillate from this layer contained less than 0.1% acetic acid and thus could be recycled directly to the extractor. The balance of the acetone was stripped from the extract layer in a short column, and the water was distilled azeotropically (*2*) from the approximately 50% acid to give an anhydrous mixture of acetic and formic acids; these acids then were separated by a second azeotropic distillation (*4*).

With aqueous solutions of acetic acid of lower solids content—usually less than about 50%—acetone has a poor selectivity (i.e., it extracts more water with the acid). As this solid's concentration is reduced, the solvent layer disappears and the acetone becomes completely miscible, usually when the solids concentration falls below 50%; sometimes it does not disappear until total solids are as low as 40%.

Industrial solutions containing acetate salts usually are evaporated to a higher concentration before sulfuric acid is added to free the acetic acid. With the higher percentage of solids in the solution, addition of acetone gives a separate layer. This method, however, is not always the most desirable or economical. Instead, one canadd, in amounts of 10-40% of the acetone, another solvent that is much less miscible with water than acetone. This cosolvent improves acetone's selectivity for dissolving acetic acid with a minimum of water, but invariably it will reduce the partition coefficient, and usually it increases the tendency of emulsification in the extractor.

Isopropyl ether is a desirable cosolvent (*2*) and has, with acetone, a constant boiling mixture, boiling at 53.3°C, containing 56.5% acetone and 43.5% isopropyl ether. This cosolvent helps in distilling acetone from the extract layer by increasing its relative volatility with respect to both acetic acid and water.

Liquors from Neutral Semichemical Pulping and from Kraft Pulping. In making neutral sulfite semichemical pulp, the black liquors may have about 10 parts of water to 1 part of total solids, of which about one-third is sodium acetate and sodium formate. After evaporation to about 1 part solids to 1 part water, sulfuric acid is added to the concentrate to free the acetic and formic acids. When the concentrate is extracted with acetone, the mixed acids are obtained, the acetone is separated for recycle, and the acids are concentrated and refined. The raffinate is stripped and is passed to the usual furnace to be burned for recovery of the inorganic salt values. This process gives a smelt of sodium sulfate, which may be used in the kraft process as chemical makeup. The loss of the fuel value of the acetic and formic acids is practically negligible.

Liquors of the kraft process itself contain a large amount of free alkali, which must be neutralized by sulfuric acid before the acetic and formic acids are freed. The corresponding amount of sodium sulfate produced from the free alkali and from the acetate and formate decomposition may be more than can be reused in the recycle liquor system. On the other hand, since the makeup of liquors for a kraft process requires the addition of fresh sodium sulfate as well as the sulfate radical from sulfur burning, a significant part of this sodium sulfate can be used, up to a chemical equivalence of the makeup amount.

Hence, a part of the black liquors (10–50%) of a kraft plant may be processed to free, and subsequently to recover, the acetic acid as described by the addition of sulfuric acid and the formation of sodium sulfate. This new sodium sulfate from the sulfuric acid is present in the raffinate from the extractor; it is thus the makeup for the balance of the liquors (50–90%) that can be processed conventionally without recovery of the acetic and formic acids. The liquors from the raffinate are added to the other liquors before they go to the drier and the furnace. This part of the liquors (10–50%) recovers that much of the volatile acids that would otherwise be lost in the burner as salts.

Thus, with black liquors from kraft pulping, the addition of sulfuric acid for recovering acetic acid from a part of the liquors replaces the sulfur and sodium sulfate makeup, which would otherwise be used. However, since the cost of sulfuric acid is greater than that of the equivalent amount of sulfur and sodium sulfate makeup normally used (because of the cost of making sulfuric acid from sulfur vs. the cost of making sulfur oxides), the difference is charged against the cost of recovering acetic and formic acids.

Economics. Because of a wide range of conditions, the weight of acetic and formic acid recovered per ton of neutral sulfite semichemical pulp produced can vary between approximately 80 and 150 pounds, with formic acid representing 5–30% of this total. The possible revenue would thus be \$20–\$35/ton of pulp produced, and after capital and operating costs are subtracted, the profit would be about half of the revenues. In addition, one would have the value of the sodium sulfate, which will have its normal use and value in the kraft process. These figures do not include the value of this processing in pollution abatement—the effluent liquors from a 300 tons per day semichemical plant are equal in BOD to that from the municipal sewage of a city of 300,000 people, and a large part of this BOD comes from the acids that would be recovered.

Material and cost balances for a kraft plant are more intricately tied in with its specifics but also offer a major opportunity for additional revenue and elimination of problems in water pollution.

Literature Cited

1. Kirk–Othmer, "Encyclopedia of Chemical Technology," 2nd ed., Vol. 19, p. 151–220, Wiley-Interscience, New York, 1969.
2. Othmer, Donald F., U.S. and Foreign Patents and Applications.
3. Othmer, Donald F., Luley, Arthur H., *Sugar* (1949) **44**(6).
4. Othmer, Donald F., Conti, James J., U.S. Patent **3,024,170** (March 1962).

RECEIVED January 9, 1978.

2

Salt Effect in Vapor–Liquid Equilibrium at Fixed Liquid Composition

JOHN A. BURNS[1] and WILLIAM F. FURTER

Department of Chemical Engineering, Royal Military College of Canada, Kingston, Ontario, Canada, K7L 2W3

The effect of salt concentration on equilibrium vapor composition at atmospheric pressure was studied in four ethanol–water–salt systems at the boiling point, with mixed-solvent composition held constant. The salts investigated were potassium iodide, ammonium bromide, sodium acetate, and potassium acetate. As expected, the two inorganic salts were in good agreement with the Furter equation. In the case of the two salts of organic acids the agreement became less and less valid as salt concentration increased. The failure of the equation with the two acetate salts was attributed to the presence of significant organic ion–alcohol interaction, a situation which does not normally become significant with inorganic ions in aqueous alcohol solution except at high alcohol concentrations.

The original equation for salt effect in vapor–liquid equilibrium, proposed by Furter in 1958 (*1, 2*), predicts the effect of a salt dissolved in a mixed solvent on vapor–liquid equilibrium when the composition of the solvent is held constant:

$$\ln(\alpha_s/\alpha) = kz. \qquad (1)$$

In the equation, equilibrium vapor composition expressed in the form of an improvement factor, α_s/α, is related to salt concentration z in the liquid phase by the salt effect parameter, k. According to the theoretical

[1] Current address: P.O. Box 74, Kandahar, Saskatchewan, Canada, S0A 1T0

0-8412-0428-4/79/33-177-011$05.00/1

background from which the equation was derived, k remains constant when the relative proportions of the two volatile components forming the mixed solvent are held constant. The background, range of applicability, previous usage, limitations, and significance of this equation have been discussed recently in detail (*3*).

Jaques and Furter (*3*) tested Equation 1 with isobaric vapor–liquid equilibrium data at atmospheric pressure, under conditions of fixed liquid composition, for four systems of the alcohol–water–inorganic salt type. The systems tested were ethanol–water with each of sodium bromide, ammonium chloride, and sodium chloride, and methanol–water–calcium chloride, over a range of salt concentrations up to saturation. For all four systems agreement with the equation was very good, in all cases yielding average absolute deviations between predicted and experimental values of one-half percent or less. However, Jaques and Furter noted that solubilities and boiling point elevations were not large at the liquid composition values in the systems investigated. They suggested testing the equation where these are known to be large as in the ethanol–water–potassium acetate system.

Burns and Furter (*4*) tested Equation 1 with two more ethanol–water–inorganic salt systems under conditions of fixed-liquid composition, the salts in this case being the potassium and ammonium bromides. Again, very close agreement with the equation was observed, although some tailing-off seemed to occur close to saturation. (It is not clear whether the tailing-off was real or whether saturation had already been reached: the exact saturation point with certain salts can be difficult to observe.) They then extended their study to include four organic-cation salts of the tetraalkylammonium (TAA) bromide series, testing ethanol–water at fixed-liquid composition with the methyl, ethyl, *n*-propyl, and *n*-butyl members of the TAA bromide series. The methyl salt, which exhibits mainly electrostrictive behavior in ethanol–water solutions, and the *n*-propyl and *n*-butyl salts, which exhibit mainly hydrophobic behavior (*4*), all yielded results in close agreement with Equation 1. However, tetraethylammonium bromide, the member of the TAA bromide series in which the crossover from salting-out to salting-in of the alcohol, and hence of electrostrictive to hydrophobic behavior, was observed in ethanol–water solution, did not yield data in good agreement with the equation. The reason was postulated to be the location of this salt, which was in the TAA bromide series with ethanol–water, at or close to the transition point between competing solute–solvent interactions.

The purpose of the present investigation was to extend the study of Equation 1 at fixed-liquid composition to salts in which the anion, rather than the cation, is organic. For this purpose, the sodium and potassium acetates were chosen, both having very high solubilities in ethanol–water solution at the liquid compositions used, and hence being capable of causing large boiling point elevations. For comparison, two inorganic salts, potassium iodide and ammonium bromide, were studied also, all in ethanol–water at atmospheric pressure. Potassium iodide and sodium acetate were investigated at a single mixed-solvent composition (0.309 mole fraction ethanol), and potassium acetate, the most soluble salt, at three fixed-liquid compositions. Ammonium bromide, which had been studied previously at three liquid values (*4*), was studied at six more in the present investigation in order to gain insight into the influence of mixed-solvent composition over a wide range. With the exception of ammonium bromide, the fixed-liquid composition values were selected from the range in which the relative volatility of the ethanol–water system is highest, and non-ideal behavior is at a maximum. All of the systems of the present investigation have been studied previously, but none (with the exception of ammonium bromide as noted above) under the conditions of fixed-liquid composition. For references to previous studies the reader is referred to two literature reviews of salt effect in vapor–liquid equilibrium, the first covering the work up to 1965 (*5*), and the second from 1965–1975 (*6*). A recent review of the complex topic of correlation and prediction of salt effect in vapor–liquid has been published also (*7*).

Experimental

The apparatus used—the improved Othmer recirculation still modified for salt effect studies—was described previously (*1, 7*), as were the analytical techniques, specifications on the purity and treatment of the solvents, and details of surface tension and solubility measurements (*4*).

The ammonium bromide, potassium iodide, and sodium acetate were British Drug Houses Analar analytical reagent grade. The potassium acetate was Baker Analyzed crystal reagent grade from J. T. Baker Chemical Co. Owing to the extreme hygroscopicity of some of these salts, they all were dried for at least 72 hours at 120°C immediately prior to each series of experiments. The dried salts then were stored under vacuum over phosphoric pentoxide between each experimental measurement.

Results

Isobaric vapor–liquid equilibrium data at atmospheric pressure are reported for the four systems of the present investigation in Tables I–VI. Salt concentrations are reported as mole fraction salt in the solution, while mixed-solvent compositions are given on a salt-free basis. A single fixed-liquid composition was used for potassium iodide and sodium acetate; potassium acetate used three—all chosen from the region of ethanol–water composition where relative volatility is highest. In the

Table I. Isobaric Vapor–Liquid Equilibrium Data for the Potassium Iodide–Ethanol–Water System at $x = 0.309$ (758 ± 3 torr)

z	y	t	$log_{10} \frac{\alpha_s}{\alpha}$
0	0.5837	82.0	0.0000
0	0.5844	82.0	0.0005
0.0100	0.5983	82.1	0.0263
0.0113	0.6044	82.2	0.0373
0.0218	0.6158	82.2	0.0581
0.0275	0.6233	82.2	0.0719
0.0344	0.6316	82.3	0.0873
0.0438	0.6437	82.6	0.1100
0.0488	0.6508	82.7	0.1235
0.0608	0.6634	82.9	0.1479
0.0687	0.6732	82.9	0.1670
0.0768	0.6839	83.0	0.1885
0.0823	0.6882	83.2	0.1971
0.0881	0.6943	83.5	0.2094
0.0992	0.7013	83.6	0.2238
0.1033	0.7112	83.8	0.2446
0.1187 (saturated)	0.7213	84.0	0.2662

case of ammonium bromide, six fixed-liquid composition values were used, which, when combined with previous data for three other values (*4*), provided data for a wide range of mixed-solvent composition in this particular system. With the exception of the potassium acetate–ethanol–water system, the boiling point elevations were generally small.

Instead of applying Equation 1 directly to the experimental data, the expression was modified to read $\log_{10}(\alpha_s/\alpha) = k'z$, where $k' = k/2.303$

Figure 1 is a graphical presentation of the data obtained for potassium iodide–ethanol–water at $x = 0.309$ mol fraction ethanol in the mixed solvent, and Figure 2 is an example of the results for ammonium bromide–ethanol–water, in this case at $x = 0.246$. In both cases the linearity predicted by Equation 1 is noted, although some slight tailing-off appears to occur close to saturation. (As previously mentioned, the tailing-off may or may not be real; in the latter case the saturation would have been reached already and no further vapor composition change was possible.)

Because of the wide range of fixed x values for which data had been taken, it was possible to use interpolated data from Table II to construct a family of vapor–liquid equilibrium curves for the ammonium bromide–ethanol–water system at various constant salt concentration values—the condition most closely representing that existing from tray to tray in a

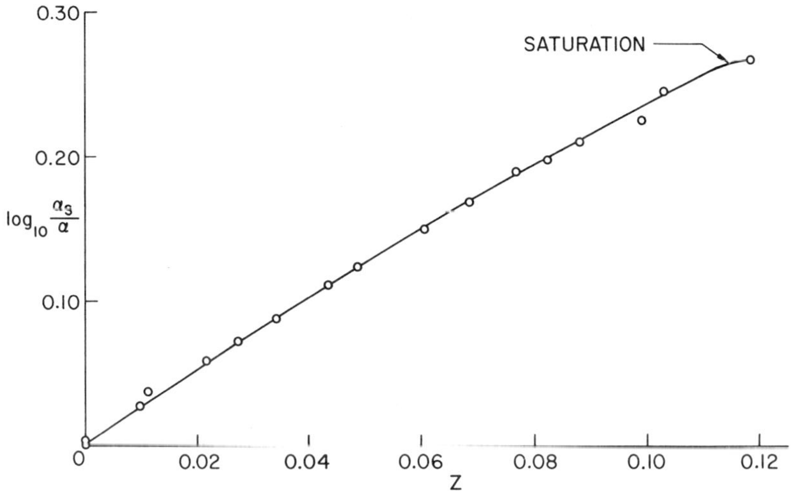

Figure 1. Salt effect of potassium iodide on the ethanol–water system at x = *0.309*

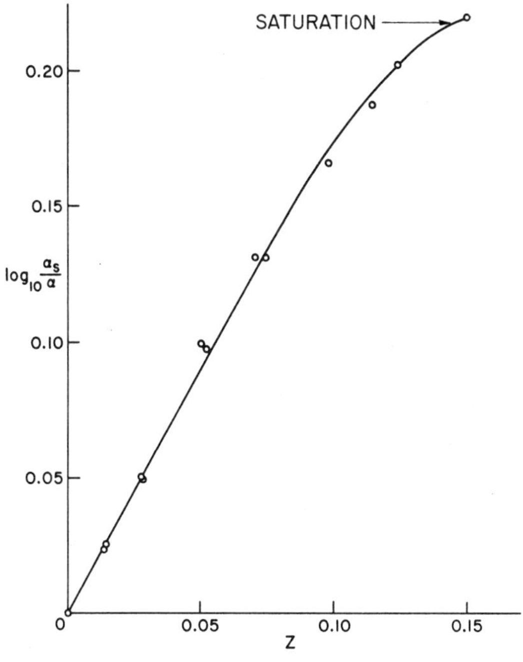

Figure 2. Salt effect of ammonium bromide on the ethanol–water system at x = *0.246*

Table II. Isobaric Vapor–Liquid Equilibrium Data for the Ammonium Bromide–Ethanol–Water System (759 ± 8 torr)

x	z	y	t	$\log_{10} \frac{\alpha_s}{\alpha}$
0.1533	0	0.5040	84.8	0.0000
	0.0248	0.5660	85.0	0.1083
	0.0482	0.5860	85.0	0.1439
	0.0705	0.6054	85.2	0.1790
	0.0925	0.6164	85.0	0.1991
	0.1139	0.6378	85.0	0.2388
	0.1344	0.6536	85.5	0.2687
	0.1534	0.6697	86.1	0.3000
	0.1755 (saturated)	0.6765	86.4	0.3134
0.2461	0	0.5828	83.0	0.0000
	0.0139	0.5956	83.0	0.0230
	0.0143	0.5968	83.0	0.0251
	0.0279	0.6106	83.0	0.0502
	0.0284	0.6100	83.0	0.0491
	0.0502	0.6372	83.1	0.0994
	0.0523	0.6360	83.2	0.0971
	0.0722	0.6536	83.5	0.1305
	0.0750	0.6536	83.6	0.1305
	0.0984	0.6715	84.0	0.1653
	0.1151	0.6822	84.0	0.1866
	0.1245	0.6899	84.0	0.2021
	0.1510 (saturated)	0.6982	84.0	0.2191
0.2778	0	0.5616	83.0	0.0000
	0.0324	0.6312	83.6	0.1258
	0.0619	0.6566	83.5	0.1739
	0.0896	0.6815	83.5	0.2229
0.37777	0	0.6013	81.5	0.0000
	0.0367	0.6622	81.9	0.1139
	0.0693	0.6944	82.5	0.1781
	0.1009	0.7246	83.0	0.2417
0.5198	0	0.6596	80.0	0.0000
	0.0306	0.7009	80.5	0.0826
	0.0667	0.7434	81.9	0.1748
	0.0838	0.7572	82.0	0.2068
0.6624	0	0.7326	79.5	0.0000
	0.0336	0.7762	80.0	0.1025
	0.0612	0.7934	80.8	0.1467
	0.0828 (saturated)	0.7999	80.9	0.1641

Table III. Isobaric Vapor–Liquid Equilibrium Data for the Sodium Acetate–Ethanol–Water System at $x = 0.309 \pm 0.001$ (758 ± 3 torr)

z	y	t	$\log_{10} \frac{\alpha_s}{\alpha}$
0	0.581	82.0	0.0000
0	0.581	82.0	0.0000
0.008	0.600	82.2	0.033
0.013	0.609	82.2	0.049
0.043	0.652	82.3	0.131
0.045	0.658	82.3	0.142
0.072	0.688	82.5	0.202
0.081	0.699	82.5	0.223
0.108	0.725	82.7	0.277
0.118	0.733	82.7	0.296
0.130	0.742	82.8	0.317
0.142	0.744	82.9	0.321
0.144	0.752	83.2	0.339
0.160	0.760	83.4	0.358
0.180	0.773	83.5	0.390
0.200	0.777	83.9	0.399
0.204	0.783	84.0	0.414
0.224	0.787	84.0	0.424
0.242	0.782	83.9	0.421

fractional distillation column. The family of constant salt concentration curves so constructed is shown in Figure 3. For further reference to distillation applications, the reader is referred to a recent review of the topic (*8*). The k′ salt effect parameter values measured at the six fixed x values of the present investigation and the three values of the previous study (*4*) are summarized in Table VII and plotted in Figure 4 as functions of x, with the standard deviation shown in each case. (Relative deviations tend to be larger in mixed-solvent composition regions where relative volatility is lower, and vice versa).

The results for the two organic anion salts of the present investigation are demonstrated in Figures 5 and 6. Figure 5 presents the data for sodium acetate–ethanol–water at $x = 0.309$; and Figure 6 for potassium acetate–ethanol–water at $x = 0.245$ and 0.311, two of the three values used . Equation 1 is unable to correlate data in these two systems; in each case a linear regression yields a large deviation of the experimental points from the best straight line, which in any event would fail to pass through the origin.

The values of k′ reported in Tables VII and VIII for the systems studied were determined by applying a linear least-squares calculation to the experimental data. The average absolute deviation (A.A.D.) of $\log_{10}(\alpha_s/\alpha)$ was determined at fixed intervals of z from the individual

Table IV. Isobaric Vapor–Liquid Equilibrium Data for the Potassium Acetate–Ethanol–Water System at $x = 0.245$ (758 ± 3 torr)

z	y	t	$\log_{10} \frac{\alpha_s}{\alpha}$
0	0.5875	84.0	0.0000
0.0267	0.6331	83.0	0.0832
0.0522	0.6610	83.0	0.1364
0.0762	0.6879	82.5	0.1896
0.0990	0.7064	82.0	0.2276
0.1207	0.7281	82.0	0.2740
0.1413	0.7418	81.5	0.3048
0.1611	0.7531	81.0	0.3309
0.1799	0.7647	81.0	0.3582

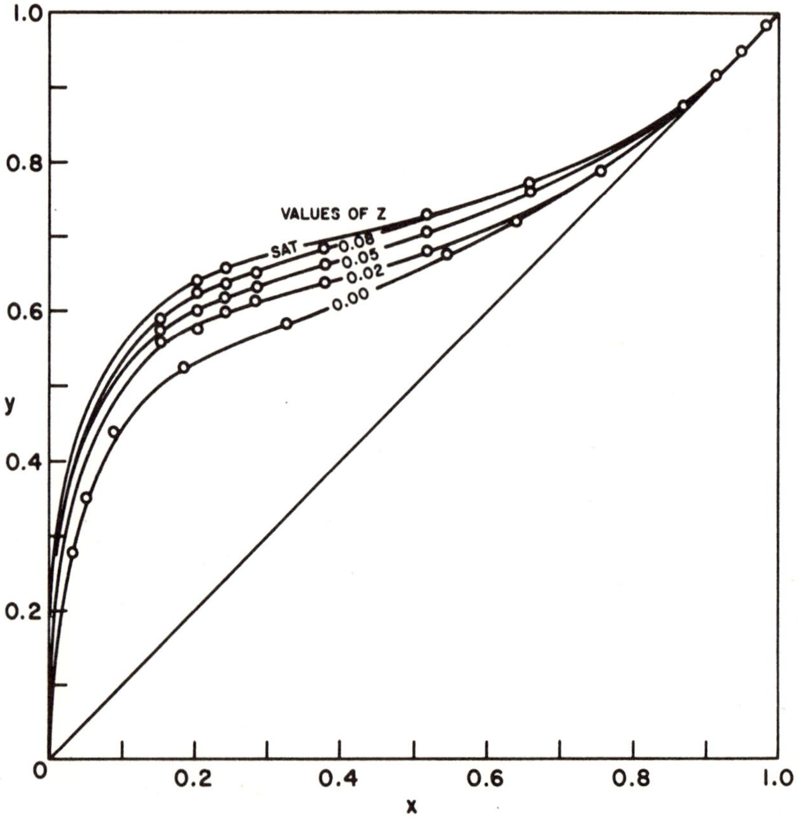

Figure 3. The ammonium bromide–ethanol–water system at various fixed-salt concentrations and saturation

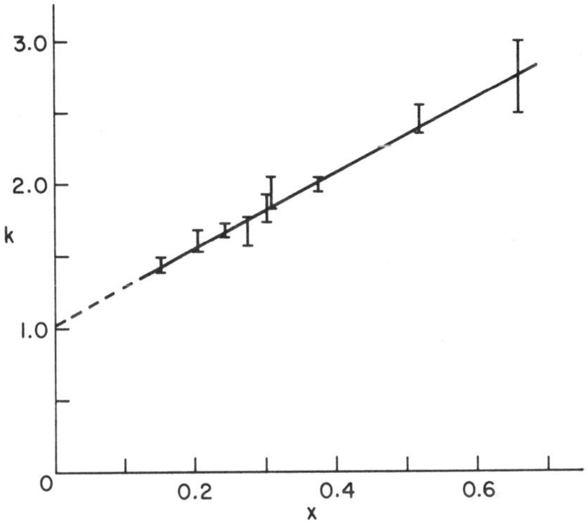

Figure 4. Salt effect parameter as a function of the liquid composition ratio for the ammonium bromide–ethanol–water system

Table V. Isobaric Vapor–Liquid Equilibrium Data for the Potassium Acetate–Ethanol–Water System at $x = 0.311$ (758 ± 3 torr)

z	y	t	$log_{10} \frac{\alpha_s}{\alpha}$
0	0.5839	82.0	0.0000
0.0308	0.6413	82.0	0.1053
0.0543	0.6741	82.0	0.1686
0.0794	0.7061	82.1	0.2335
0.0892	0.7186	82.2	0.2601
0.1055	0.7371	82.4	0.3007
0.1302	0.7586	82.8	0.3503
0.1532	0.7780	83.4	0.3975
0.1746	0.7947	84.1	0.4407
0.1792	0.7982	84.3	0.4501
0.1959	0.8076	84.5	0.4759
0.2160	0.8194	85.8	0.5096
0.2354	0.8286	86.5	0.5373
0.2565	0.8377	87.4	0.5657
0.2702	0.8422	88.0	0.5801

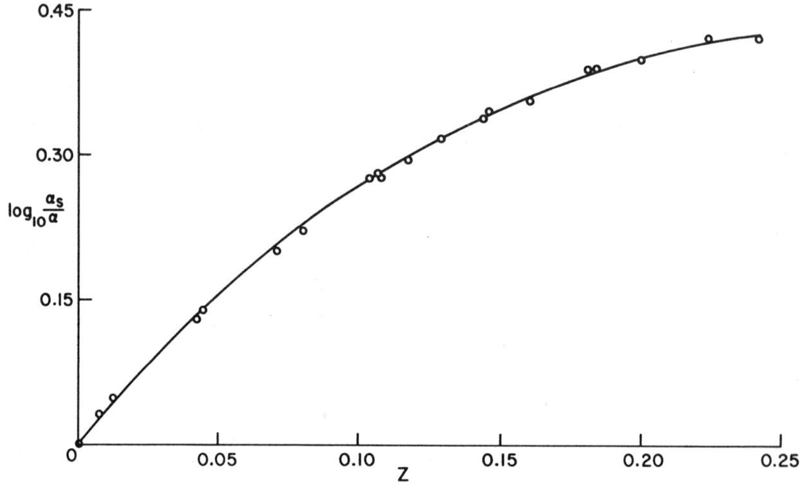

Figure 5. Salt effect of sodium acetate on the ethanol–water system at x = *0.309*

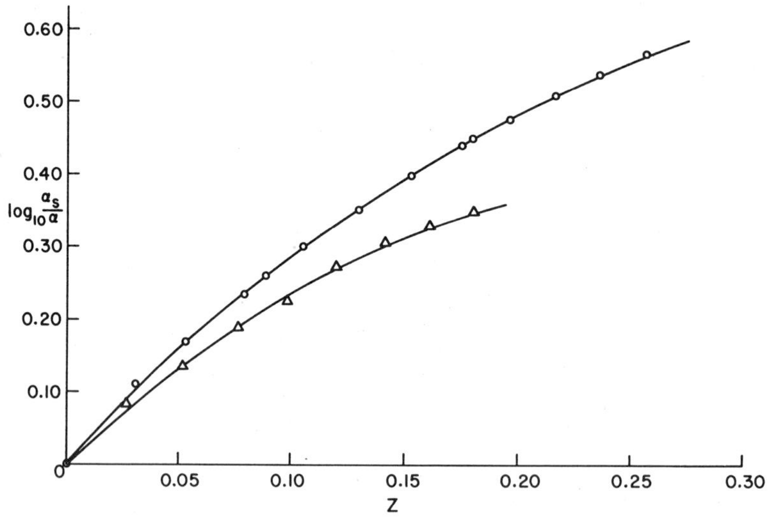

Figure 6. Salt effects on the potassium acetate–ethanol–water system at x = *0.245,* △*; and* x = *0.311,* ○

Table VI. Isobaric Vapor–Liquid Equilibrium Data for the Potassium Acetate–Ethanol–Water System at $x = 0.313$ (753 ± 3 torr)

z	y	t	$log_{10} \frac{\alpha_s}{\alpha}$
0	0.5989	82.0	0.0000
0.0155	0.6186	82.0	0.0360
0.0295	0.6420	82.0	0.0796
0.0578	0.6799	82.3	0.1530
0.0876	0.7167	82.5	0.2290
0.1098	0.7414	82.6	0.2834
0.1217	0.7524	83.0	0.3087
0.1504	0.7768	83.4	0.3676
0.1782	0.7956	84.0	0.4161
0.1913	0.8059	85.0	0.4440
0.2036	0.8116	85.5	0.4603
0.2151	0.8169	86.0	0.4753
0.2679	0.8403	87.7	0.5470
0.3417	0.8612	92.0	0.6185
0.4023 (saturated)	0.8740	94.2	0.6670

Table VII. Values of the Salt Effect Parameter at Various Values of x for the Ammonium Bromide–Ethanol–Water System

x	k'
0.154	1.44 ± 0.05
0.206[a]	1.61 ± 0.07
0.246	1.67 ± 0.06
0.278	1.67 ± 0.10
0.305[a]	1.83 ± 0.10
0.309[a]	1.94 ± 0.11
0.378	1.99 ± 0.05
0.520	2.45 ± 0.10
0.662	2.75 ± 0.25

[a] Data from Burns and Furter (*4*).

Table VIII. Salt Effect Parameters and Reliability of Equation 1 to Predict Salt Effects for Some Inorganic Salts in Ethanol–Water Mixtures at Various Values of *x*

Salt	x	*k′*	*R.A.A.D.*[a] %	*Reference*
KBr	0.206	2.89 ± 0.04	1.3	*4*
	0.311	3.02 ± 0.09	2.5	*4*
KI	0.309	2.33 ± 0.09	4.0	this work
NaOAc	0.309	2.05 ± 0.20	13.8	this work
KOAc	0.245	2.11 ± 0.18	8.5	this work
	0.311	2.31 ± 0.20	8.3	this work
	0.313	2.20 ± 0.15	7.7	this work
NH_4Br	0.206	1.61 ± 0.07	3.9	*4*
	0.246	1.67 ± 0.06	3.0	this work
	0.305	1.83 ± 0.10	5.8	*4*
	0.309	1.94 ± 0.11	4.5	*4*
NaCl	0.240	3.46 ± 0.11		*3*
	0.248	3.62		*3*
NaBr	0.248	2.87		*3*
NH_4Cl	0.143	2.50		*3*
	0.223	2.20		*3*
	0.313	2.19		*3*
	0.367	2.29		*3*
	0.451	2.35		*3*
	0.527	2.23		*3*

[a] The relative average absolute deviation, % (*see* Results).

deviations of the smoothed curve from the linear regression passing through the experimental points. From this A.A.D. value its magnitude, relative to the experimental value of $\log_{10}(\alpha_s/\alpha)$ at the mean *z*, was obtained, giving the R.(relative)A.A.D. values listed in Table VIII. An assessment of the capability of Equation 1 to correlate the salt effects can be made more effectively from these R.A.A.D.

The significance and magnitude of the errors involved in the experimental procedures and analytical methods, and the precision of the solubility and surface tension data listed in Table IX have been discussed fully elsewhere (*4*).

Discussion

The k′ standard deviation data and R.A.A.D. values listed in Table VIII for the data of the present investigation and certain data from previous studies (*3, 4*) demonstrate clearly that Equation 1 can satisfactorily correlate the salt effects of the inorganic salts tested at fixed-liquid composition in ethanol–water, at all salt concentrations up to saturation. The R.A.A.D. yields a critical evaluation of the applicability of the equation since it removes the complicating factor of experimental data scatter. This treatment accentuates the validity of Equation 1 in correlating accurately the effects of the inorganic salts; however, it also accentuates its failure for the salts containing organic anions, especially if the organic ions have a propensity to promote both a hydrophilic and hydrophobic effect on the water molecules, the net result of which is a balance between these two types of solute–solvent interactions with one or the other slightly predominating. In the case of the acetate ions, the hydrophilic interactions are more evident. A similar observation was made previously by the present authors for tetraethylammonium bromide (*4*). However, it was found that when hydrophobic interactions predominated, then the salt effects again could be correlated using Equation 1, but a salting-in of ethanol was noted instead of the normal salting-out. Figures 5 and 6 reveal that a diminishing return appears to occur as the

Table IX. Physical Properties of Saturated Solutions of Some Inorganic Salts in Water, Ethanol, and 0.206 Mole Fraction Ethanol–Water at 25°C

Solvent	*Salt*	*Surface Tension Pa m × 10³*	*Solubility* *m mol Kg⁻¹*	z
H_2O		72.0		
	KBr	73.4	5.6 ± 0.1	0.092
	KBr		5.71[a]	0.093
	NH_4Br	60.7	8.0 ± 0.2	0.126
	NH_4Br		7.92[a]	0.125
	KI	69.6	8.92[a]	0.138
	NaOAc	57.5	15.2[a]	0.215
	KOAc	66.8	27.4[a]	0.331
C_2H_5OH		22.3		
	KBr	22.3	< 0.1	< 0.0004
	KBr		0.113[a]	0.0004
	NH_4Br	22.4	0.33 ± 0.02	0.015
	NH_4Br		0.35[a]	0.016
	KI	22.5	0.118[a]	0.005
	NaOAc	22.5		
	KOAc	23.3		
0.206 mole fraction C_2H_5OH–H_2O		30.5		
	KBr	29.5	1.9 ± 0.1	0.043
	NH_4Br	31.3	3.5 ± 0.2	0.073
	KI	31.3	4.52[a]	0.097
	NaOAc	28.1		
	KOAc	28.6		

[a] Data from Seidell (*13*).

salt concentration is increased; this may be caused by either ion reassociation occurring at high salt concentrations, or saturation effects with respect to the hydration spheres; that is, the added ions and their solvent spheres may find it difficult to fit into spaces among other ions and their solvent spheres. In addition, surface effects could, at least partially account for the inability of Equation 1 to correlate the salt effects in the case of organic ions where the hydration spheres at the surface are disrupted severely compared with those within the homogeneous solution. This would result in a large degree of heterogeneity occurring at the surface, and the water molecules finding it easier to escape than if there were no salt present.

A comparison of the dependence of the improvement factor on the salt concentration for the ammonium bromide–ethanol–water system at $x = 0.246$ (Figure 3) and that observed by Jaques and Furter (*3*) for ammonium chloride–ethanol–water at $x = 0.223$ reveals many similarities between these two systems. Their k′ values are similar, they both yield good correlation with Equation 1, and in both systems an inflection can be detected in the experimental plot of improvement factor vs. salt concentration at ~ 0.05 mole fraction salt. The inflection probably can be attributed to a change in the solvent structure owing to the presence of salt.

Although Johnson and Furter (*1, 2*), among others, observed a surprising insensitivity of k′ to mixed-solvent composition in many alcohol–water–inorganic salt systems, such does not appear to be the case with ammonium bromide–ethanol–water. A linear dependence of k′ with x was observed and is demonstrated in Figure 4. The slope of this dependence is 2.63 and the intercept with the y-axis occurs at approximately a value of unity. This extrapolated salt effect when $x = 0$, that is, with water as the single solvent, is consistent with Raoult's Law in that the vapor pressure of the aqueous salt solution depends directly on the salt concentration. However the same behavior has not been observed for the ammonium chloride–ethanol–water system (*3*); as seen in Table VIII its salt effect parameter shows essentially no dependence on the liquid composition. Therefore the two systems differ in this respect.

In the case of the acetate salts, Meranda and Furter (*9*) found that k′ changed relatively little with x for sodium acetate–ethanol–water, but it underwent a major variation with potassium acetate in the same binary solvent. They also found (*10*) that at x values below 0.5, the improvement factor in the potassium acetate–ethanol–water system fell off with decreasing x, even though the study was conducted at saturation and hence salt concentration was increasing with decreasing x. The same phenomenon is observed in the two curves of Figure 6 where, for the same salt concentration, the larger effect occurs at the higher value of x (both values being below 0.5).

Table X. Solubility of Various Inorganic Salts in Boiling Solutions of Water, Ethanol, and Ethanol–Water Mixtures

Solvent	*Salt*	*Solubility*[a] *m* *mol kg*$^{-1}$	*Solubility*[a] z
H_2O	NaOAc	23.8	0.30
	KOAc	54.2	0.494
	KBr	8.89	0.138
	KI	12.50	0.183
	NH_4Br	13.8	0.200
C_2H_5OH	NaOAc	0.22	0.01
	KOAc	2.58	0.106
	KBr	0	0
	NH_4Br	0.48	0.022
0.25 m.f.[b] EtOH–H_2O	NaOAc	17.1	0.30
	KOAc	28.6	0.417
	KBr	3.24	0.075
0.31 m.f.[b] EtOH–H_2O	NaOAc	1.61	0.28
	KOAc	23.4	0.40
	KBr	2.39	0.06

[a] Data from Seidell (*13*).
[b] m.f. = mole fraction.

Bedrossian and Cheh (*11*) studied vapor–liquid equilibrium in the sodium acetate–ethanol–water system, using constant lower values of salt concentration rather than the saturation values used by Meranda and Furter (9). Analysis of their data using Equation 1 indicates a larger variation of k′ with x than that observed at saturation by Meranda and Furter. Bedrossian and Cheh concluded that hydration as well as hydrotropism of ions plays a major role in this particular system.

The surface tension and solubility data listed in Tables IX and X demonstrate that the sodium and potassium acetates have much more dramatic effects in aqueous solution than do the inorganic salts for which data are given.

All salts of the present investigation salted-out ethanol from aqueous solution; that is, they raised its concentration in the equilibrium vapor. Also, all are more soluble in water than in ethanol. While there are exceptions to this general parallel (*12*), in most systems it is more or less applicable (7).

Acknowledgments

The research presented in this chapter was supported financially through Grant No. 3610–457 from the Defence Research Board of Canada. J. S. Labbé obtained some of the experimental results for the potassium acetate–ethanol–water system at $x = 0.245$.

Nomenclature

k = salt effect parameter, as defined by Equation 1.

k′ = salt effect parameter; $k' = \frac{k}{2.303}$

t = boiling point, °C

x = mole fraction of ethanol in liquid phase; calculated on a salt-free basis

$$= \frac{\text{mole ethanol}}{\text{mole ethanol} + \text{mole water}}$$

y = mole fraction of ethanol in the vapor phase

z = mole fraction of salt in the liquid phase

$$= \frac{\text{mole salt}}{\text{mole ethanol} + \text{mole water} + \text{mole salt}}$$

α = relative volatility in absence of salt

$$= \frac{y(1-x)}{x(1-y)}$$

α_s = relative volatility in presence of salt; calculated using liquid compositions on a salt-free basis.

Literature Cited

1. Furter, W. F., Ph.D. thesis, University of Toronto (1958).
2. Johnson, A. I., Furter, W. F., *Can. J. Chem. Eng.* (1960) **38,** 78.
3. Jaques, D., Furter, W. F., *Ind. Eng. Chem. Fundam.* (1974) **13,** 238.
4. Burns, J. A., Rurter, W. F., Adv. Chem. Ser. (1976) **155,** 99.
5. Furter, W. F., Cook, R. A., *Int. J. Heat Mass Transfer* (1967) **10,** 23.
6. Furter, W. F., *Can. J. Chem. Eng.* (1977) **55,** 229.
7. Johnson, A. I., Furter, W. F., *Can. J. Technol.* (1957) **34,** 413.
8. Furter, W. F., Adv. Chem. Ser. (1972) **115,** 35.
9. Meranda, D., Furter, W. F., *AIChE J.* (1971) **17,** 38.
10. Meranda, D., Furter, W. F., *Can. J. Chem. Eng.* (1966) **44,** 298.
11. Bedrossian, A. A., Cheh, H. Y., *AIChE Symp. Ser.* (1974) **70**(140), 102.
12. Meranda, D., Furter, W. F., *AIChE J.* (1974) **20,** 103.
13. Seidell, A., Linke, W. F., "Solubilities of Inorganic and Metal-Organic Compounds," Vol. II, 4th ed., American Chemical Society, Washington, D.C., 1965.

Received January 16, 1978.

3

Prediction of Salt Effect on Vapor–Liquid Equilibrium: A Method Based on Solvation II

SHUZO OHE

Ishikawajima–Harima Heavy Ind. Co. Ltd., Research Institute,
1 Shinnakahara Isogoku, Yokohama 235, Japan

The preferential solvation formed between salt and solvent molecules causes a salt effect on vapor–liquid equilibria. A method of prediction of salt effect based on the preferential solvation number was reported previously for the case in which salt was solved below the saturation level. The idea introduced in this chapter applies for salt solved in saturation. The alcohol–ester–calcium chloride system for which the preferential solvation was thought to be formed was examined. Specifically, calcium chloride dissolves readily in alcohol but only sparingly in ester. Thus, when calcium chloride is solved into alcohol–ester mixed solvent, the calcium chloride will form a preferential solvation with alcohol only. Methanol–methyl acetate, butanol–butyl acetate, and methanol–ethyl acetate systems were selected for the mixed-solvent systems.

When salt is added to a volatile solvent mixture, there is a salt effect—a change in the vapor–liquid equilibrium relation. This salt effect occurs because salt forms a preferential solvate with a particular component of the solvent mixture, causing a drop in partial pressure of the particular component which forms the preferential solvate. Results of the studies conducted based on this idea are reported by the author in References *1* and *2*. In the past studies, the vapor–liquid equilbrium relation of the system for which formation of preferential solvate had been expected was observed, preferential solvation number was calculated based on the actually observed values, and further, salt effect was predicted based on the preferential solvate number. The author has

0-8412-0428-4/79/33-177-027$05.00/1

developed this study further by expanding the concentration of the salt to the saturation level (in his former studies, it was limited to the non-saturated concentration). This chapter reports on this most recent study.

Preferential Solvate System

The author selected the system containing salt which is not dissolved with other components but only with a particular component of a solvent mixture as a system with which the phenomenon of preferential solvate can be understood easily. Calcium chloride is dissolved with alcohol but it is not dissolved well with ester. Thus, calcium chloride forms a preferential solvate with alcohol and does not with ester. For the component system which consists of calcium chloride, alcohol, and ester, the author selected the following three systems for which vapor–liquid equibrium relations have been measured: methanol–ethyl acetate–calcium chloride (*1*); methyl acetate–methanol–calcium chloride (*3*); and *n*-butyl acetate–*n*-butanol–calcium chloride (*3*).

Solubility of Salt into Solvent Mixture

The solubility of the salt into a solvent mixture is decided by the concentration of a particular component in that mixture when salt is readily dissolved only with the particular component in the solvent mixture. Figures 1, 2, and 3 show solubilities of calcium chloride for the above mentioned three systems. Figure 1 shows the solubility of calcium

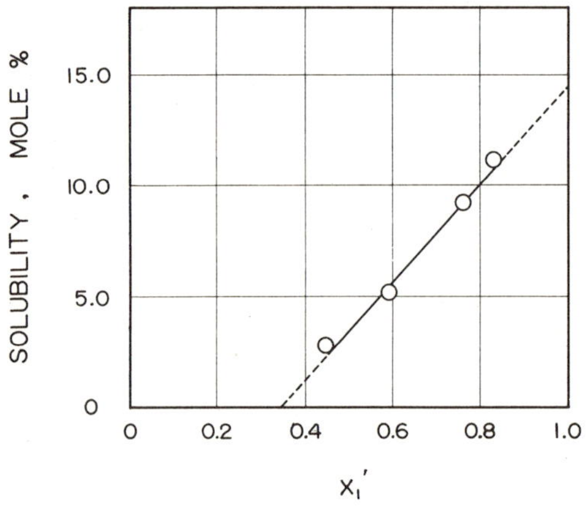

Journal of Chemical Engineering, of Japan

Figure 1. Solubility of calcium chloride in boiling methanol–ethyl acetate mixture at 1 atm (3)

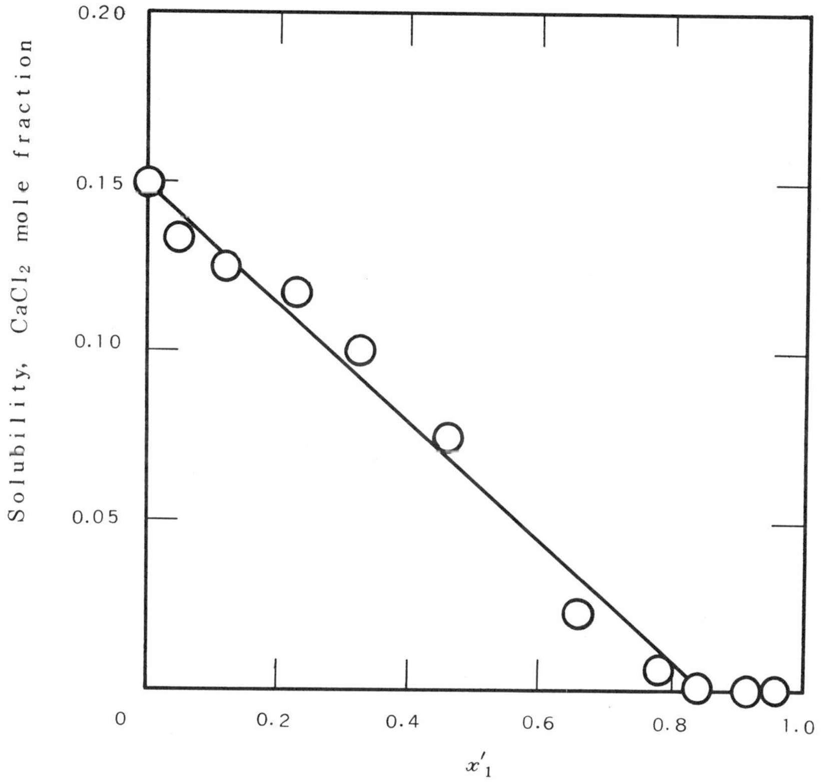

Figure 2. Solubility of calcium chloride in boiling methyl acetate–methanol mixture at 1 atm

chloride into the methanol–ethyl acetate system. From 0–0.333 mole fraction of methanol, the solubility is almost zero. These solubility data indicate that if calcium chloride is dissolved by only the methanol contained in the methanol–ethyl acetate solution, both solvents exist in the form of clustered molecules comprised of one methanol molecule and two ethyl acetate molecules (*2*). In methanol concentrations greater than 0.333 mole fraction, free molecules forming nonclustered molecules are present in the system, so that the salt is dissolved in the free molecules of methanol. From the extrapolated solubility (the mole ratio of calcium chloride to methanol is —1:6), calcium chloride and methanol are believed to form a solvate of $CaCl_2 \cdot 6CH_3OH$. Figure 2 shows the solubility of calcium chloride into the methyl acetate–methanol system. Figure 2 was drawn by the author based on Hashitani's report (*3*). This figure indicates that the solubility changes linearly against the concentration of methanol in the same manner as that depicted in Figure 1. Solubility is almost zero in the range where the mole fraction of methanol is between 0 and 0.15. The molecules of methanol and methyl acetate form a cluster of some kind and because of this cluster, no free methanol molecule exists. Figure 3 shows the solubility of calcium chloride in the butyl acetate–butanol system. This figure was drawn also by the author based on Hashitani's report (3). This figure shows that the solubility of calcium chloride reduces linearly against the reduction of the mole fraction of butanol and that the existence of butanol in the solvent mixture contributes to dissolving calcium chloride. Similarly, in this system also, the solubility of calcium chloride is almost zero at the range where the mole fraction of butanol is between 0 and 0.2. This occurs because four molecules of butyl acetate form a cluster against one molecule of butanol

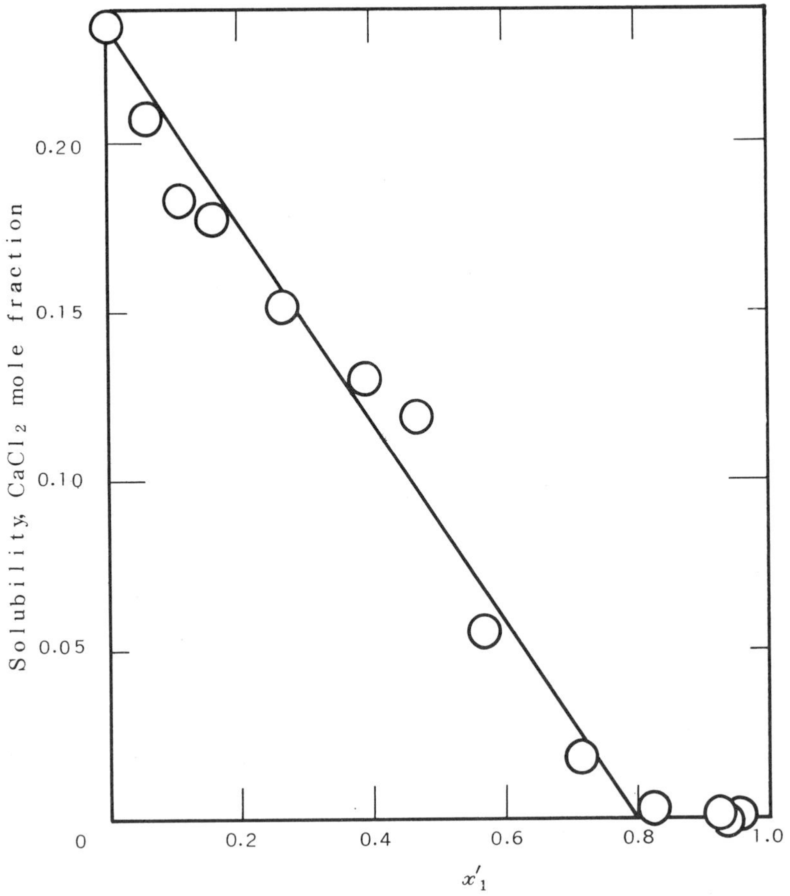

Figure 3. Solubility of calcium chloride in boiling butyl acetate–butanol mixture at 1 atm (3)

resulting in the inability of a butanol molecule to exist. It is obvious from the above described three examples that calcium chloride is dissolved only by an alcohol molecule in a solvent mixture. This indicates that calcium chloride is related only to alcohol molecules in the solvent mixture, suggesting the existence of a preferential solvate.

Preferential Solvation Number

When a preferential solvate is formed across salt and a particular component in a solvent mixture, the preferentially solvated component is assumed to be nonvolatile. Hence, the essential concentration of the preferentially solvated component in the solvent mixture is reduced as much as the solvated component. The vapor–liquid equilibrium relation obtained under the addition of a salt may well be considered to be the same as the vapor–liquid equilibrium without the salt for liquid-phase composition from which the solvents forming solvates are excluded. Based on this idea, the essential concentration at the time when salt forms a preferential solvate with the primary component is given by Equation 1. Then we can obtain the preferential solvation number from the observed values of the salt effect. As the concentration of solvent is decreased by the number of solvated molecules, the actual solvent composition participating in the vapor–liquid equilibrium is changed. Assuming that a salt forms the solvate with the first component, the actual composition x_{1a} is given by

$$x_{1a} = \frac{x_1 - Sx_3}{(x_1 - Sx_3) + x_2}. \tag{1}$$

Since $x_1 = x_1'(1 - x_3)$, $x_2 = x_2'(1 - x_3)$, and $x_1' + x_2' = 1$, Equation 2 is rewritten as follows:

$$x_{1a}' = \frac{x_1'(1 - x_3) - Sx_3}{(1 - x_3) - Sx_3}. \tag{2}$$

Solving Equation 3 for S, we obtain

$$S = \frac{1 - x_3}{x_3} \frac{x_1' - x_{1a}'}{1 - x_{1a}'}. \tag{3}$$

Therefore, the solvation number can be calculated by determining x_{1a}' from the measured values using the vapor–liquid equilibrium relation obtained without adding a salt. When a salt forms the solvation with the second component, the following three equations can be derived in a similar manner.

$$x_{1a} = \frac{x_1}{x_1 + (x_2 - Sx_3)} \tag{4}$$

$$x_{1a}' = \frac{(1 - x_3)x_1'}{(1 - x_3) - Sx_3} \tag{5}$$

$$S = \frac{1 - x_3}{x_3} \frac{x_{1a}' - x_1'}{x_{1a}'} \tag{6}$$

Figures 4, 5, and 6 indicate caluculated results of the preferential solvation numbers for the three systems. As shown by each figure, preferential solvation numbers are almost constant against compositions of the solvent. On the other hand, the concentration of salt increases linearly against an increase in the concentration of alcohol in the solvent as indicated in Figures 1, 2, and 3. This fact denotes that for an increase of solvent which forms a preferential solvate in a solvent mixture, the salt required to form a certain solvation number with that solvent is dissolved. For essential concentration x_{1a}' in Equations 3 and 4, which are required in calculating solvation numbers, the data observed by the author et al. (*1*) were used for the methanol–ethyl acetate system;

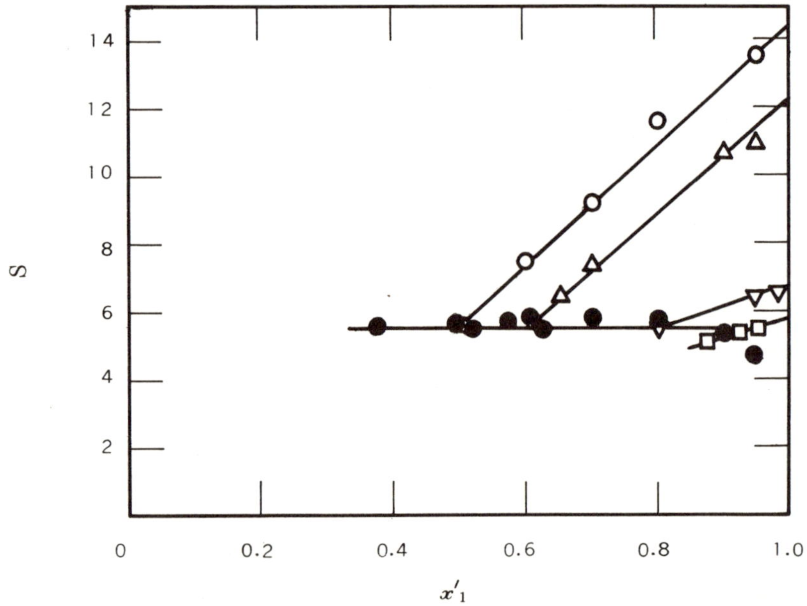

Journal of Chemical and Engineering Data

Figure 4. Preferential solvation number in the methanol–ethyl acetate system at 1 atm: (○), $CaCl_2$: 5 wt %; (△), $CaCl_2$: 10 wt %; (▽), $CaCl_2$: 20 wt %; (□), $CaCl_2$: 25 wt %; (●), $CaCl_2$: saturated (1).

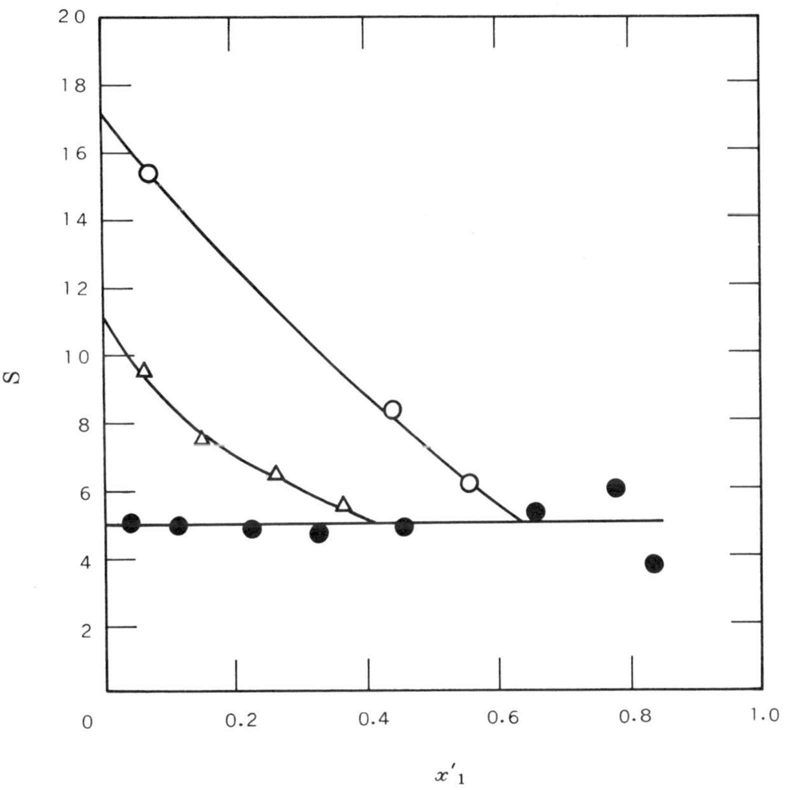

Figure 5. Preferential solvation number in the methyl acetate–methanol–calcium chloride system at 1 atm (○), $CaCl_2$: 6 wt %; (△), $CaCl_2$: 15 wt %; (●), $CaCl_2$: saturated (3).

Nagata's data (4) were used for the methyl acetate–methanol system; and for the butyl acetate–butanol system, data observed by Brunjes et al. (5) were used. Figure 4 shows that the preferential solvation number is about 5.5 in the methanol–ethyl acetate–calcium chloride system, and it is constant at the range where the mole fraction of methanol is between 0.333 and 1.000. At the range where the mole fraction of methanol is less than 0.333, the concentration of calcium chloride is zero and consequently the preferential solvation number is also zero. Figure 5 indicates that for the methyl acetate–methanol–calcium chloride system, the preferential solvation number is about 5, and it is constant at the range where the mole fraction of methyl acetate is between 0 and 0.85. At the range where the mole fraction of methyl acetate is 0.85 to 1.0, the concentration of calcium chloride is zero, and consequently, the preferential solvation number is also zero. For the butyl acetate–butanol–calcium chloride sys-

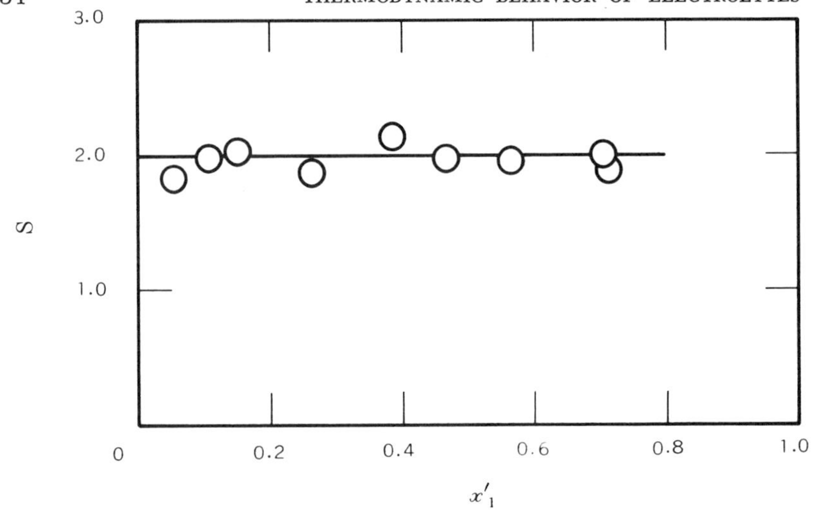

Figure 6. Preferential solvation number in the butyl acetate–butanol–calcium chloride system at 1 atm

tem, the preferential solvation number is about 2 and it is constant at the range where the mole fraction of butyl acetate is between 0 and 0.8 as shown in Figure 6. At the range where the mole fraction of butyl acetate is 0.8 to 1.0, the concentration of calcium chloride is zero, and the preferential solvation number is also zero.

Prediction of Salt Effect

A method to predict salt effect on vapor–liquid equilibrium in which salt is dissolved in a saturated state is introduced. In this method, salt effect is predicted by using preferential solvation numbers, the concentration of the salt, and the vapor–liquid equilibrium data for which salt is not involved. It is possible to predict salt effect completely without using actually measured data if the preferential solvation number can be predicted. Presently, however, it is impossible to completely predict preferential solvation number. Hence, the preferential solvation numbers are obtained through actual measurements, and these numbers are used for the prediction. If preferential solvation number can be predicted independently in the future, this method will be an extremely hopeful one. The salt effect prediction method is entirely in reverse sequence of that used to obtain preferential solvation number. Specifically, it is carried out in the following sequence.

(1) The components which form preferential solvate in a mixed-solvent system are determined.

(2) The preferential solvation number across salt and solvent is determined.

(3) Solubility of salt into the solvent mixture is determined.

(4) The essential composition (x_{1a}') is calculated by applying Equation 2 or 5.

(5) Vapor composition for the x_{1a}' is read out by using the vapor–liquid equilibrium data for which salt is not involved, and this vapor composition is used as the vapor composition for the actual concentration (x_1').

Figures 7, 8, and 9 indicate the prediction results for the following three systems: methanol–ethyl acetate, methyl acetate–methanol, and butyl acetate–butanol with saturated calcium chloride, respectively. The absolute value of mean errors $|\Delta y|$ were 0.018 and 0.014 for each system, while the maximum and minimum errors were 0.047 and 0, 0.039 and 0.005, and 0.039 and 0.005, rcspcctivcly.

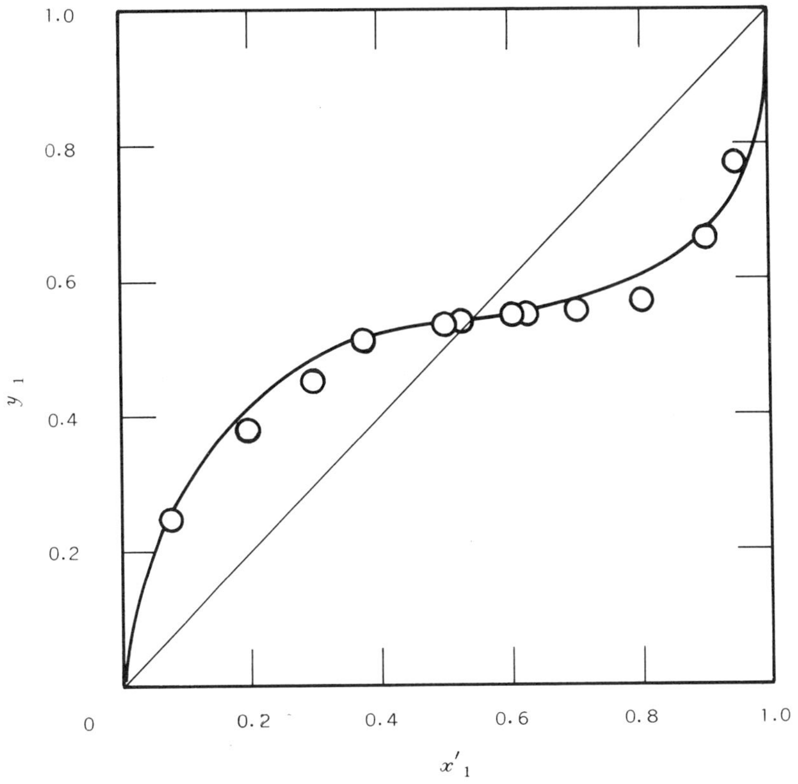

Figure 7. Result of prediction for methanol–ethyl acetate–calcium chloride system at 1 atm: (○), observed; (—), calculated.

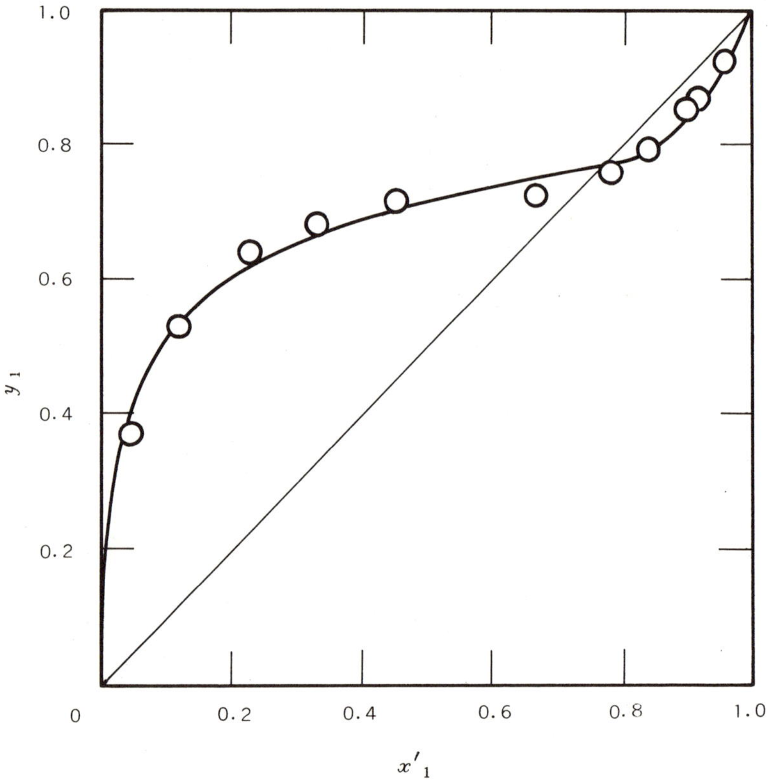

Figure 8. Result of prediction for methyl acetate–methanol–calcium chloride system at 1 atm: (○), observed; (—), calculated.

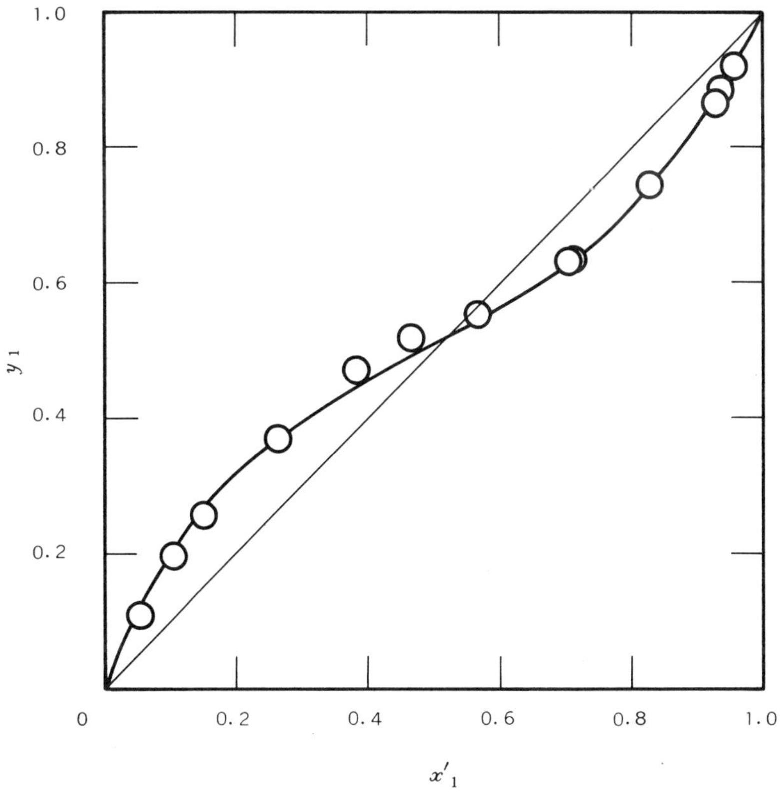

Figure 9. Result of prediction for butyl acetate–butanol–calcium chloride system at 1 atm: (○), observed; (—), calculated.

Conclusion

As a cause of salt effect, the existence of a preferential solvate formed across salt and particular component in a solvent mixture is considered. Preferential solvation number was calculated for the alcohol–ester–calcium chloride system in which formation of preferential solvate was believed to exist. The preferential solvation number was found to be constant regardless of compositions of a solvent mixture. On the other hand, the solubility of salt into mixed solvent increased linearly against an increase of composition of the component which dissolved salt. This fact indicates that the entire dissolved salt contributes to the formation of solvation with the particular component in the mixed solvent. The author feels that the study should be continued in order to make the salt effect clear so that it will be possible to apply the above described idea to other systems also.

Glossary of Symbols

S = preferential solvation number (—)
x = liquid-phase composition (mole fraction)
y = vapor-phase composition (mole fraction)

Superscript

$'$ = salt free

Subscripts

1 = first component
2 = second component
3 = third component
a = free solvent molecule not solvated

Literature Cited

1. Ohe, S., Yokoyama, K., Nakamura, S., *J. Chem. Eng. Data* (1971) **16**(1), 70.
2. Ohe, S., *Adv. Chem. Ser.* (1976) **155**, 53.
3. Hashitani, M., Mirata, M., *J. Chem. Eng. Jpn.* (1969) **1**(2), 116.
4. Nagata, I., *J. Chem. Eng. Data* (1969) **14**(4), 418.
5. Brunjes, A. S., Furnas, C. C., *Ind. Eng. Chem.* (1935) **27**, 396.

RECEIVED March 1, 1978.

4

Error Analysis of Isobaric Liquid–Vapor Equilibrium Data for Mixed Solvents Containing Salts at Saturation

DEREK JAQUES

Department of Applied Chemistry, Royal Melbourne Institute of Technology, Melbourne, Victoria, 3000, Australia

In the calculation of total pressure and vapor composition from boiling point data using the indirect method, the greatest source of error lies in the liquid-phase composition. We have attempted to characterize the frequency distribution of the error in the calculated vapor composition by the standard statistical methods and this has given a satisfactory result for the methanol–water system saturated with sodium chloride when the following estimates of the standard deviation were used: x, *0.003;* y, *0.006;* T, *0.1°C; and* π, *2 mm Hg. This work indicates that in the design of future experiments more data points are required and, for each variable, a reliable estimate of the standard deviation is highly desirable.*

Recently there has been considerable interest in the chemical literature on the subject of thermodynamic consistency, evaluation of data, and in error analysis of salt-free data. These authors (*1, 2, 3, 4*), for reasons of simplicity, chose isothermal data where the isothermal–isobaric form of the Gibbs–Duhem equation can be used with only a very small error. We are interested in isobaric data containing salts at saturation because most salt data are measured under these conditions. Also the model we wish to use is based upon Barker's method (*5*) which predicts vapor composition from boiling point data. This approach has been discussed previously in some detail by the present author (*6*) and for alcohol/water systems it was preferable to the correlation of excess free energy which incorporates the redundant *y*-values and their associated errors. Further-

0-8412-0428-4/79/33-177-039$05.00/1

more, a suggestion was made (*7*) that as Othmer liquid–vapor equilibrium stills have been used extensively there is uncertainty in the determination of temperature. Hence we ask what is the relative importance of the errors in each variable. If the importance of the T-error was much greater than the y-error the Barker method might not be the best approach.

There are three sources of error in the calculated vapor composition when these are calculated from boiling point data: random error in each experimental observation; systematic error in one or more of the observations; and the model is imperfect (this is particularly true for isobaric data because use is made of the Gibbs–Duhem equation which was derived for constant temperature and pressure). In the present work we shall assume that the only error in the data is caused by randomness.

The purpose of this work is to attempt to analyze the random errors in each independent variable and assess which one contributes the greatest error in the calculated quantities when use is made of the indirect method.

Correlation Procedure

This procedure has been given in detail elsewhere (*6*) so it will only be described here briefly for the sake of completeness. The function $\Sigma(\pi - \pi_c)^2$ is minimized where the total pressure is given by the equation:

$$\pi_c = x p_1' \gamma_1 \Phi_1 + (1 - x)\, p_2' \gamma_2 \Phi_2 \tag{1}$$

The vapor pressure of each solvent is replaced by the vapor pressure of the solvent saturated with salt at the observed temperature. The Wilson Equation (*8*) is used to relate the activity coefficient on a salt-free basis:

$$\ln \gamma_1 = -\ln(1 - A_{21}(1 - x)) + (1 - x)\left\{\frac{(1 - x)A_{12}}{1 - A_{12}x} - \frac{xA_{21}}{1 - A_{21}(1 - x)}\right\} \tag{2}$$

$$\ln \gamma_2 = -\ln(1 - A_{12}x) - x\left\{\frac{(1 - x)A_{12}}{1 - A_{12}x} - \frac{xA_{21}}{1 - A_{21}(1 - x)}\right\} \tag{3}$$

where

$$A_{21} = 1 - \frac{V_2}{V_1} \exp\,(-Z_1/RT) \tag{4}$$

and

$$A_{12} = 1 - \frac{V_1}{V_2} \exp(-Z_2/RT) \tag{5}$$

One of two procedures can be used now. The equilibrium vapor composition is evaluated using the observed temperatures and Equation 6:

$$\ln y = \ln(xp_1'\Phi_1/\pi_c) - \ln(1 - A_{21}(1-x)) + (1-x)\left\{\frac{(1-x)A_{12}}{1 - A_{12}x} - \frac{xA_{21}}{1 - A_{21}(1-x)}\right\} \tag{6}$$

or alternatively the equilibrium temperature is calculated and substituted in Equation 7:

$$\ln y = \ln(xp_1'\Phi_1/\pi) - \ln(1 - A_{21}(1-x)) + (1-x)\left\{\frac{(1-x)A_{12}}{1 - A_{12}x} - \frac{xA_{21}}{1 - A_{21}(1-x)}\right\} \tag{7}$$

In the present work the second procedure has been used.

Error Analysis

We begin by defining the error in total pressure (E_π) for each measurement as:

$$E_\pi = \pi - \pi_c = \pi - x\gamma_1 p_1'\Phi_1 - (1-x)\gamma_2 p_2'\Phi_2 \tag{8}$$

The independent variables are x, T, and π. In Equation 8 the activity coefficients are functions of x and T, the vapor pressures are functions of T, and the fugacity coefficients and molar volumes are assumed free of random error. Hence for the variance of the E_π error we have:

$$S_E^2 = \left(\frac{\partial E}{\partial x}\right)^2_{T,\pi} S_x^2 + \left(\frac{\partial E}{\partial T}\right)^2_{x,\pi} S_T^2 + \left(\frac{\partial E}{\partial \pi}\right)^2_{x,T} S_\pi^2 \tag{9}$$

Three differential terms are readily calculated at each datum but the corresponding set of x, T, and π standard deviations are strictly unknown. However we can make a reasonable estimate of these values and also assume that each is independent of composition.

The vapor compositions are calculated from the equation:

$$y = \frac{x\gamma_1 p_1'\Phi_1}{\pi} \tag{10}$$

using the calculated temperatures. Hence we have the following variables: x, T, π, Z_1, and Z_2, but they are not all independent. So we take x, π, Z_1, and Z_2 as independent variables. The problems associated with trying to assess the error in Z_1 and Z_2 will be discussed fully later. Here it is sufficient to note that the actual values of Z_1 and Z_2 depend upon the random errors associated with x, T, and π and hence the problem is complex and not amenable to a full statistical treatment. The variance of the y-error is given by:

$$S_y^2 = \left(\frac{\partial y}{\partial x}\right)^2_{\pi,Z_1,Z_2} S_x^2 + \left(\frac{\partial y}{\partial \pi}\right)^2_{x,Z_1,Z_2} S_\pi^2 + \left(\frac{\partial y}{dZ_1}\right)^2_{x,\pi,Z_1} S_{Z_1}^2 + \left(\frac{\partial y}{\partial Z_2}\right)^2_{x,\pi,Z_2} S_{Z_2}^2 \quad (11)$$

To check that the method can be used for isobaric data a set of perfect data are generated and random errors added to x, y, T, and π in turn and all together to see what effect they have on our standard procedure. For large samples we expect 68% of the sample values to lie within one standard deviation of the perfect value of the selected variable. In the case of small samples, e.g., twelve data, error bounds are calculated using binomial probabilities for each of the above variables so that, with probability of 0.95, we expect 41–95% of the sample observations to lie within one standard deviation of the perfect value of the selected variable (the normal distribution is assumed). Twelve is a common number of data points with salt-saturated solutions and this shows the desirability of taking more experimental observations.

Application of Error Analysis

In a previous evaluation of salt-saturated data, it was found (7) that the methanol–water system saturated with sodium chloride showed little or no average bias for the calculated vapor composition for both the $T - x$ fit and the $G^E/RT - x$ fit, it passed the area test quite easily and showed satisfactory values of all sample derivations. Hence this system was chosen for error analysis.

Stage 1. The $MeOH/H_2O/NaCl$ data are subjected to the correlation procedure described previously which gives values of the Wilson energy constants (Z_1 and Z_2) and a new set of data for temperature and vapor composition that are internally consistent (*see* Table I). The small values of the standard deviation and the bias indicate good quality data in the salt effect field. For the analysis of serial correlation among the residuals we use the Durbin–Watson test (9). A run of positive or negative signs in the series of residuals is some indication that the model

Table I. Experimental and Calculated Data for the $MeOH/H_2O/NaCl$ System at π = 762 mm Hg Pressure[a]

x_{obs}[a]	y_{obs}	y_{calc}[a]	T_{obs}	T_{calc}[a]
0.029	0.259	0.301	99.6	99.05
0.050	0.418	0.415	95.0	94.55
0.074	0.515	0.499	90.5	90.84
0.110	0.590	0.578	86.5	86.95
0.174	0.661	0.658	82.3	82.58
0.250	0.721	0.714	79.0	79.29
0.348	0.766	0.763	76.2	76.30
0.448	0.804	0.803	74.0	73.88
0.557	0.841	0.842	72.2	71.61
0.653	0.875	0.875	70.2	69.77
0.768	0.913	0.915	68.0	67.70
0.878	0.953	0.954	66.1	65.82

	RMS Deviation	*Bias*	*Durbin–Watson Test (D)*
π-values	10.4	−3.0	0.57
y-values	0.014	0.0	1.04
T-values	0.40	0.11	0.60

[a] New data set.

used is inadequate. In the present case for isobaric data we would not expect the model to be perfect because of the use of the Gibbs–Duhem equation which is not strictly applicable to isobaric data and so a value close to two is not to be expected. However, we shall use the test to give a relative measure of the adequacy of the model.

Stage 2. We produce 99 equally spaced x-values between 0 and 1 excluding the two extreme values. By using the values of Z_1 and Z_2 found in Stage 1 and the experimental total pressure value for the $MeOH/H_2O/NaCl$ system we calculate the corresponding values of y and T. Next we introduce normally distributed random errors of zero mean for each variable by specifying the standard deviation of x, y, T, and π, respectively, and add them in turn and then all together to the generated data. The following values of the standard deviation were selected: x, 0.003; y, 0.006; T, 0.1°C; and π, 2 mm Hg. The latter was high because Johnson and Furter (*10*) did not connect a monostat to their equilibrium still. The average variation of atmospheric pressure quoted for their data set is ±2 mm Hg. The choice of the other values was determined by the requirement that 68% of the differences between the generated data plus random error and the generated data must lie within the specified confidence levels based on Equation 11. Figure 1 shows S_E

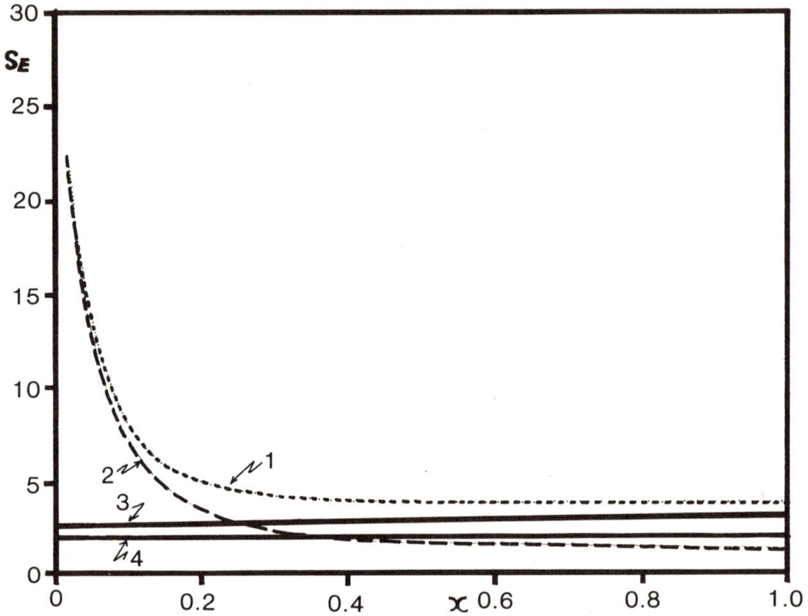

Figure 1. Effect of errors in x, T, *and* π *on the calculated standard deviation of* E *in Equation 9: (1) 3 simultaneous errors; (2)* x*-error; (3)* T*-error; (4)* π*-error*

plotted against x (Equation 9) where the effect of error in each variable separately and then all together is shown. Note that the effect of x is always appreciable and is the dominant variable at low x-values.

Figure 2 shows S_y plotted against x (Equation 11) for all four variables (i.e., x, π, Z_1, and Z_2) separately and then together. Again x is the dominant variable at low x and has an appreciable effect over the remaining concentration range.

It is perhaps worth mentioning that in Equation 10 y is a function of x, π, T, Z_1, and Z_2 but they are not independent variables, because if x, π, Z_1, and Z_2 are known T can be calculated. Hence the error in the measured T is included in the errors associated with the energy parameters. The fact that they also contain x and π errors complicates the statistical treatment. The level of error associated with the Wilson energy parameters is difficult to quantify. The problem arises because the values of the parameters are governed by the errors in x, T, and π through the use of Equation 8. We examined the sum of squares (Equation 8) for a range of values of the two parameters to see if they are robust, i.e., to see if slight changes in value caused large changes in the sum of squares and found this not to be so. We assumed an error level of 2% for each energy parameter as being reasonable.

Table II gives the standard deviations of pressure, vapor composition and temperature, and the corresponding bias and D-value as each variable is changed randomly and then as all four are changed simultaneously. We see that the random error of x contributes ca. 75% of the induced error in the value of the standard deviation of both the pressure and temperature while the random error of T and π only contribute about 12% each. On the other hand the random errors of x and y contribute equally to the induced-vapor composition standard deviation with the pressure making a negligible contribution. The bias values are negligibly small except for the pressure standard deviations where they are still not large. The final column has D-values at least equal to two and this gives one confidence in the model and suggests it is adequate for good quality data as in this particular case the only source of error is caused by random behavior.

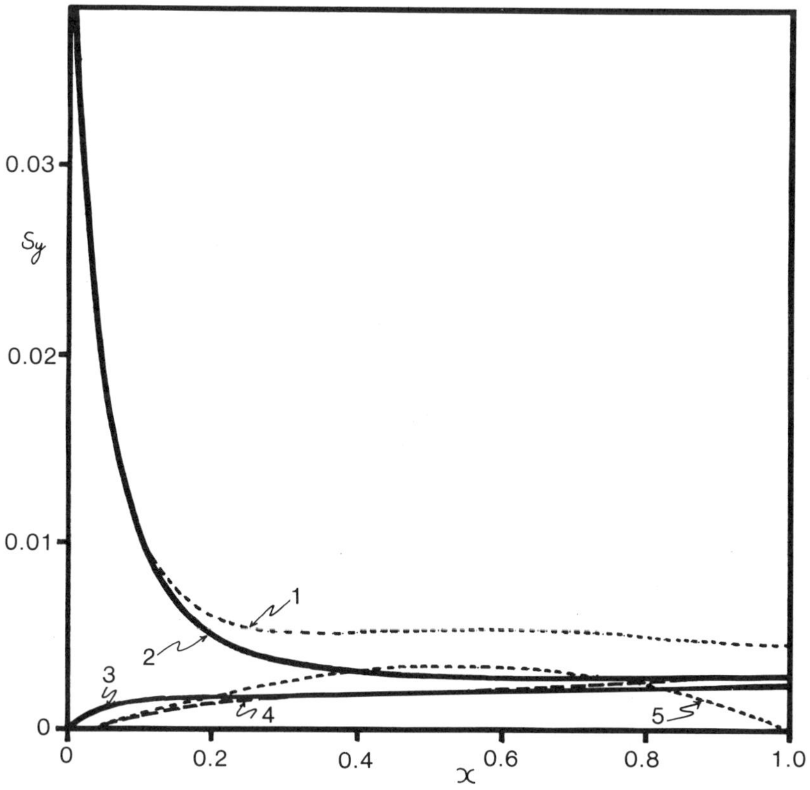

Figure 2. Effect of errors in x, T, Z_1, *and* Z_2 *on the calculated standard deviation of the* y*-error in Equation 11: (1) 4 simultaneous errors; (2)* x*-error; (3)* π*-error; (4)* Z_1*-error; (5)* Z_2*-error*

Table II. Effect of Random Error on 99

	Standard Deviation				
	x	y	T	π	*All 4*
π-Value	5.7	—	2.6	2.4	6.8
y-Value	0.006	0.006	—	0	0.008
T-Value	0.23	—	0.09	0.08	0.27

Stage 3. The twelve calculated data from Table I are processed by adding normally distributed random errors of zero mean to each variable in turn and then all together. The results are shown on Figures 3 and 4 for $\Delta\pi$ and Δ_y, respectively. The confidence regions also are shown and we observe that 92% and 58% of the calculated differences, respectively, fall within these limits. For 12 data points there is a 95% probability that between 41% and 95% of the calculated values should lie within the confidence limits. This wide range for a small number of data points again

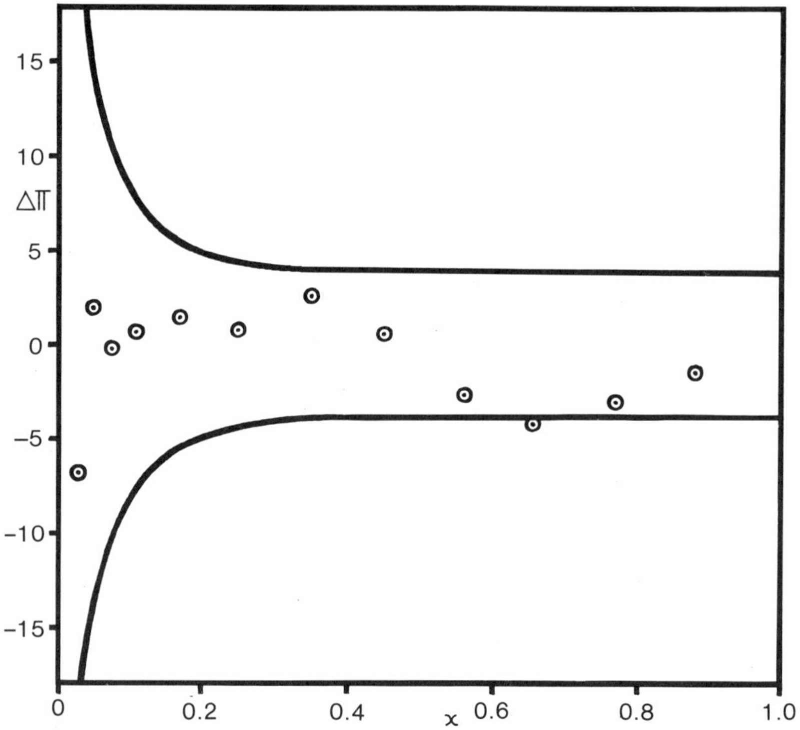

Figure 3. Pressure difference and the 68% confidence region for the 12 calculated data of Table I

Simulated Data for MeOH/H_2O/NaCl

Bias					Durbin Watson Test (D)				
x	*y*	*T*	*π*	*All 4*	*x*	*y*	*T*	*π*	*All 4*
−0.3	—	0.16	0.06	0.11	1.4	—	1.6	2.1	2.0
0	0	—	0	0	1.4	1.7	—	1.9	2.1
0.01	—	−0.01	−0.01	−0.01	1.5	—	1.8	2.3	2.0

highlights the desirability of taking a large number of observations. Table III shows the standard deviation, bias, and *D*-value for the simultaneous addition of random errors to all variables.

Finally the original data are shown on Figures 5 and 6 together with the confidence regions. Now we see that 42% of the pressure differences lie within the confidence levels while 66% of the vapor composition differences are within the levels. Included in Table III are the standard deviations, bias, and results of the Durbin–Watson test. Comparison of the two sets of results indicates appreciably larger values for standard deviation and bias for the experimental results whereas for the *D*-test the

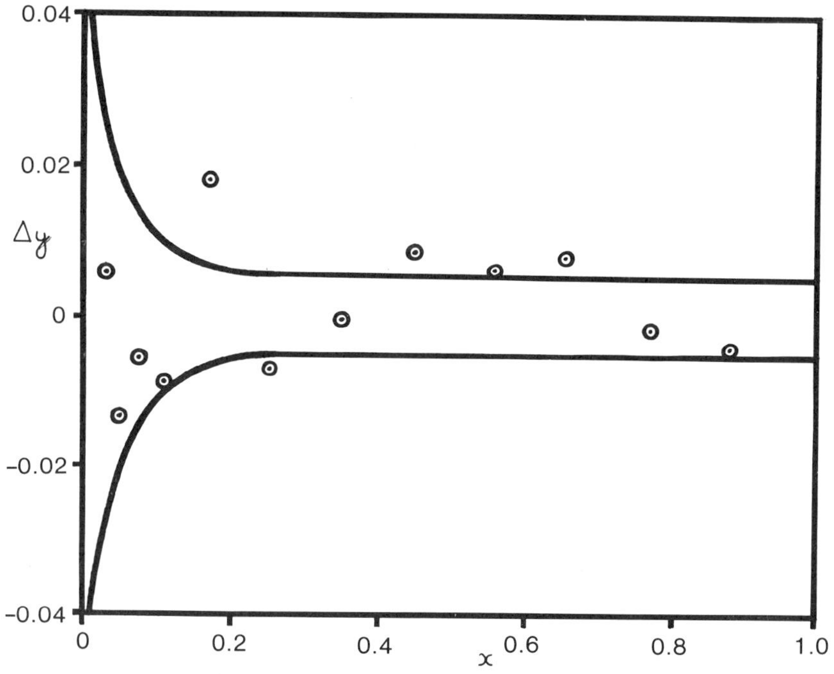

Figure 4. Vapor composition difference and the 68% confidence region for the 12 calculated data of Table I

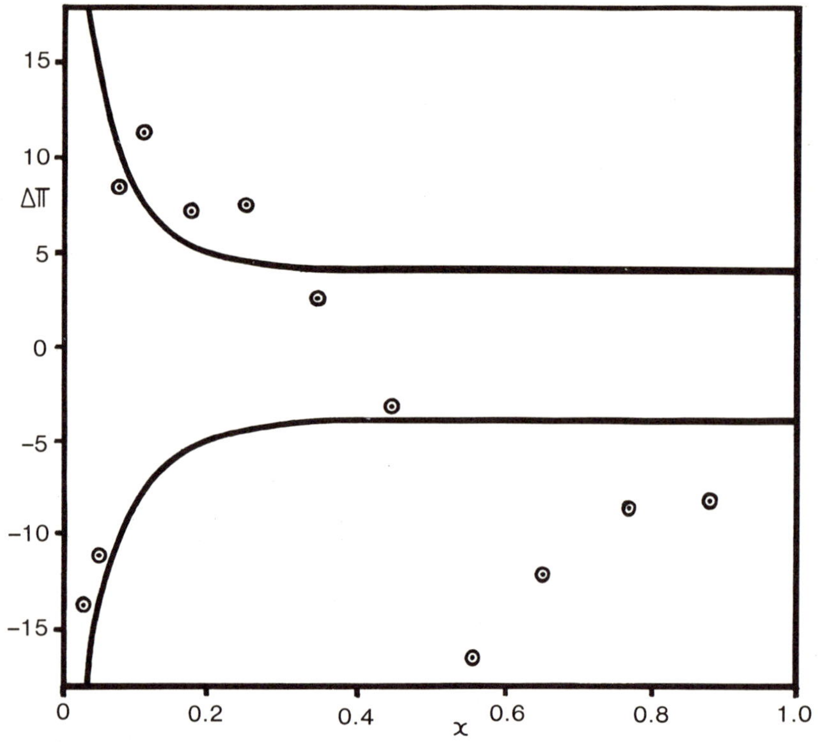

Figure 5. Pressure difference and the 68% confidence region for the methanol–water–sodium chloride system

reverse is true. Part of the explanation for this lies in the particular set of calculated data plus random errors used in Table III. One hundred sets of calculated data containing simulated random errors were processed and the average value of the pressure standard deviation and its standard deviation calculated. This was 5.0 ± 3.4. Hence the particular set used in Table III was on the low side of the mean. The other part of the explanation for the apparent discrepancy is given in the next section.

Table III. Comparison of 12 Experimental Data and 12 Simulated Data Containing Random Error

	Standard Deviation		*Bias*		*D-value*	
	Exptl.	*Calc.*	*Exptl.*	*Calc.*	*Exptl.*	*Calc.*
π-Value	10.4	3.1	−3.0	−0.8	0.57	1.14
y-Value	0.014	0.009	0	0	1.04	2.28
T-Value	0.40	0.11	0.11	0.3	0.60	1.20

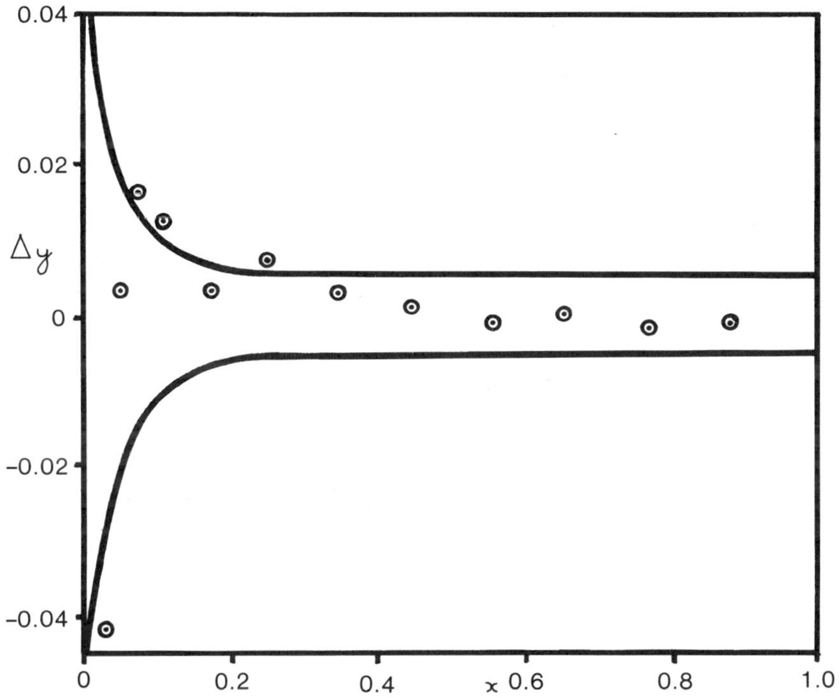

Figure 6. Vapor composition difference and the 68% confidence region for the methanol–water–sodium chloride system

Conclusions

In the analysis of the effect on the calculated quantity of random errors in measured quantities it is unfortunate that the only model susceptible to an exact statistical treatment is the linear one (*11*). Here we have attempted to characterize the frequency distribution of the error in the calculated vapor composition by the standard methods and have not included a co-variance term for each pair of dependent variables (*12*). Our approach has given a satisfactory result for the methanol–water–sodium chloride system but it has not been tested on other systems and perhaps of more importance, it has not been possible, so far, to confirm the essential correctness of the method by an independent procedure. Work is currently being undertaken on this project.

Several conclusions can be drawn from this work. First, in the calculation of total pressure and vapor composition from boiling point data the greatest source of error lies in the liquid-phase composition, particularly at low concentration. Second, the estimates of the standard deviation for vapor composition and temperature of 0.006 and 0.1°C, respectively, are quite low and suggest that the main effort is required in reducing the error in the first conclusion. Certainly with the present set of data the measurement of T-π-x data only would have given very satisfactory y-values. Third, unless a more sophisticated approach is to be used for collecting isobaric data, data determined at very low x-values are going to be subject to a very large random error and hence it would be more profitable to obtain extra data at higher x-values. Finally, in the design of future experiments we need more data points and, for each variable, a reliable estimate of the standard deviation should be determined.

Nomenclature

Subscripts:

1 = alcohol
2 = water
c = calculated

A_{21}, A_{12} = constants in Wilson equation
E_π = pressure difference in Equation 8
p_i' = vapor pressure of Component i saturated with salt
s_x, s_y, s_π, s_T = estimate of standard deviation of the error in the experimental variables x, y, π, and T, respectively
s_{Z_1}, s_{Z_2} = estimate of the standard deviation of the error in the calculated energy parameters
S_E = calculated standard deviation of E in Equation 9
S_y = calculated standard deviation of the error in y in Equation 11
T = temperature, °C
V_i = molar volume of Component i
x = mole fraction of alcohol in liquid phase, calculated on a salt-free basis
y = mole fraction of alcohol in vapor phase
Z_1, Z_2 = energy parameters in Wilson equation commonly expressed as $(\lambda_{ij} - \lambda_{ii})$
γ_i = activity coefficient of Component i
$\Delta\pi$ = $\pi - \pi_c$
Δy = $y - y_c$
π = total pressure
Φ_i = correction term for nonideality of Component i in an indeal gaseous solution and is the reciprocal of the fugacity coefficient.

Acknowledgments

The author wishes to thank I. R. I. Cox of the Department of Mathematics at R.M.I.T. for many helpful discussions and the Computer Centre at R.M.I.T. for the provision of facilities.

Literature Cited

1. Ulrichson, D. L., Stevenson, F. D., *Ind. Eng. Chem. Fundam.* (1972) **11,** 287.
2. Van Ness, H. C., Byer, S. M., Gibbs, R. E., *AIChE. J.* (1973) **19,** 238.
3. Fabries, J. F., Renon, H., *AIChE J.* (1975) **21,** 735.
4. Peneloux, A., Deyrieux, R., Canals, E., Neau, E., *J. Chem. Phys.* (1976) **73,** 706.
5. Barker, J. A., *Aust. J. Chem.* (1953) **6,** 207.
6. Jaques, D., *Ind. Eng. Chem. Process Des. Dev.* (1976) **15,** 236.
7. Jaques, D., *Ind. Eng. Chem. Process Des. Dev.* (1977) **16,** 129.
8. Wilson, G. M., *J. Am. Chem. Soc.* (1964) **86,** 127.
9. Kendall, M. G., "Time Series," p. 163, Griffin, London, 1973.
10. Johnson, A. I., Furter, W. F., *Can. J. Chem. Eng.* (1960) **38,** 78.
11. Draper, N. R., Smith, H., "Applied Regression Analysis," Wiley, New York, 1968.
12. Kempthorne, O., Folks, L., Probability, Statistics, and Data Analysis," p. 129, Iowa State University, Ames, 1971.

RECEIVED February 6, 1978.

5

The Concept of Basicity in Mixtures of Water with Organic Solvents

C. F. WELLS

Department of Chemistry, University of Birmingham, Edgbaston, P.O. Box 363, Birmingham B15 2TT, England

The concept of acidity and basicity in mixed solvents is discussed and a method for analyzing differential solvation effects is described. This enables the free energy of transfer of the proton between water and the mixed solvent, $\Delta G_t^\circ(H^+)$, *to be calculated, and thereby* $\Delta G_t^\circ(i)$ *for* $i = X^-$ *and* M^+, *using values for* $\Delta G_t^\circ(HX)$ *and* $\Delta G_t^\circ(MX)$. *The* pK_a *values for acids are combined with* $\Delta G_t^\circ(H^+)$ *to calculate proton affinities in mixed solvents, and these are used as measures of free energies of transfer of the charges on the molecular species,* $\Delta G_t^\circ(i)_e$. *Values of* $\Delta G_t^\circ(i)$ *and* $\Delta G_t^\circ(i)_e$ *are compared for a range of co-solvents and the factors influencing the way these quantities vary with solvent composition are discussed.*

The early observations of Goldschmidt and his co-workers and of others (*1–12*) using kinetic, conductiometric, and electrometric techniques suggested that water, when present in small concentrations in an organic hydroxylic solvent, appears to be more basic than the organic solvent. However, Hammett's Acidity Function (*13*), H_0, showed that the acidity of a solvent to a dissolved solute varied quite considerably with composition. Although the latter concept has proved to be very useful in interpreting the kinetics of acid-dependent reactions of solutes, it is of little value when one of the reactants is a component of the solvent system (*14, 15*). The simple Brönsted–Lewis notion of an acid–base relationship requires considerable modification in a mixed solvent with varying component ratios. The extrema found in a mixed solvent in the variation of H_0 with composition (*16, 17, 18, 19*) must mean that the

0-8412-0428-4/79/33-177-053$05.75/1

nature and the structure of the solvation of the proton varies over the whole range and that its acidity depends strongly on solvation.

This partition of the proton between the two components of the solvent has been investigated in water-rich conditions spectrophotometrically using *p*-nitroaniline B as the solute.

$$(H_2O)_xH^+_{solv} + ROH_{solv} \overset{K}{\rightleftharpoons} \{(H_2O)_{x-1}ROH\}H^+_{solv} + H_2O_{solv} \quad (1)$$

$$B_{solv} + (H_2O)_xH^+_{solv} \overset{K_1}{\rightleftharpoons} BH^+_{solv} + H_2O_{solv} \quad (2)$$

$$B_{solv} + \{H_2O)_{x-1}ROH\}H^+_{solv} \overset{K_2}{\rightleftharpoons} BH^+_{solv} + ROH_{solv} \quad (3)$$

K is the thermodynamic equilibrium constant for the solvation sorting of the proton as in Equation 1, and K_1 and K_2 are the thermodynamic equilibrium constants for Equations 2 and 3, respectively. For a range of concentrations of mineral acid at a constant added concentration of ROH, *a*, at constant ionic strength and temperature, with c_o equal to the total added concentration of *p*-nitroaniline and c and c_R equal to $[B_{solv}]$ with and without the ROH present, respectively, plots of $c\,c_R/(c_R - c)$ against $c_R/(c_o - c_R)$ are always linear (*20, 21, 22*) for a wide variation in the chemical identity of ROH. The mineral acid used is normally HCl, with the ionic strength maintained at 1.00 mol L^{-1}. If $F_1 = f_B f_P/f_{BH^+} f_{H_2O}$ and $F_2 = f_B f_{ROH_2}/f_{BH^+} f_{ROH}$, where f = activity coefficient, P = $(H_2O)_xH^+_{solv}$, and $ROH_2 = \{(H_2O)_{x-1}ROH\}H^+_{solv}$, Equation 4 can be deduced from Equations 1, 2, and 3,

$$\frac{cc_R}{c_R - c} = \frac{K_2F_2}{K_1F_1} \cdot \frac{w_Rc_o}{a} \cdot \frac{c_R}{c_o - c_R} + \frac{c_ow_R}{K_1F_1a} \quad (4)$$

provided $wF_1 = w_RF_1'$; $w_R = [H_2O]$ in the presence of ROH, $w = [H_2O]$ in the absence of ROH and $F_1' = F_1$ in the absence of ROH. This requirement is equivalent to the assumption that $K_1F_1'w^{-1} = K_1F_1w_R^{-1}$: it is supported by the observed linearity of the plots for $c\,c_R/(c_R - c)$ against $c_R/(c_o - c_R)$, the coincidence of the intercepts of these plots with $c_ow/K_1F_1'a$ using $K_1F_1'w^{-1}$ determined in the absence of ROH, and by the invariance of $K_1F_1'w^{1-}$ for a wide range of [ROH] when ROH is glycerol for which K of Equation 1 is very small (*20, 21, 22, 32, 33, 34, 35, 36*). Applying Equation 4 to the plots, their slopes are given by Equation 5,

$$\text{slope} = \frac{K_2F_2}{K_1F_1} \cdot \frac{w_Rc_o}{a} = \frac{[ROH][P]}{[\overset{+}{R}OH_2]} \cdot \frac{f_{\overset{+}{R}OH_2}\, f_{H_2O}}{f_P\, f_{ROH}} \cdot \frac{c_o}{a} = \frac{F_cc_o}{K_ca} \quad (5)$$

where $F_c = f_{ROH_2} f_{H_2O}/f_P f_{ROH}$ and $K_c = [ROH_2]/[P][ROH]$. Assuming that the symmetrical $F_c = 1.0$, $K_c = c_o/(\text{slope})a$. Now, K_2F_2 can be calculated from the slope and the intercept of these plots, and this enables $[ROH_2]$ to be calculated at individual total concentrations of mineral acid, $[H^+]_T$, at constant [ROH], ionic strength (I), and temperature (T). Thereby, $K_c = [ROH_2]/(a - [ROH_2])([H^+]_T - [ROH_2])$. Values of K_c calculated by this latter method agree well with those calculated from the slope alone, which adds further support for the above assumption (*20, 21, 22, 32, 33, 34, 35, 36*). At low [ROH], K_c depends strongly on the electron-releasing properties of R and on the chemical structure of R (*20, 21, 22*). K_c determined in this way is in good agreement with values determined from the kinetics of acid-catalyzed reactions (*24*), conductivity measurements (*25–31*), ionic transport (*25–31*), and calorimetry (*25–31*). K_c at low [ROH] varies in the range 0.1–1.0 L mol^{-1} for a wide variation in the chemical identity of ROH. This implies that an H_2O is less basic than ROH in water-rich conditions, as indeed the electron-releasing properties of R require: this can be pictured as an ROH replacing an H_2O in the solvation sheath of $H_3{}^+O$ so that the stability of $(H_3{}^+O)_{solv}$ is increased by extending the sharing of the distribution of electrons via hydrogen bonding among the solvent molecules to include an ROH, with the bonding strengthened by the electron-releasing properties of R.

When this treatment is extended to higher concentrations of organic component (i.e. 10–50% wt), similar experimental results are obtained (*20, 21, 22, 23, 32, 33, 34, 35, 36*): linear plots for $c\,c_R/(c_R - c)$ against $c_R/(c_o - c_R)$ and agreement between K_c via the two methods. Although the solvated aqueous proton is commonly assumed to be $(H_2O)_4H_{aq}{}^+$, i.e. $\{H_3{}^+O(H_2O)_3\}_{aq}$, with a trigonal–pyramidal arrangement of H_2O molecules around $H_3{}^+O$, it is more convenient when discussing the transfer of the proton between water and a mixed-aqueous solvent to consider the solvated aqueous proton as approximating to a sphere with an additional H_2O molecule at the apex of the tetrahedron. This structure, $\{H_3{}^+O$-$(H_2O)_4\}_{aq}$, has a tetrahedral arrangement of H_2O molecules in the immediate coordination sphere around the central $H_3{}^+O$: this is consistent with the tetrahedral arrangement of orbitals around the central O atom with each orbital involved in hydrogen bonding. The arrangement of O atoms for these two structures is shown in Figure 1 with O–O indicating a hydrogen bond. The free energy change for Equation 1 will include contributions from the replacement of an H_2O in the coordination sphere of $H_3{}^+O$ in Figure 1b by an ROH molecule (as discussed above) and from changes in the interactions with the solvent molecules outside the coordination sphere. This free energy change, $\Delta G°(ROH_2)$, is given by Equation 6,

$$\Delta G°(\overset{+}{R}OH_2) = -RT \ln K = -RT \ln \{K_c(55{,}345 - a)\} \qquad (6)$$

Upon transferring $\{H_3^+O(H_2O)_4\}$ from water into the mixture, water + ROH, the free energy of transfer, $\Delta G_t°(H^+)$, will consist of a component derived from the transfer of the spherical $\{H_3^+O(H_2O)_4\}$ between the two dielectric surroundings plus the contribution from Equation 6 owing to the changes represented by Equation 1: this is shown in Equation 7,

$$\Delta G_t°(H^+) = \{Ne^2(D_s^{-1} - D_w^{-1})/(6\, r_{H_2O})\}$$
$$-[\overset{+}{R}OH_{2\,solv}]\, RT \ln \{K_c(55{,}345 - a)\} \qquad (7)$$

where N is Avogadro's number, e is the electronic charge, r_{H_2O} is the radius of the water molecule, and D_w and D_s are the dielectric constants for water and the mixed solvent respectively. $[ROH_{2\,solv}] = \{(H_2O_{x-1}$-$ROH\}H^+_{solv}$ in the mixed solvent is calculated from Equation 8 which is derived using Equation 9,

$$[\overset{+}{R}OH_{2\,solv}] = 0.5\,\{A - (A^2 - 4a)^{\frac{1}{2}}\} \qquad (8)$$

$$[\overset{+}{R}OH_{2\,solv}] + [(H_2O)_xH^+_{solv}] = 1 \qquad (9)$$

where A is as defined in Equation 10,

$$A = a + 1 + K_c^{-1} \qquad (10)$$

The free energy of transfer at 25°C for a species MX between water and the mixed solvent can be calculated from Equation 11,

$$\Delta G_t°(MX) = 96.5\,(E_w° - E_s°)\ \mathrm{kJ\ mol^{-1}} \qquad (11)$$

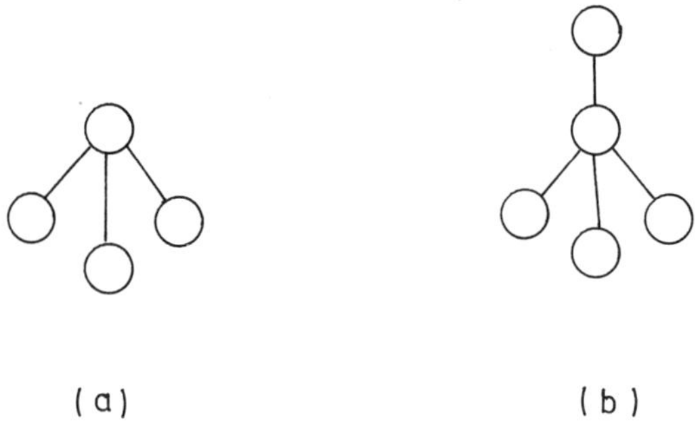

Figure 1. Structures for the hydrated proton in water: (a) trigonal pyramidal; (b) tetrahedral

where $E_w°$ and $E_s°$ are the standard electrode potentials in water and the mixed solvents, respectively, for an appropriate cell (*32, 33, 34, 35, 36*). $\Delta G_t°(HX)$ can be determined from Equation 11 with M = H, and when these values are combined with the values for $\Delta G_t°(H^+)$ calculated as described above, values for $\Delta G_t°(X^-)$ can be obtained; then, from $\Delta G_t°(MX)$, values for $\Delta G_t°(M^+)$ can be calculated. Alternatively, for a sparingly soluble salt MX, with a solubility product K_p, $\Delta G_t°(MX)$ is obtained from Equation 12,

$$\Delta G_t°(MX) = RT \ln (K_p^w/K_p^s) \qquad (12)$$

where superscripts w and s indicate water and the mixed solvent respectively. For $M = H^+$ and $X = OH^-$, Equation 13 can be used to derive values for $\Delta G_t°(HOH)$,

$$\Delta G_t°(HOH) = RT \ln (K_{ip}^w/K_{ip}^s) + RT \ln \{(a_{H_2O}^s)^2/m_w m_s\} \qquad (13)$$

where K_{ip} is the ionic product as designated by the appropriate superscript, $a_{H_2O}^s$ is the activity of water in the mixed solvent on the molality scale, and m_w and m_s are the molalities of water in pure water and in the mixed solvent respectively. $\Delta G_t°(OH^-)$ then can be calculated using $\Delta G_t°(H^+)$ for the appropriate mixture.

Values for $\Delta G_t°(H^+)$, $\Delta G_t°(M^+)$, and $\Delta G_t°(X^-)$ have been published (*32, 33, 34, 35*) for the mixtures water + methanol, water + acetone, water + isopropanol, water + ethylene glycol, water + glycerol, and water + *tert*-butanol. As values are now available for water + ethanol, water + dioxan, and water + dimethyl sulfoxide, it is interesting to compare free energies of transfer for these new mixtures with those already available in the above mixtures. $E°$ values for MX with these new co-solvents have been derived from the published data (*37–61*): values for K_c have been determined experimentally from linear plots of $c\, c_R/(c_R - c)$ against $c_R/(c_o - c_R)$ for the appropriate solvent mixtures using HCl and *p*-nitroaniline at a constant ionic strength of 1.0 mol L^{-1} maintained by the addition of NaCl and at a constant temperature of 25°C. If there are no interactions of the ions with the solvent, then $\Delta G_t°$ obtained using the Born Expression (as used above for the dielectric contribution to $\Delta G_t°(H^+)$) should be always positive for both M^+ and X^- for co-solvents like those listed above where $D_s < D_w$. Figures 2 and 3 show that H^+ and K^+ are more stable in the mixtures than in water, except for K^+ in water + methanol. In general, $\Delta G_t°$ is negative for all cations, including quaternary ammonium cations and the ferrocinium cation, for all solvent mixtures including water + methanol. The curves for $X^- = Cl^-$ in Figure 4 are typical of those for all simple anions, with positive values for $\Delta G_t°(X^-)$. However, the values for $\Delta G_t°(X^-)$ are much greater

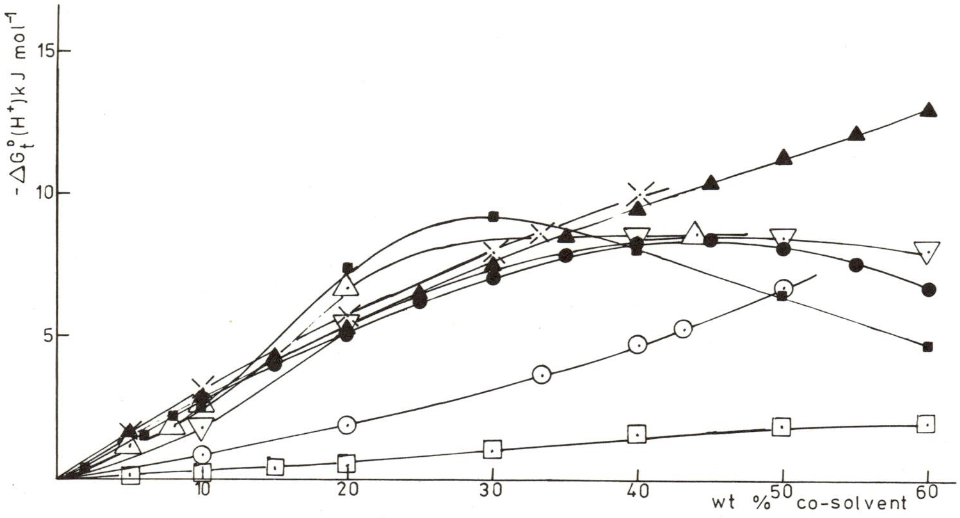

Figure 2. Plots of $\Delta G_t°(H^+)$ for water + co-solvent at 25°C against composition of the mixture with the following solvents: (⊙) methanol; (⊡) ethylene glycol; (△) isopropyl alcohol; (▽) ethanol; (✕) acetone; (●) dioxan; (■) tert-butanol; (▲) dimethyl sulfoxide.

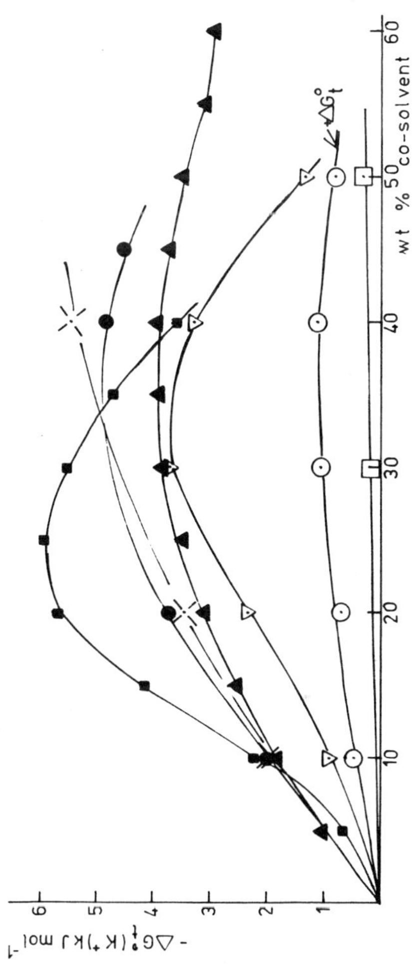

Figure 3. Plots of $\Delta G_t°(K^+)$ for water + co-solvent at 25°C against composition of the mixture with the following co-solvents: (⊙) *methanol;* (⊡) *ethylene glycol;* (▽) *ethanol;* (⁘) *acetone;* (●) *dioxan;* (■) tert-*butanol;* (▲) *dimethyl sulfoxide.*

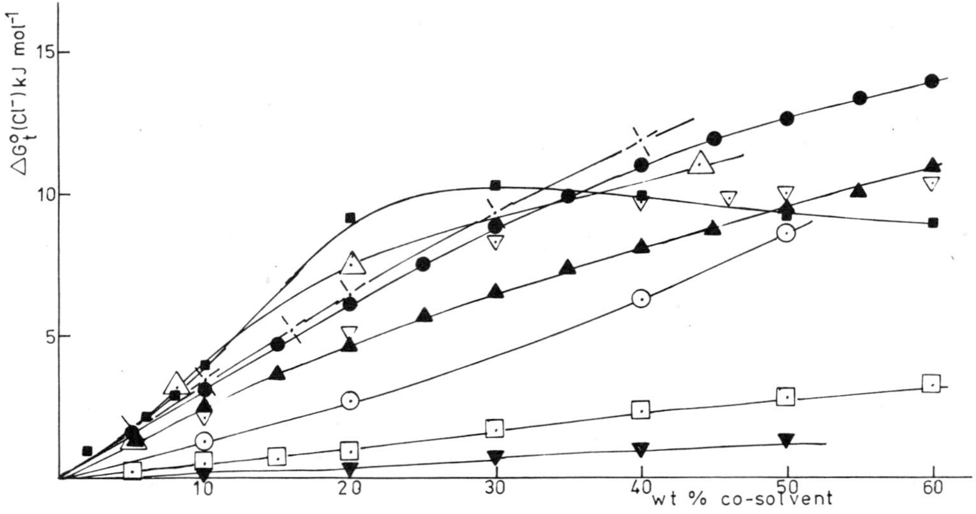

Figure4. Plots of $\Delta G_t°(Cl^-)$ *for water + co-solvent at 25°C against composition of the mixture with the following co-solvents:* (⊙) *methanol;* (⊡) *ethylene glycol;* (△) *isopropyl alcohol;* (▽) *ethanol;* (-·-) *acetone;* (●) *dioxan;* (■) tert-*butanol;* (▲) *dimethyl sulfoxide;* (▼) *glycerol.*

than those expected from the simple Born Expression. The order of $\Delta G_t°(X^-)$ among the halide ions is always $Cl^- > Br^- > I^-$; $\Delta G_t°(F^-)$ is available in water + methanol, and there the order is $F^- > Cl^- > Br^- > I^-$ (*32, 66*). When data are available for ClO_4^-, the order in $\Delta G_t°(X^-)$ is $Cl^- > Br^- > I^- > ClO_4^-$ ($F^- > Cl^- > Br^- > I^- > ClO_4^-$ in water + methanol). Thus, although the values of $\Delta G_t°(X^-)$ are much bigger than the values expected using the Born Expression, the order of $\Delta G_t°(X^-)$ is still in the order expected from this expression as indicated by the order of their ionic radii: however, the above order is also the reverse order of the structure-breaking effects of the anions on the solvent (*62, 63, 64, 65*). The position of OH^- in the series of $\Delta G_t°(X^-)$ depends on the identity of the co-solvent. For co-solvents where an H atom can ionize to give a solvent anion, $\Delta G_t°$ is always low, e.g. methanol and ethylene glycol, becoming negative for ethylene glycol (Figure 5); clearly, some stabilization in the mixed solvent arises from the production of the co-solvent anion by proton exchange. Only with co-solvents not possessing such an ionizable proton does OH^- sit in its expected position on the structure breaking/forming order, $OH^- > Cl^- > Br^- > I^- > ClO_4^-$; this is exemplified by the curves for OH^- with the co-solvents dioxan and dimethyl sulfoxide in Figure 5. Alkyl and aryl groups, especially when forming the major bulk of the molecule, are known to confer a structure-forming capacity on the molecule. Thus, $\Delta G_t°(M^+)$ for $M^+ = R_4N^+$ with R = alkyl tend to be more negative than values for M^+ = a unipositive cation like Na^+ or K^+. Therefore, it is not surprising that BPh_4^- behaves so differently from the other anions—having large negative values for $\Delta G_t°$ (Figure 6).

For larger cations and anions that eject or associate with, respectively, a proton, this analysis can be taken a stage farther (*66*). It is possible to calculate $\Delta G_t°$ which derives solely from the transfer of the charge on the species, eliminating the contribution to $\Delta G_t°$ which arises as a result of its size and chemical identity. For an acid–base system as in Equation 14,

$$A_{solv} \overset{K_a}{\rightleftharpoons} B_{solv} + H^+_{solv} \qquad (14)$$

the change in the proton affinity P_a of B on transferring the system from water to water + co-solvent at 25°C is given by Equation 15,

$$\Delta P_a = 5.70\ (pK_a{}^w - pK_a{}^s) + \Delta G_t°(H^+)\ kJ\ mol^{-1} \qquad (15)$$

where superscripts w and s refer to water and the mixture, respectively. When A and B are large, the chemical difference between A and B is minimal; so for a cationic acid A^+ producing a neutral conjugate base B, $\Delta G_t°(B)$ equals $\Delta G_t°$ for A arising solely from its chemical bulk, i.e. it

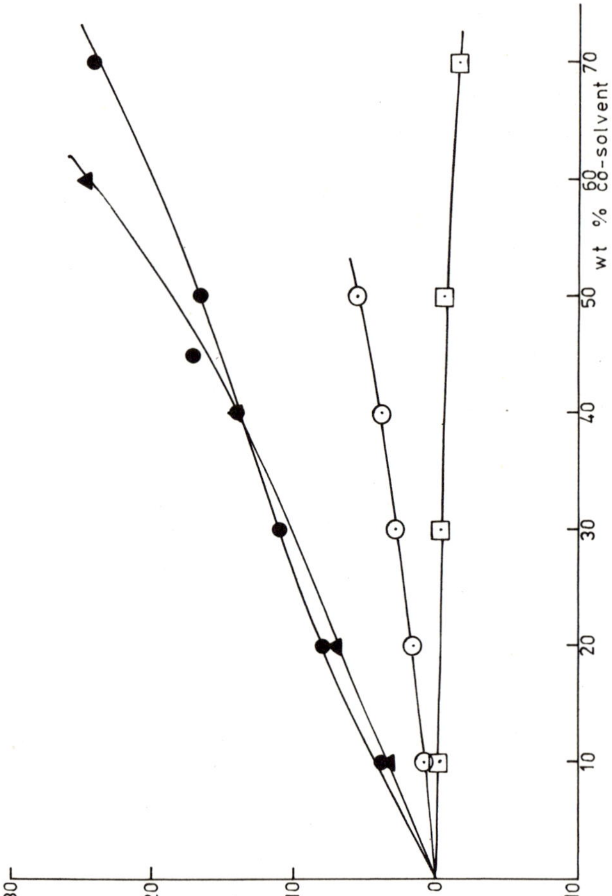

Figure 5. Plots of $\Delta G_t^\circ(OH^-)$ for water + co-solvent at 25°C against composition of the mixture with the following co-solvents: (⊙) methanol; (□) ethylene glycol; (●) dioxan; (▲) dimethyl sulfoxide.

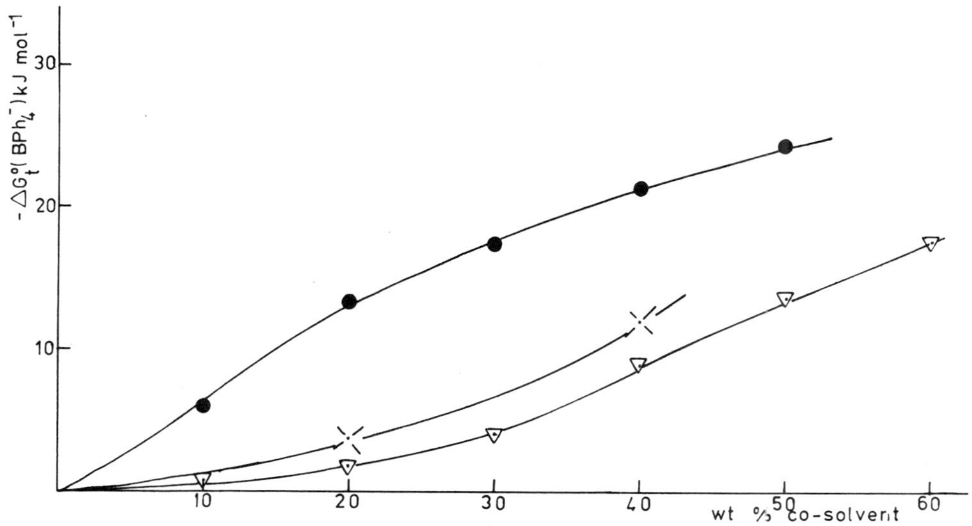

Figure 6. Plots of $\Delta G_t^\circ(BPh_4^-)$ *for water + co-solvent at 25°C against composition of the mixture with the following co-solvents:* (▽) *ethanol;* (✕) *acetone;* (●) *dioxan.*

equals the neutral component of $\Delta G_t°(A^+)$, designated as $\Delta G_t°(A^+)_n$, and therefore ΔP_a equals $\Delta G_t°$ of A^+ arising solely from the effect of its charge, designated as $\Delta G_t°(A^+)_e$. Similarly, for the "bulk" component for an anionic B^- produced from a neutral A, $\Delta G_t°(B^-)_n$ is equal to $\Delta G_t°(A)$, and consequently $\Delta G_t°(B^-)_e$ is equal to $- \Delta P_a$. Positive values are found always for $\Delta G_t°(B^-)_e$, and are similar to those found for $\Delta G_t°(X^-)$ for simple anions. However, as Figure 7 shows for B^- = acetate ion, $\Delta G_t°(B^-)_e$ is not even nearly related linearly to the reciprocal of the dielectric constant, as the Born Expression predicts for the transfer of the charge between two dielectric media. Other examples of $\Delta G_t°(B^-)_e$ are shown in Figures 9, 10, and 11. Figures 12 and 13 show that $\Delta G_t°(A^+)_e$ is negative at low-mole fractions of co-solvent x_2, comparable with the variation of $\Delta G_t°(M^+)$ with x_2 for simple cations. The pK_a data used in the derivation of these free energies were taken from Refs. 67–89.

Looking at Figures 8, 9, 10, and 11, one notices the special place which *tert*-butanol appears to occupy among the co-solvents: there is always an extremum at low x_2 in the variation of $\Delta G_t°$ with x_2 for *tert*-butanol. To a lesser extent, this also occurs with ethanol. This extremum

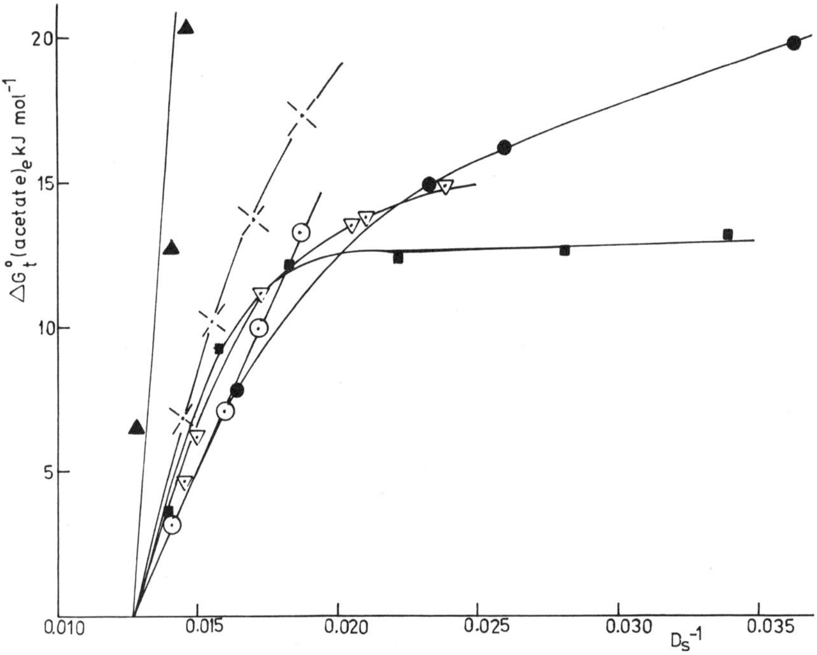

Figure 7. Plots of $\Delta G_t°(CH_3COO^-)_e$ for water + co-solvent at 25°C against reciprocal of the dielectric constant with the following co-solvents: (⊙) methanol; (▽) ethanol; (✕) acetone; (●) dioxan; (■) tert-butanol; (▲) dimethyl sulfoxide

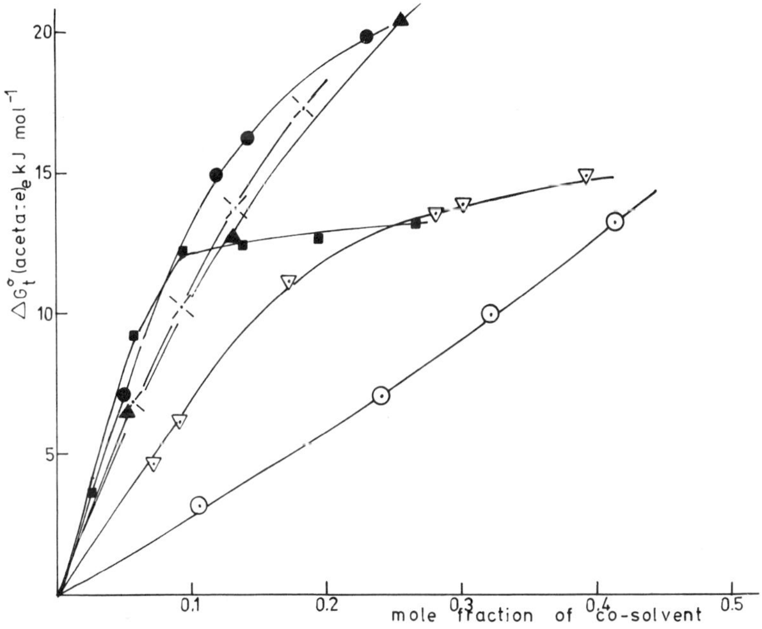

Figure 8. Plots of ΔG_t° $(CH_3COO^-)_e$ *for water + co-solvent at 25°C against mole fraction of co-solvent with the following co-solvents: (⊙) methanol; (▽) ethanol; (⋋⋌) acetone; (●) dioxan; (■)* tert-*butanol; (▲) dimethyl sulfoxide*

with *tert*-butanol is emphasized by the variation of $\Delta G_t^\circ(\text{anion})_e$ with x_2 for several other anions in Figure 14. Using properties of solvent mixtures such as changes in the partial molar volume of the co-solvent, $\Delta vol_2 = vol_2 - vol_2^\circ$ (*90*), changes in the structural contribution of the co-solvent to the variation in maximum density, ΔT^E (*91, 92*), and changes in the ultrasonic absorption (*93*) with x_2. This sharp change in ΔG_t° always occurs in the region where there are considerable changes in structure in solution. This behavior can be contrasted with the absence of an extremum in ΔG_t° at low x_2 for solvents like dioxan and dimethyl sulfoxide where, if extrema occur in the other properties of the mixed solvent, they are found at higher x_2 (*91, 92, 94, 95, 96, 97, 98*). Therefore, in addition to the classification of the types of variation in ΔG_t° according to the structure-breaking or -forming capacities of the ions themselves, there is also a pronounced effect depending on the structural effect of the co-solvent in the mixture.

For a species *i*, $\Delta G_t^\circ(i)$ must be related to changes in the excess free energy of mixing the two solvents, $\Delta G_M{}^E$, produced by *i*, $\Delta\Delta G_M{}^E$. This latter quantity is in turn related to the differential effects on the excess

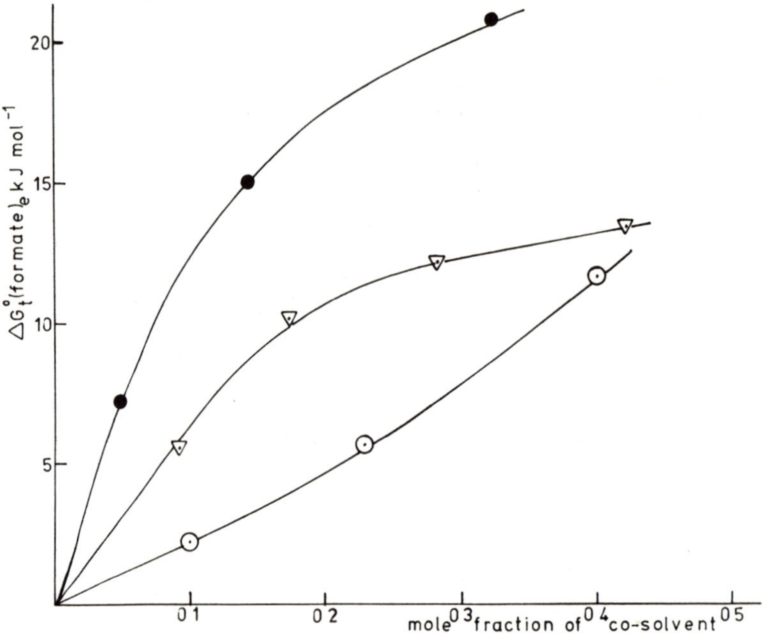

Figure 9. Plots of $\Delta G_t^\circ(HCCO^-)_e$ for water + co-solvent at 25°C against mole fraction of co-solvent with the following co-solvents: (⊙) methanol; (▽) ethanol; (●) dioxan

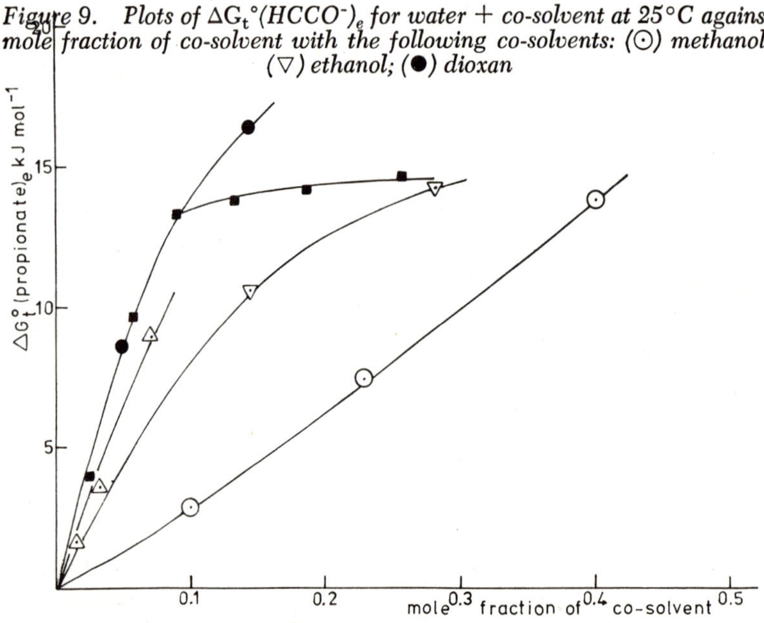

Figure 10. Plots of $\Delta G_t^\circ(CH_3CH_2COO^-)_e$ for water + co-solvent at 25°C against mole fraction of co-solvent with the following co-solvents: (⊙) methanol; (△) isopropyl alcohol; (▽) ethonol; (●) dioxan; (■) tert-butanol

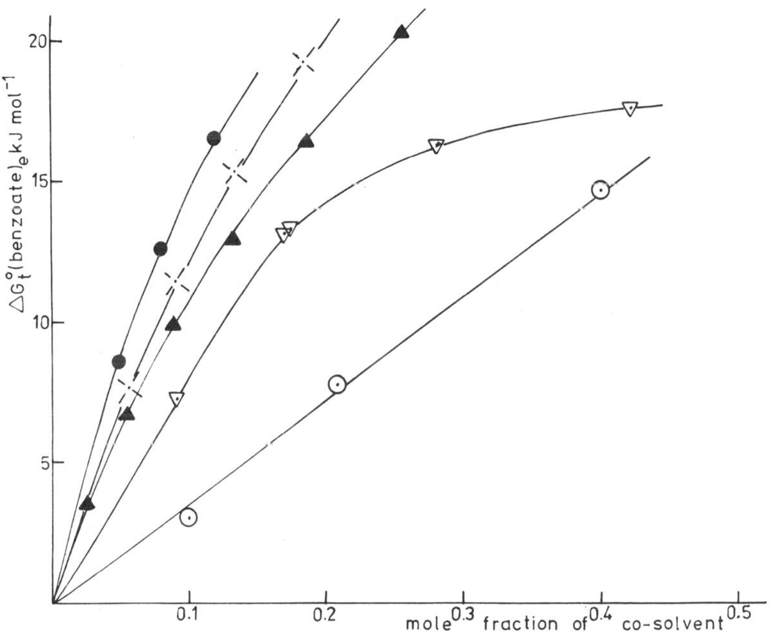

Figure 11. Plots of $\Delta G_t°(C_6H_5COO^-)_e$ for water + co-solvent at 25°C against mole fraction of co-solvent with the following co-solvents: (⊙) methanol; (▽) ethanol; (✕) acetone; (●) dioxan; (▲) dimethyl sulfoxide

enthalpy of mixing, ΔH_M^E, and the excess entropy of mixing, ΔS_M^E, of the solvents, $\Delta\Delta H_M^E$ and $\Delta\Delta S_M^E$, respectively, as expressed in Equation 16,

$$\Delta G_t°(i) = \Delta\Delta H_M^E - T\Delta\Delta S_M^E \qquad (16)$$

The sign of $\Delta G_t°(i)$ will depend on the relative magnitudes of $\Delta\Delta H_M^E$ and $T\Delta\Delta S_M^E$, i.e. on whether the energetic or configurational effect dominates. These are classified in Table I according to the structure-forming or -breaking effect of the ion. The proton is always in Category 1, except for glycerol where $\Delta G_t°(H^+)$ is small and positive (33) and Category 2 applies (not shown in Figure 2). K^+ is in Category 1, except

Table I. (34)

Sign of $\Delta G_t°(i)$	*Structure-forming* i: $\Delta\Delta H_M^E$ *and* $\Delta\Delta S_M^E$ *(both negative)*	*Structure-breaking* i: $\Delta\Delta H_M^E$ *and* $\Delta\Delta S_M^E$ *(both positive)*
Negative	$\lvert\Delta\Delta H_M^E\rvert > \lvert T\Delta\Delta S_M^E\rvert$ (1)	$\lvert T\Delta\Delta S_M^E\rvert > \lvert\Delta\Delta H_M^E\rvert$ (3)
Positive	$\lvert T\Delta\Delta S_M^E\rvert > \lvert\Delta\Delta H_M^E\rvert$ (2)	$\lvert\Delta\Delta H_M^E\rvert > \lvert T\Delta\Delta S_M^E\rvert$ (4)

Journal of the Chemical Society, Faraday Transactions 1

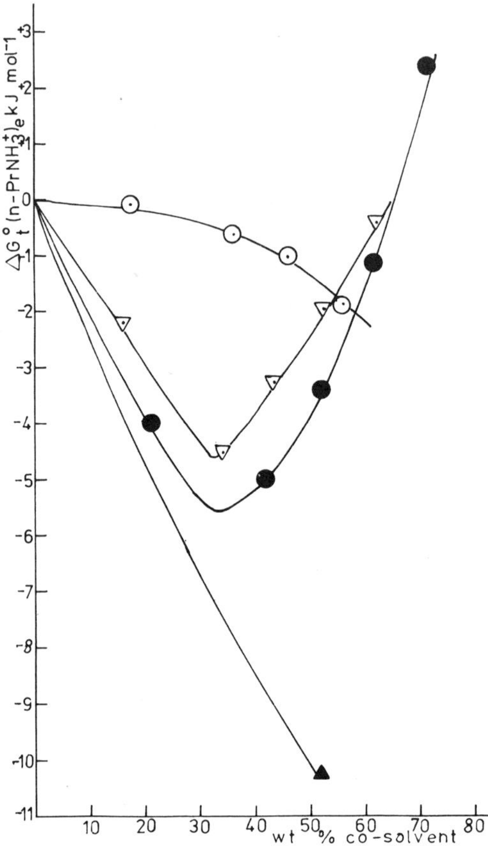

Figure 12. Plots of ΔG_t° (n-*propylammonium*$^+$)$_e$ *for water + co-solvent at 25°C against composition of the mixture with the following co-solvents: (⊙) methanol; (▽) ethanol; (●) dioxan; (▲) dimethyl sulfoxide*

for methanol where Category 2 operates (Figure 3). Cl^- is always in Category 4, as are all the other halide ions and ClO_4^- for all co-solvents. In contrast, where the proton exchange with co-solvent is absent, the structure fromer OH^- is in Category 2 along with F^- and the structure former BPh_4^- is always in Category 1 (Figure 6). The difference between OH^- and F^- on the one hand and BPh_4^- on the other may lie in the two different types of structure-forming effect: OH^- and F^- presumably are involved in hydrogen bonding with solvent molecules, whereas BPh_4^- is involved in hydrophobic bonding (99). The anilinium ion and the *n*-propylammonium ion, as structure formers, are always in Category 1 at low x_2, but they soon change to Category 2 as x_2 rises: here, too, hydrophobic bonding will also occur, but now accompanied by the structure-forming effect of the positive charge. The carboxylate anions are more difficult to classify, but assuming that the normal effect of a negative charge is structure breaking, then they always fall into Category 4. For structure-forming positive ions, generally $-\Delta G_t°(i)$ increases with increasing structure-forming capacity, e.g. increase in the positive charge on i for the simple cations or increase in the size and branching of R in R_4N^+: these effects cause the difference between $T\,\Delta\Delta S_M^E$ and $\Delta\Delta H_M^E$ in Category 1 to increase.

The behavior of the proton in the mixed solvent is very similar to that of the other cations: in general they are stabilized in the mixed

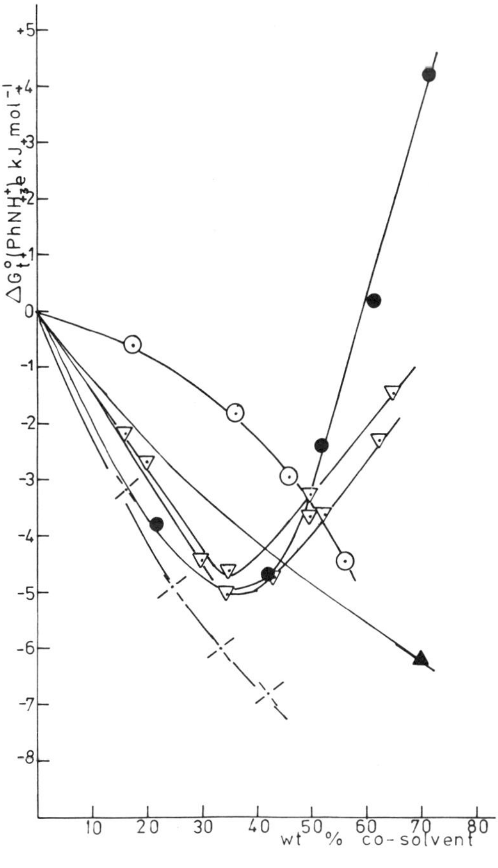

Figure 13. Plots of $\Delta G_t°(anilinium^+)_e$ for water + co-solvent at 25°C against composition of the mixture with the following co-solvents: (⊙) methanol; (▽) ethanol; (⋋) acetone; (●) dioxan; (▲) dimethyl sulfoxide

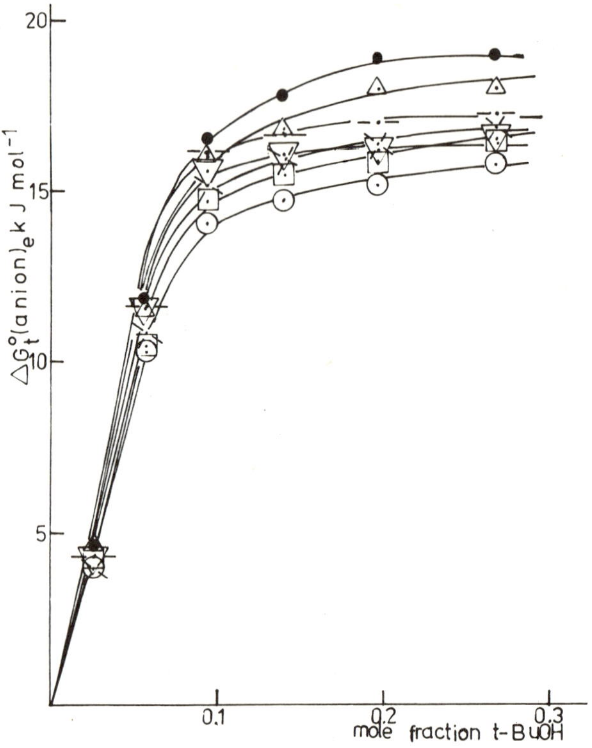

Figure 14. Plots of ΔG_t°*(carboxylate)*$_e$ *for water +* tert-*butanol at 25°C against mole fraction of* tert-*butanol with the following carboxylate anions: (⊙) butyrate; (⊡) 1-butyrate; (⋋⋌) 1-valerate; (△) pivalate; (▽) caproate; (●) diethylacetate; (– · –) cyclohexanecarboxylate*

solvent. For the proton, the structural effects are presumably a combination of the penetration of the inner coordination sphere around H_3^+O by the co-solvent through its electron-donating power being higher than that of H_2O and of orientation effects outside this coordination sphere i.e. general stabilization of the "cluster" of solvent molecules around the ion, with a greater effect in the mixed solvent than in pure water. Similar effects will operate with other cations, except that for a co-solvent molecule with a lower dipole moment than that of water with inorganic cations not possessing d orbitals, these effects in the cluster must be largely outside the inner coordination shell. For inorganic anions, $\Delta G_t^{\circ}(i)$ is always positive except for some anions in water + glycerol (*32, 33, 34, 35, 36*), but the values are much greater than those required by the Born Expression, even though the order in anion, $OH^- > Cl^- > Br^- > I^- >$

ClO_4^- in a solvent where OH^- cannot produce a co-solvent anion ($F^- >$ $Cl^- > Br^- > I^- > ClO_4^-$ in water + methanol), is roughly related to the order of the reciprocal of size, as required by the Born Expression. As pointed out above, this is also the reverse of their order of structure-forming capacity; but OH^- and F^-, as net structure formers, require a different balance between energetic and configurational effects than those of the other anions. In general, as the dipole moments of the co-solvent molecules are less than that of H_2O (*34*), replacement of H_2O in the inner coordination sphere of an anion by a co-solvent molecule is unlikely, except for ethylene glycol and glycerol which have dipole moments greater than that of H_2O (*34*). Perhaps it is best to discuss possible effects in water + dioxan, where dioxan has a dipole moment of zero and is the least likely to appear contiguous with a halide anion and the structural effects are most likely to be produced outside the coordination sphere. The introduction of an anion except OH^- and F^- induces structure breaking with energetic changes dominating configurational effects (Category 4). The reverse order in $\Delta G_t°(i)$ with respect to the order of the increasing structure-breaking effect of the anion may arise from an enhancement of the polarization of the anion electron density by the positive end of the H_2O dipole caused by quenching of the vibrational and rocking motions of the latter (*32, 33, 34*) in the more rigid conditions prevailing in the mixed solvents at low x_2. Although, in water + alcohol mixtures, this increase in rigidity over that in water at low x_2 probably largely results from the stabilization of the "iceberg-like" structures of water caused by the alkyl groups (*63, 64, 65*), the physical properties of water + dioxan suggest (*36*) that this effect is relatively small with dioxan, although the electron-releasing properties should cause dioxan to encourage hydrogen-bond formation in the mixture extending to higher x_2. If the change in $-\Delta H$ from this effect is greater than the accompanying change in $-\Delta S$ and it increases in the same order as the polarizability of the anions, $Cl^- < Br^- < I^- < ClO_4^-$ (*33*), the resulting difference between $\Delta\Delta H_M^E$ and T $\Delta\Delta S_M^E$ could decrease in the latter order. With OH^- or F^-, the dioxan molecule's capability for hydrogen bonding may allow a dioxan molecule to become contiguous with the ion, and this structure formation, together with that outside the coordination sphere, results in Category 2 and a change in the balance of energetic and configurational effects in favor of the latter.

Glossary of Symbols

H_o = Hammett's Acidity Function
B = *p*-nitroaniline
BH^+ = protonated *p*-nitroaniline

K, K_1, and K_2 = thermodynamic equilibrium constants for Reactions 1, 2, and 3
ROH = co-solvent
a = added molar concentration of ROH
c_o = total added molar concentration of B
c = molar concentration of B with HCl present
c_R = molar concentration of B with both HCl and ROH present
P = $(H_2O)_xH^+_{solv}$
F_1, F_1', and F_2 = activity coefficient quotients
f_y = activity coefficient on molar scale for species y
w = molar concentration of water in the absence of ROH
w_R = molar concentration of water in the presence of ROH
$[H^+]_T$ = total added molar concentration of mineral acid
K_c = concentration quotient $[ROH_2]/[P][ROH]$
I = ionic strength
T = temperature in Kelvin
R = gas constant
$\Delta G°(ROH_2)$ = standard free energy change for the formation of ROH_2 from H_{aq}^+
$\Delta G_t°(i)$ = standard free energy change for the transfer of i from water into the mixed solvent
N = Avogadro's number
e = electronic charge
D_w = dielectric constant of water
D_s = dielectric constant of the mixed solvent
r_{H_2O} = radius of the water molecule
$E_w°$ = standard electrode potential in water
$E_s°$ = standard electrode potential in the mixed solvent
M = cation
X = anion
K_p^w = solubility product in water
K_p^s = solubility product in the mixed solvent
K_{ip}^w = ionic product in water
K_{ip}^s = ionic product in the mixed solvent
$a_{H_2O}^s$ = activity of water in the mixed solvent on the molality scale
m_w = molality of water in pure water
m_s = molality of water in the mixed solvent
A_{solv} = acid molecule
B_{solv} = conjugate base of A_{solv}
K_a^w = acidic dissociation constant in water
K_a^s = acidic dissociation constant in the mixed solvent

P_a = proton affinity

ΔP_a = difference in proton affinity between the mixed solvent and pure water

$\Delta G_t^\circ(i)_n$ = component of $\Delta G_t^\circ(i)$ arising from the chemical bulk of i

$\Delta G_t^\circ(i)_e$ = component of $\Delta G_t^\circ(i)$ arising from the charge on i

x_2 = mole fraction of co-solvent

vol_2 = partial molar volume of the co-solvent molecule

T^E = structural contribution to the temperature of maximum density

$\Delta G_M{}^E$ = excess free energy of mixing of water with the co-solvent

$\Delta H_M{}^E$ = excess enthalpy of mixing water with the co-solvent

$\Delta S_M{}^E$ = excess entropy of mixing water with the co-solvent

Literature Cited

1. Goldschmidt, H., Udby, O., *Z. Phys. Chem.* (1907) **60**, 728.
2. Goldschmidt, H., *Z. Elektrochem.* (1909) **15**, 4.
3. Goldschmidt, H., Theusen, A., *Z. Phys. Chem.* (1912) **81**, 30.
4. Goldschmidt, H., *Z. Elektrochem.* (1914) **20**, 473.
5. Goldschmidt, H., *Z. Phys. Chem.* (1915) **89**, 129.
6. Ibid. (1916) **91**, 46.
7. Goldschmidt, H., Dahll, P., *Z. Phys. Chem.* (1924) **108**, 121,
8. Ibid. (1925) **114**, 1.
9. Goldschmidt, H., Meybye, R. S., *Z. Phys. Chem.* (1929) **143**, 139.
10. Goldschmidt, H., Haaland, H., Melbye, R. S., *Z. Phys. Chem.* (1929) **143**, 278.
11. Thomas, L., Marum, E., *Z. Phys. Chem.* (1929) **143**, 191.
12. Bezman, I. I., Verhoek, F. H., *J. Am. Chem. Soc.* (1945) **67**, 1330.
13. Hammett, L. P., "Physical Organic Chemistry," p. 260, McGraw-Hill, New York, 1940.
14. Wells, C. F., *Discuss. Faraday Soc.* (1965) **39**, 132.
15. Wells, C. F., "Hydrogen-Bonded Solvent Systems," A. K. Covington, P. Jones, Eds., pp. 323–334, Taylor and Francis, London, 1968.
16. Braude, E. A., Stern, E. S., *J. Chem. Soc.* (1948) 1976.
17. Bates, R. G., Schwarzenbach, G., *Helv. Chim. Acta* (1955) **38**, 699.
18. Salomoa, P., *Acta Chem. Scand.* (1957) **11**, 125.
19. Strehlow, H., *Z. Phys. Chem.* (1960) **24**, 240.
20. Wells, C. F., *Trans. Faraday Soc.* (1965) **61**, 2194.
21. Ibid. (1966) **62**, 2815.
22. Ibid. (1967) **63**, 147.
23. Wells, C. F., *J. Phys. Chem.* (1973) **77**, 1994.
24. Ibid., 1997.
25. Wells, C. F., *J. Chem. Soc., Faraday Trans. I* (1972) **68**, 993.
26. Palm, V. A., Haldna, U. L., Talvik, A. I., "Chemistry of the Carbonyl Group," S. Pattai, Ed., p. 429, Interscience, New York, 1966.
27. Palm, V. A., Haldna, U. L., *Dokl. Akad. Nauk. S. S. S. R.* (1960) **135**, 667.
28. Haldna, U. L., Ploom, L., Maroos, A., *Zap. Tartu Gos. Univ.* (1962) **127**, 65.
29. Haldna, U. L., Püss, R. K., *Russ. J. Phys. Chem. (Engl. Transl.)* (1964) **38**, 1529.
30. Haldna, U. L., *Org. React. (U.S.S.R.)* (1964) **1**, 184.

31. Ibid. (1965) **2,** 381.
32. Wells, C. F., *J. Chem. Soc., Faraday Trans. 1* (1973) **69,** 984.
33. Ibid. (1974) **70,** 694.
34. Ibid. (1975) **71,** 1868.
35. Ibid. (1976) **72,** 601.
36. Ibid. (1978) **74,** 1569.
37. Harned, H. S., Morrison, J. O., *J. Am. Chem. Soc.* (1936) **58,** 1908.
38. Harned, H. S., Walker, F., Calmon, C., *J. Am. Chem. Soc.* (1939) **61,** 44.
39. Mussini, T., Massarani-Formaro, C., Andrigo, P., *J. Electroanal. Chem.* (1971) **33,** 189.
40. Smits, R., Massart, D. L., Juillard, J., Morel, J.-P., *Electrochim. Acta* (1976) **21,** 431.
41. Das, P. K., Mishra, U. C., *Electrochim. Acta* (1977) **22,** 59.
42. Schwabe K., Schwenke W., *Z. Elektrochem.* (1959) **63,** 441.
43. Feakins, D., Turner, D. J., *J. Chem. Soc.* (1965) 4986.
44. Bax, D., Alfenaar, M., De Ligny, C. L., *Recl. Trav. Chim.* (1971) **90,** 1002.
45. Bax, D., De Ligny, C. L., Remijnse, A. S., *Recl. Trav. Chim.* (1973) **92,** 374.
46. Harned, H. S., Fallon, L. D., *J. Am. Chem. Soc.* (1939) **61,** 2374.
47. Villermaux, S., Baudet, V., Delpuech, J.-J., *Bull. Soc. Chim. Fr.* (1974), 1781.
48. Rat, J. C., Villermaux, S., Delpuech, J.-J., *Bull. Soc. Chim. Fr.* (1974), 815.
49. Luca, C., Enea, O., *Electrochim. Acta* (1970) **15,** 1305.
50. Harned, H. S., Allen, D. S., *J. Phys. Chem.* (1954) **58,** 191.
51. Seguela, P., Paraiud, J.-C., *C. R. C* (1961) **253C,** 1565.
52. Schwabe, K., Kunz, M., *Z. Elektrochem.* (1960) **64,** 1188.
53. Bax, D., De Ligny, C. L., Remijnse, A. G., *Recl. Trav. Chim.* (1972) **91,** 965.
54. Schwabe, K., Urlass, R., Ferse, A., *Ber. Bunsenges. Phys. Chem.* (1964) **68,** 46.
55. Popovych, O., Dill, A. J., *Anal. Chem.* (1969) **41,** 456.
56. Morel, J.-P., *Bull. Soc. Chim. Fr.* (1967) 1405.
57. Khoo, K. H., *J .Chem. Soc. A* (1971) 1177.
58. Ibid., 2932.
59. Das, A. K., Kundu, K. K., *J. Chem. Soc., Faraday Trans 1* (1973) **69,** 730.
60. Ibid., 1983.
61. Das, A. K., Kundu, K. K., *J. Chem. Soc., Faraday Trans. 1* (1974) **70,** 1452.
62. Frank, H. S., Evans, M. W., *J. Chem. Phys.* (1945) **13,** 507.
63. Frank, H. S., Wen, W.-Y., *Discuss. Faraday Soc.* (1957) **24,** 133.
64. Nemethy, G., Sheraga, H. A., *J. Chem. Phys.* (1962) **36,** 3382.
65. Ibid., 3401.
66. Wells, C. F., *J. Chem. Soc., Faraday Trans. 1* (1978) **74,** 636.
67. Harned, H. S., *J. Phys. Chem.* (1938) **43,** 275.
68. Harned, H. S., Kazanjian, G. L., *J. Am. Chem. Soc.* (1936) **58,** 1912.
69. Harned, H. S., Fallon, L. D., *J. Am. Chem. Soc.* (1939) **61,** 2377.
70. Douhéret, G., *Bull. Soc. Chim. Fr.* (1967) 1412.
71. Bacarella, A. L., Grunwald, E., Marshall, H. P., Purlee, E. L., *J. Org. Chem.* (1955) **20,** 747.
72. Reynaud, R., *Bull. Soc. Chim. Fr.* (1967) 4597.
73. Harned, H. S., Nestler, F. H. M., *J. Am. Chem. Soc.* (1946) **68,** 966.
74. Grunwald, E., Berkowitz, B. J., *J. Am. Chem. Soc.* (1951) **73,** 4939.
75. Gelsema, W. J., De Ligny, C. L., Visserman, G. F., *Recl. Trav. Chim.* (1965) **84,** 1129.
76. Morel, J.-P., *Bull. Soc. Chim. Fr.* (1967) 1465.
77. Harned, H. S., Dedell, T. R., *J. Am. Chem. Soc.* (1941) **63,** 3308.
78. Harned, H. S., Done, R. S., *J. Am. Chem. Soc.* (1941) **63,** 2579.
79. Dunsmore, H. S., Speakman, J. C., *Trans. Faraday Soc.* (1954) **50,** 236.

80. Juillard, J., *Bull. Soc. Chim. Fr.* (1964) 3069.
81. Juillard, J., Simonet, N., *Bull. Soc. Chim. Fr.* (1968), 1884.
82. Patterson, A., Felsing, W. A., *J. Am. Chem. Soc.* (1942) **64,** 1480.
83. Moore, R. E., Felsing, W. A., *J. Am. Chem. Soc.* (1947) **69,** 2420.
84. Morel, J.-P., Fauve, J., Avédikian, L., Juillard, J., *J. Solution Chem.* (1974) **3,** 403.
85. Auriacombe, J., Morel, J.-P., *C. R. C* (1968) **267C,** 377.
86. De Ligny, C. L., *Recl. Trav. Chim.* (1960) **79,** 731.
87. Gutbezahl, B., Grunwald, E. *J. Am. Chem. Soc.* (1953) **75,** 559.
88. Pawlak, Z., Bates, R. G., *J. Solution Chem.* (1975) **4,** 817.
89. Yates, K., Welch, G., *Can. J. Chem.* (1972) **50,** 474.
90. Nakanishi, K., *Bull. Chem. Soc. Jpn.* (1960) **33,** 793.
91. Wada, G., Umeda, S., *Bull. Chem. Soc. Jpn.* (1962) **35,** 646.
92. Ibid., 1797.
93. Blandamer, M., "Introduction to Chemical Ultrasonics," Chap. 11, Academic, London, 1973.
94. Hoverka, F., Schaefer, R. A., Dreisbach, D., *J. Am. Chem. Soc.* (1936) **58,** 2264.
95. Kenttämaa, J., Lindberg, J. J., *Suom. Kemistil. B* (1960) **33,** 32.
96. Schott, H., *J. Chem. Eng. Data* (1961) **6,** 19.
97. Hammes, G. G., Knoche, W., *J. Chem. Phys.* (1966) **45,** 4041.
98. Arakawa, K., Takenaka, N., *Bull. Chem. Soc. Jpn.* (1969) **42,** 5.
99. Sarma, T. S., Ahluwalia, J. C., *Chem. Soc. Rev.* (1973) **2,** 203.

RECEIVED February 14, 1978.

6

Conductance and Ionic Association of Several Electrolytes in Binary Mixtures Involving Sulfolane (TMS) and Protic Solvents

GIUSEPPE PETRELLA[1], ANTONIO SACCO, and MAURIZIO CASTAGNOLO

Institute of Physical Chemistry, University of Bari, Via Amendola, 173–70126 Bari, Italy

Conductometric and spectrophotometric behavior of several electrolytes in binary mixtures of sulfolane with water, methanol, ethanol, and tert-*butanol was studied. In water–sulfolane, ionic Walden products are discussed in terms of solvent structural effects and ion–solvent interactions. In these mixtures alkali chlorides and hydrochloric acid show ionic association despite the high value of dielectric constants. Association of LiCl, very high in sulfolane, decreases when methanol is added although the dielectric constant decreases. Picric acid in ethanol–sulfolane and* tert-*butanol–sulfolane behaves similarly. These findings were interpreted by assuming that ionic association is mainly affected by solute–solvent interactions rather than by electrostatics. Hydrochloric and picric acids in sulfolane form complex species* HCl_2^- *and* $Pi(HPi)_2^-$.

In recent years, in organic syntheses studies, dipolar aprotic solvents or mixtures of these with protic solvents have been used more and more frequently as media in which to carry out the reaction, because the rate of reaction is much higher in these solvent systems than in protic ones. These findings aroused the interest of research workers, so that electro-

[1] To whom correspondence should be addressed.

0-8412-0428-4/79/33-177-077$05.50/1

chemical studies multiplied with the aim of obtaining useful information on ion–solvent and ion–ion interactions in aprotic and mixed protic–aprotic solvents.

In this regard a systematic investigation has been planned in our laboratory to study the behavior of several electrolytes in sulfolane (TMS) and in its mixtures with different protic solvents. TMS possesses a high value of the dipolar moment ($\mu^{30} = 4.65$ D.U.) (*1*) and a dielectric constant of intermediate value ($\epsilon^{30} = 43.33$) (*2*). The small changes of these two quantities over a wide temperature range (*1*) shows that TMS is a scarcely structured solvent, even though it has low ΔH and ΔS fusion values (2.84 cal/g (*3*) and 1.1 eu (*4*), respectively), which are connected to the fact that it first solidifies into plastic crystals of a mesomorphic phase at 28.45°C and then at 15.43°C into a nonrotational crystalline phase with transition ΔS higher than that of fusion (*8* eu (*4*)). TMS has a low autoprotolysis constant (pK = 25.45) (*5*), and, in spite of its high dipole moment, it has very weak acidic and basic properties ($pK_{BH^+} = -12.88$ (*6*) and $pK_a > 31$ (*7*). TMS has been chosen for our studies because it appears anomalous among nonaqueous solvents since in this medium several ions have very high Walden products, even higher than in water (*8, 9, 10*). Moreover, ionic association constants greater than expected on the basis of its dielectric constant were reported in literature for some salts (*11*). These findings led us to believe that research on ionic mobility and association to ion pairs extended to TMS mixtures with protic solvents and might provide us with some interesting results.

Experimental Data

Ionic mobilities under consideration here are based on the results of conductometric measurements carried out on diluted solutions ($10^{-3} \leqslant c \leqslant 7.10^{-3}$ mol/L) of Bu_4NCl, Bu_4NBr, Bu_4NI, Bu_4NClO_4, iAm_3BuNI (TABI), $NaBPh_4$, and NaI in water–TMS at 30°C (*2, 12*). Experimental

Table I. Limiting

			Λ_o	
x_2	$\eta(cP)$	ϵ	Bu_4NCl	Bu_4NBr
0	0.8004	76.77	105.29	107.57
0.0208	0.9304	73.95	90.782	91.277
0.0744	1.282	68.32	66.83	66.297
0.1586	1.869	62.51	45.59	47.109
0.3098	2.860	55.79	30.36	31.298
0.6080	4.988	48.40	17.69	19.12
1	10.29	43.33	12.27	11.722

[a] For Bu_4NCl, Bu_4NBr, Bu_4NI, Bu_4NClO_4, iAm_3BuNI, $NaBPh_4$, and NaI in water–TMS mixtures at 30°C.

data were analyzed according to Fuoss–Onsager–Skinner treatment (*13*) and the values of derived limiting equivalent conductances Λ_0 are summarized in Table I together with TMS mole fraction x_2, viscosity η, and dielectric constant ϵ of solvent mixtures. Limiting ionic equivalent conductances have been obtained on the basis of the hypothesis suggested by Coplan and Fuoss (*14*) $\lambda_0^+ (iAm_3BuN^+) = \lambda_0^- (BPh_4^-) = [\Lambda_0(iAm_3BuNBPh_4)]/2$. Moreover, since $iAm_3BuNBPh_4$ ($TABBPh_4$) was not soluble in water–sulfolane mixtures, the value of its limiting equivalent conductance has been calculated by the equation:

$$\Lambda_0\ (TABBPh_4) = \Lambda_0\ (TABI) + \Lambda_0\ (NaBPh_4) - \Lambda_0\ (NaI) \qquad (1)$$

given that the salts TABI, $NaBPh_4$ and NaI are soluble in water–TMS mixtures.

Information on ionic association phenomena have been obtained conductometrically in water–TMS at 35°C for diluted solutions of LiCl (*15*), NaCl (*16*), KCl (*17*), HCl (*18*), and $NaClO_4$ (*19*). The study of association to ion pairs has been extended conductometrically to diluted solutions of LiCl in methanol–TMS at 35°C (*20*), and spectrophotometrically to picric acid (HPi) in solutions of ethanol–TMS, and *tert*-butanol–TMS at 30°C (*21*).

Also, in this case, conductivity data were analyzed by Fuoss–Onsager–Skinner equations and limiting equivalent conductances Λ_0 and association constants K_A are collected in Table II together with physical properties of solvent mixtures. Furthermore, Table III shows ethanol and *tert*-butanol concentration in the mixture [ROH], the relevant dielectric constant ϵ, and the pK of picric acid.

Discussion

Ionic Walden Products. Fundamental work by Kay and Evans (*22*) has shown that a correct interpretation of the conductometric behavior of ions in water cannot be made without considering both the complex three-dimensional structure of water and the structure-breaking, structure-making properties of ions. On the other hand, if water is an atypical

Equivalent Conductances[a]

		Λ_0		
Bu_4NI	Bu_4NClO_4	iAm_3BuNI	$NaBPh_4$	NaI
106.14	95.84	107.43	77.63	140.23
87.46	78.76	86.09	67.30	118.33
60.854	52.635	59.86	49.93	85.67
43.205	37.249	42.29	36.46	61.56
29.744	26.34	29.29	24.85	41.31
18.703	16.771	18.47	14.19	24.25
9.996	9.481	9.80[b]	—	10.87[c]

[b] Value obtained from the data of A. P. Zipp in Ref. *10*.
[c] Value obtained from the data of R. Fernandez-Prini and J. E. Prue in Ref. *11*.

Table II. Limiting Equivalent Conductances and Association Constants[a]

x_2	$\eta(cP)$	ϵ	Λ_o	K_A
		Water–TMS		
		LiCl		
0	0.7194	74.64	140.22[b]	—
0.0190	0.8225	72.43	122.83	—
0.0633	1.093	68.05	96.19	—
0.1522	1.620	61.82	66.61	2
0.3115	2.574	54.87	42.45	10
0.4445	3.398	51.08	31.51	13
0.6259	4.543	47.33	24.33	36
1	9.033	42.71	15.92	14595
		NaCl		
0	0.7194	74.64	153.84[c]	—
0.0206	0.8304	72.27	133.76	—
0.0731	1.150	67.20	101.90	—
0.1585	1.665	61.42	71.85	—
0.3119	2.602	54.69	45.93	3
0.4323	3.338	51.34	35.53	7
0.6248	4.537	47.34	26.10	19
0.7443	5.466	45.60	21.89	25
0.8298	6.117	44.54	19.67	40
		KCl		
0	0.7194	74.64	180.53[b]	—
0.0200	0.8271	72.33	157.93	—
0.1584	1.665	61.43	81.21	—
0.2735	2.428	56.03	55.29	—
0.4335	3.347	51.31	37.58	3
0.6282	4.555	47.30	26.56	7
		HCl		
0	0.7194	74.64	489.92	—
0.0206	0.8304	72.27	426.80	—
0.0731	1.150	67.18	310.33	—
0.1583	1.663	61.42	199.13	26
0.3116	2.602	54.69	99.30	51
0.6248	4.537	47.34	32.63	76
1	0.033	42.71	—	10^9

Table II. Continued

x_2	$\eta(cP)$	ϵ	Λ_o	K_A
		$NaClO_4$		
0	0.7194	74.64	142.15	—
0.0205	0.8287	72.30	120.91	—
0.0731	1.145	67.24	87.22	—
0.1573	1.658	61.47	61.79	—
0.3050	2.545	55.09	42.99	—
0.6111	4.441	47.57	25.52	—
1	9.033	42.71	11.74	—
		Methanol–TMS		
		LiCl		
0	0.4813	30.71	104.63	—
0.0440	0.5178	32.37	99.16	—
0.1566	0.6832	34.52	84.19	16
0.2960	1.010	36.90	64.08	25
0.4973	1.790	39.27	42.36	32
0.7927	4.408	41.55	24.30	121
1	9.033	42.71	15.92	14595

[a] For LiCl, NaCl, KCl, HCl, and $NaClO_4$ in water–TMS mixtures and LiCl in methanol–TMS mixtures at 35°C.
[b] Ref. *49*.
[c] Ref. *50*.

solvent because of its structural characteristics, we can wholly agree with F. Franks when he says, "If there is one way of producing a solvent even more eccentric and anomalous than pure water, then this is achieved by the addition of small amounts of some organic compounds" (*23*). Numerous conductometric studies on electrolytes in mixtures of water with organic solvents have been reported in the literature; but it is only recently, thanks to Kay and Broadwater, that ion–solvent interactions in these binary mixtures have been studied, particularly with respect to what happens in water-rich mixtures. They use *tert*-butanol (*tert*-BuOH) (*24*), ethanol (EtOH) (*25*), and dioxane (Diox) (*26*) as co-solvents. Among these solvents, the alcohols, when added in small amounts, are able to generate mixtures which are more structured than pure water. This effect is greater for *tert*-BuOH than for EtOH. However, dioxane does not enhance the long-range order in water (*26*). Kay and Broadwater observe that in the case of alcoholic mixtures, in water-rich regions, alkali and halide ions show maxima in Walden products. This could be explained by the fact that, in these regions of the mixtures, solvent structure is higher than that of pure water, so that the mobility of structure-

Table III. The pK Values of Picric Acid in Ethanol–TMS and *tert*-Butanol–TMS Mixtures at 30°C

[ROH](mol/L)	ϵ	pK_{HPi}
	Ethanol–TMS	
16.95	23.8	3.66
15.36	25.2	3.62
12.07	28.2	3.72
8.10	31.8	3.72
5.31	36.0	4.40
1.93	40.5	5.54
0	43.3	7.6
	tert-Butanol–TMS	
10.46	11.5	4.60
9.44	13.3	4.82
7.39	18.3	4.56
5.33	24.6	4.22
3.26	31.6	4.82
1.12	39.3	5.90

breaking ions is enhanced. However, the mobility increase is greater for smaller ions, and this does not agree with what would be expected if the maxima were caused solely by structure-breaking effects, given that these effects are more marked for larger ions. Thus Kay and Broadwater suggest that maxima in Walden products can be explained by supposing that the interactions between the components of the solvent mixture and the ions are mainly of the acid–base type. Therefore, as a result of ionic charge density and the great differences in acid–base properties between water and the organic co-solvent, the ion cosphere is enriched with water to a greater degree than the bulk mixture. Consequently, the ions have a higher mobility since the viscosity around them is less than the bulk viscosity, as can be seen from the trend of the viscosity of the two solvent mixtures (*24, 25*). Sorting effect is obviously greater for smaller ions and this agrees with experimental results.

Regarding water–dioxane mixtures, the maxima observed in Walden products are smaller than in water–alcohol mixtures and in these mixtures too the ions having a lower structure-breaking capacity show a higher increase in Walden products than in water.

In the case of water–TMS mixtures there is evidence in literature that TMS breaks down water structure in water-rich mixtures. This has been proved by studies concerning the influence of small additions of TMS on the temperature of the maximum density of water (*27*), the heat of mixing, and the vapor pressure of water–TMS mixtures (*28*). Figure 1 reports ionic Walden products normalized to their values in water R = $[\lambda_0^{\pm}\eta)s]/[\lambda_0^{\pm}\eta)w]$ as a function of mole percent TMS for Na^+, Cl^-, Br^-, I^-, ClO_4^-, Bu_4N^+, TAB^+, and BPh_4^-.

Let us now consider inorganic ionic behavior. Na^+ shows R values greater than unity throughout almost the entire range of the solvent composition with a maximum at about 30 mol % TMS. Cl^- and Br^-, up to 60 mol % in TMS, possess nearly constant values, and are roughly equal to those in water, while I^- and ClO_4^-, which are the best structure-breaking ions in water, show a minimum in Walden products at about 10 mol % TMS. Therefore Na^+, contrary to anions, behaves in water–TMS as it does in the mixtures studied by Kay and Broadwater.

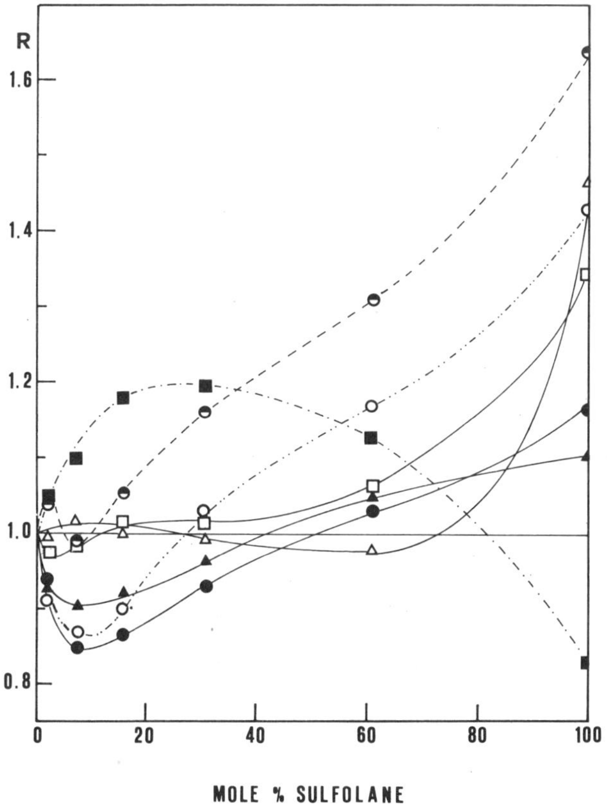

Journal of Solution Chemistry

Figure 1. Ionic Walden products normalized to their values in water as a function of mole percent TMS at 30°C: (○), $TAB^+ = BPh_4^-$; (◓), Bu_4N^+; (■), Na^+; (△), Cl^-; (□), Br^-; (▲), I^-; (●), ClO_4^- (12).

A mobility increase shown by Na^+ with respect to that in water cannot be interpreted by supposing that its structure-breaking ability is enhanced in water–TMS mixtures, because they are less structured than pure water. Therefore, the behavior of Na^+ ion may be attributed to the sorting of water in its cosphere. In fact, TMS possesses very low acid–base properties, and viscosities in the mixtures steadily increase in passing from water to TMS, as shown in Table I.

On the contrary the behavior of Cl^-, Br^-, I^-, and ClO_4^- certainly can not be explained by the sorting effect since in that case one would expect a mobility increase and thus an R maxima. On the other hand, the observed order in R values at about 10 mol % TMS ($Cl^- > Br^- > I^- > ClO_4^-$) agrees with the increasing structure-breaking ability of ions. Because the addition of TMS breaks down water structure, the behavior of these anions may be explained by supposing that their mobility is affected mainly by structural effects.

With regard to anions, it is interesting to compare our data with those reported by Kay and Broadwater. In Figure 2, R values in H_2O–*tert*-BuOH, H_2O–EtOH, H_2O–Diox, and H_2O–TMS are plotted against mole percent of nonaqueous solvent in water-rich regions. As can be seen, for each solvent mixture, mobility increases with a decrease in ion size, in agreement with the sorting effect. At the same time, however, R values for each ion decrease in passing from one mixture to another with a decrease in the organic solvent's ability to enhance long-range order in water-rich mixtures, as expected on the basis of structural effects. We believe that these findings suggest that sorting and structural effects exist at the same time in solution and that both affect the behavior of ions in solution.

It is difficult to explain the different behavior shown in water–TMS by Na^+ and halides and ClO_4^- ions; moreover, it was not found in the mixtures studied by Kay and Broadwater. This difference probably arose either from the differences between cation–TMS and anion–TMS interactions or from the fact that the sorting effect prevails over structural effects for Na^+, contrary to what happens to anions.

Let us now consider the organic ions. As far as the Bu_4N^+ ion is concerned, Figure 1 shows that R values, after some deviations from the unity which is probably caused by experimental errors, increase with increasing organic solvent percentage, as Kay and Broadwater found for this ion in the mixtures they studied. This trend is caused by a reduction of hydrophobic effects characteristic of Bu_4N^+ in pure water. It appears very difficult to explain the behavior of TAB^+ and BPh_4^- ions which show a minimum at about 30 mol % TMS. On the other hand, the lack of conductometric data for these ions in other aqueous mixtures precludes any useful comparison.

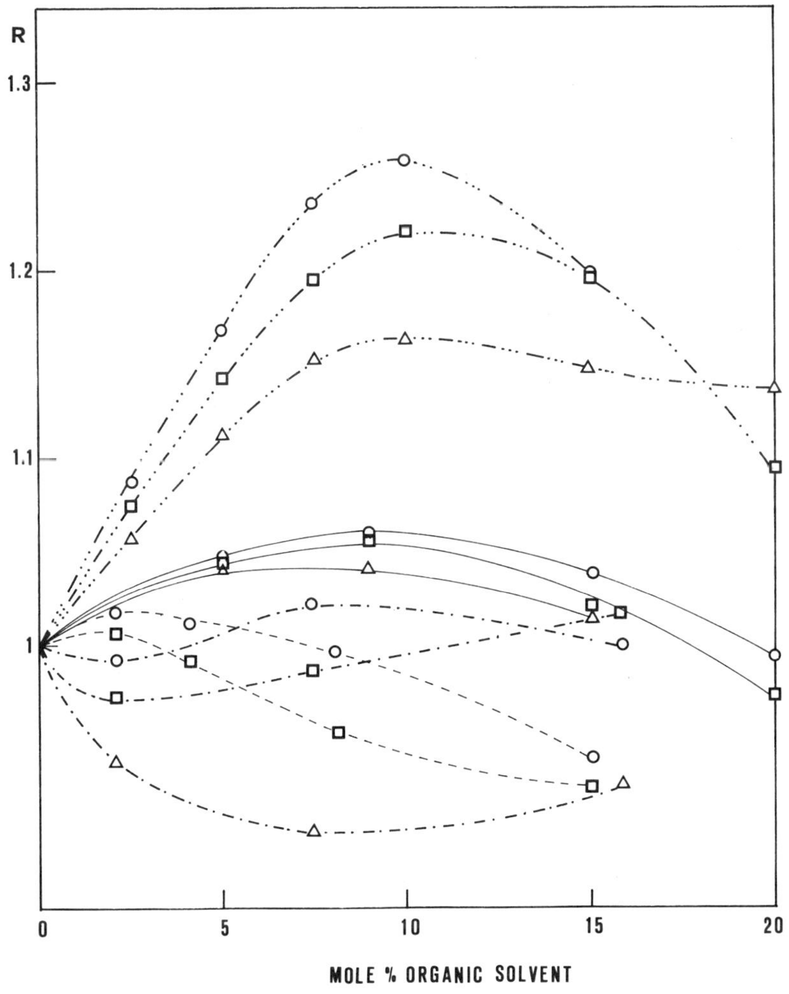

Figure 2. Ionic Walden products normalized to their values in water as a function of mole percent organic solvents: (- · · · -), tert-BuOH–H_2O; (——), EtOH–H_2O; (- - -), Diox–H_2O; (- · · -), TMS–H_2O; (○), Cl^-; (□), Br; (△), I^-.

Association Phenomena According to the theoretical model of spheres in a dielectric continuum the ions are represented as rigid, charged spheres that do not interact with solvent, which is considered to be a medium without any kind of structure. The only interaction is that which occurs between the ions, and the formation of ion pairs is controlled only by electrostatic forces. On these bases, the association constant may be expressed by the Fuoss equation (29):

$$K_A = \frac{4\pi N a^3}{3000} \exp\left(\frac{e^2}{a \epsilon kT}\right) \qquad (2)$$

Thus, according to this equation, the association constant increases with a decrease of distance of closest approach of ions a and the dielectric constant ϵ. A graph of log K_A vs $1/\epsilon$ should be linear. Nevertheless, several data in the literature show that association to the ion pair is not expressed adequately by Equation 2, and that solute–solvent interactions are fundamentally important in determining the existence and magnitude of association phenomena. This was observed also by us in the mixtures of TMS and some protic solvents. Association behavior of electrolytes in these mixtures can be understood better by first examining this phenomena in pure TMS.

Pure TMS. Given the intermediate value of the dielectric constant of TMS, no noticeable ionic association on the basis of the Fuoss equation should be expected. Moreover, much higher association constants than expected by Equation 2 have been observed for LiCl ($K_A^{35°C}$ = 14595 L/mol) (*15*) while the theoretical value, calculated for a = 2.413, the sum of crystallographic radii of Li^+ and Cl^- ions, is 7 L/mol). This high association also was observed at 30°C by Prini and Prue ($K_A^{30°C}$ = 13860 L/mol) (*9*) who also remarked on the inadequacy of the theory and explained it in terms of dielectric saturation. Furthermore, Garnsey and Prue gave apparent polymerization numbers for LiCl (*3*). We believe that this high association constant value also may be explained by keeping in mind that TMS, like other dipolar aprotic solvents, scarcely solvates anions, particularly the smallest ones such as Cl^- (*30*). Thus chloride ion is particularly reactive in TMS and this increases the intensity of interaction with Li^+.

The failure of the Fuoss equation to reproduce experimental data appears particularly evident by conductometric measurements of hydrochloric acid (*18*) and by spectrophotometric measurements of picric acid (*21*) which we have carried out in TMS at 35° and 30°C, respectively.

Regarding hydrochloric acid, in a concentration range of 30.10^{-4} to 300.10^{-4} mol/L, equivalent conductance assumes an extremely low and constant value of 0.03 S cm^2/mol, as seen in Figure 3. This behavior certainly cannot be explained on the basis of simple dissociation phenomena. Thus we have interpreted these results on the basis of theoretical work by Caruso and co-workers (*31*) who consider the conductometric, potentiometric, and spectrophotometric behavior of weak acids and bases in nonaqueous solvents. In these solvents a weak acid, HA, besides undergoing simple ionic dissociation, also may undergo conjugation phenomena by the H^+ and A^- ions which lead to the formation of ionic complex species $A(HA)_j^-$ or $H(HA)_j^+$. Caruso shows that the

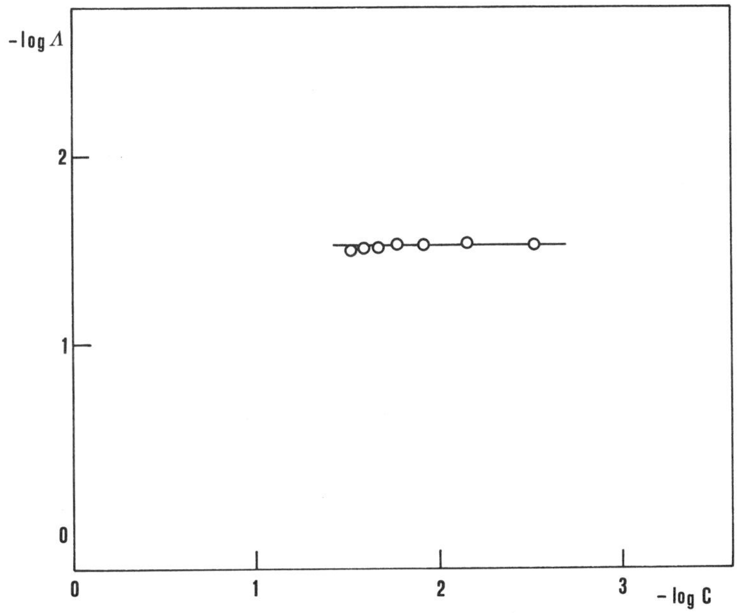

Electrochimica Acta

Figure 3. Conductance of HCl in pure TMS at 35°C (18)

existence of these species in solution reflects on the shape of $-\log \Lambda$ vs. $-\log c$, E vs. $-\log$ c, and $-\log[A^-]$ vs. $-\log c$ plots in conductance, potentiometry, and spectrophotometry, respectively. Particularly regarding conductometric measurements, if there is unilateral conjugation by the H^+ or A^- ions with a neutral molecule of the acid HA, the curve $-\log \Lambda$ vs. $-\log c$ assumes a slope equal to zero, which approaches -0.5 only for very low values of acid concentration. This behavior is observed for HCl in TMS and we suppose that it is caused by the HCl_2^- complex species, so that the equilibria established in solution are:

$$HCl \rightleftarrows H^+ + Cl^- \qquad K_1 = \frac{a_{H^+}\, a_{Cl^-}}{a_{HCl}} \qquad (3)$$

$$Cl^- + HCl \rightleftarrows HCl_2^- \qquad K_2 = \frac{a_{HCl_2^-}}{a_{Cl} - a_{HCl}} \qquad (4)$$

The hypothesis of the formation of the HCl_2^- ion agrees with the properties which TMS shows in favoring the formation of large hydrogen dihalide ions rather than small halide ions, as found by Benoit and coworkers (*32, 33*). Moreover, acid–anion complexes also were observed for 2.6-dihydroxylbenzoic acid in TMS (*34, 35*) and for phenol and phenolate ion in 3-methyl-sulfolane (*36*), which has acid–base properties similar to those of TMS.

The theoretical equation found by Caruso for the intercept of the linear portion of the plot —log Λ vs. —log *c* enables us to evaluate the constant K_1:

$$\text{intercept} = -\frac{1}{2}\log 4\, K_1K_2\, \Lambda_o\ (H^+\, HCl_2^-) \qquad (5)$$

In Equation 5, inserting the values of 1.52 for the intercept, 350 for K_2 (*32, 33*), and 13 for Λ_o (H^+ HCl_2^-), calculated approximately, on the data from the literature (*8, 35*), afirst approximation value of 10^{-9} mol/L was obtained for K_1.

More complex equilibria than those concerned with simple association to ion pair have been observed spectrophotometrically in TMS in the case of picric acid. Figure 4 shows that $\sum_{J=0}^{n}[Pi(HPi)_j^-]$ vs. — log c

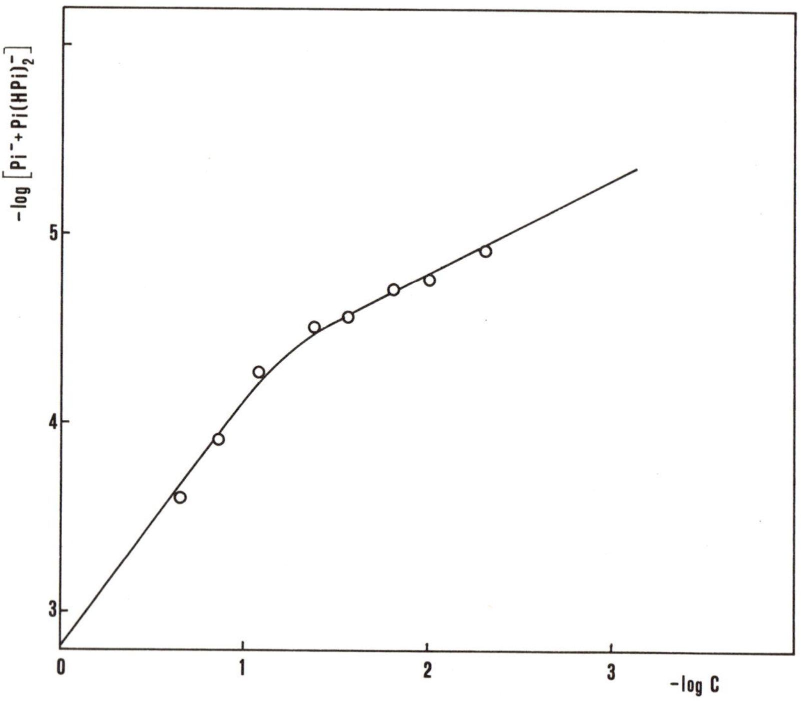

Canadian Journal of Chemistry

Figure 4. Spectrophotometric data of picric acid in TMS. The line is the theoretical curve for equilibria 6–8 ($K_{HPi} = 10^{-7.6}$, $K_{(HPi)_2} = 10^{-1.9}$, and $K_{Pi(HPi)_2^-} = 10^4$) (21).

($\sum_{j=0}^{n}$[Pi(HPi($_j^-$] is the sum of anionic picrate species, and c is the analytical concentration of HPi). As may be seen, there are two linear portions with slopes of 0.46 for $c < 0.08M$ and 1.56 for $c > 0.08M$. A very similar graph has been observed by Kolthoff and co-workers for HPi in acetonitrile (*37*) with slopes of 0.48 and 1.45 for concentrations below and above 0.1*M*, respectively. The analysis carried out by Caruso and co-workers (*31*) on these data confirms the hypothesis made by Kolthoff and co-workers that the predominating species in acetonitrile at higher concentrations is the ion $Pi(HPi)_2^-$. Moreover, conductometric measurements on HPi in TMS have been interpreted by Eller and Caruso presuming that HPi polymers are formed in solutions (*38*). On these bases we have hypothesized that picric acid in TMS too forms the complex species $Pi(HPi)_2^-$ according to the following reactions:

$$HPi \rightleftharpoons H^+ + Pi^- \qquad K_{HPi} = \frac{[H^+][Pi^-]f^2}{[HPi]} \qquad (6)$$

$$2HPi \rightleftharpoons (HPi)_2 \qquad K_{(HPi)_2} = \frac{[(HPi)_2]}{[HPi]^2} \qquad (7)$$

$$Pi^- + (HPi)_2 \rightleftharpoons Pi(MPi)_2^- \qquad K_{Pi(HPi)_2^-} = \frac{[Pi(HPi)_2^-]}{[Pi^-][(HPi)_2]} \qquad (8)$$

Spectrophotometric data have been analyzed mathematically as suggested by Caruso and the theoretical curve which fits the experimental data very well was obtained for $K_{HPi} = 10^{-7.6}$, $K_{(HPi)_2} = 10^{-1.9}$, and $K_{Pi(HPi)_2^-} = 10^4$. Comparison of pK_A values in TMS for $HClO_4$, HSO_3F, and $H_2S_2O_4$ (2.7, 3.3, and 5, respectively (*39*)) with the ones obtained for HCl (9) and HPi (7.6) shows that the weakest acid in this solvent is hydrochloric acid. This is further proof that in TMS large anions are stabilized by solvation more than little ones.

The results obtained for LiCl, HCl, and HPi in TMS show the fundamental importance of the effect of ion–solvent interactions on association phenomena. These effects also have been evidenced when studying TMS–protic solvent mixtures where ion association is conditioned mainly by the presence of TMS.

Water–TMS Mixtures. Conductometric studies on Li, Na, and K chlorides and hydrochloric acid in water–TMS have shown association constants higher than the Fuoss equation predicts in these mixtures too. In the case of HCl, K_A values equal to 26 ± 5, 51 ± 9, and 76 ± 4 corresponded to dielectric constant values of 61.42, 54.69, and 47.34, respectively. On the contrary, K_A values for the same systems calculated on the basis of the Fuoss equation using the reasonable value of 4 Å for the distance of closest approach of ions a are 1.5, 1.9, and 2.8, respectively. The difference between experimental and theoretical trends of log K_A vs. $100/\epsilon$ is shown in Figure 5.

In the case of LiCl, NaCl, and KCl, Figure 6 enables experimental values to be compared with theoretical ones calculated by the Fuoss equation, substituting for the a parameter the lowest value with a physical meaning ($a = 2.413$, the sum of Li^+ and Cl^- crystallographic radii), in order to get larger theoretical values for K_A. As seen from Figures 5 and 6, the observed trends in almost the whole range of solvent composition are linear but with different slopes from the theoretical one in the

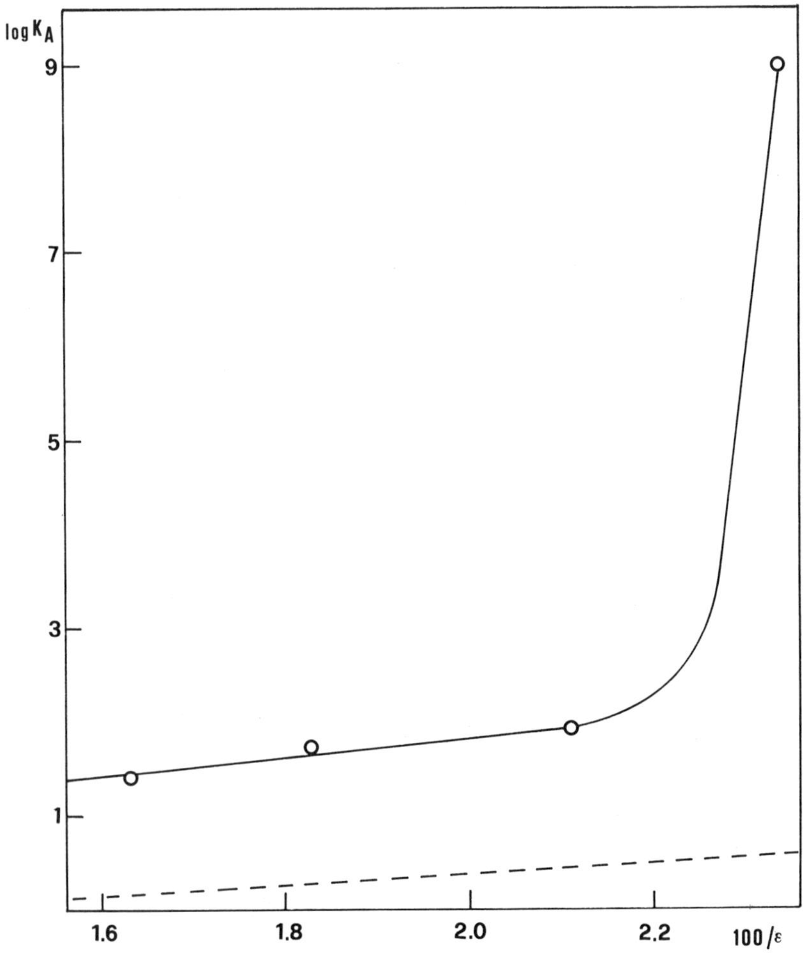

Figure 5. Dependence of association constants on the dielectric constant for HCl in water–TMS at 35°C: (– – –), association constants calculated from the Fuoss equation (a = 4 Å).

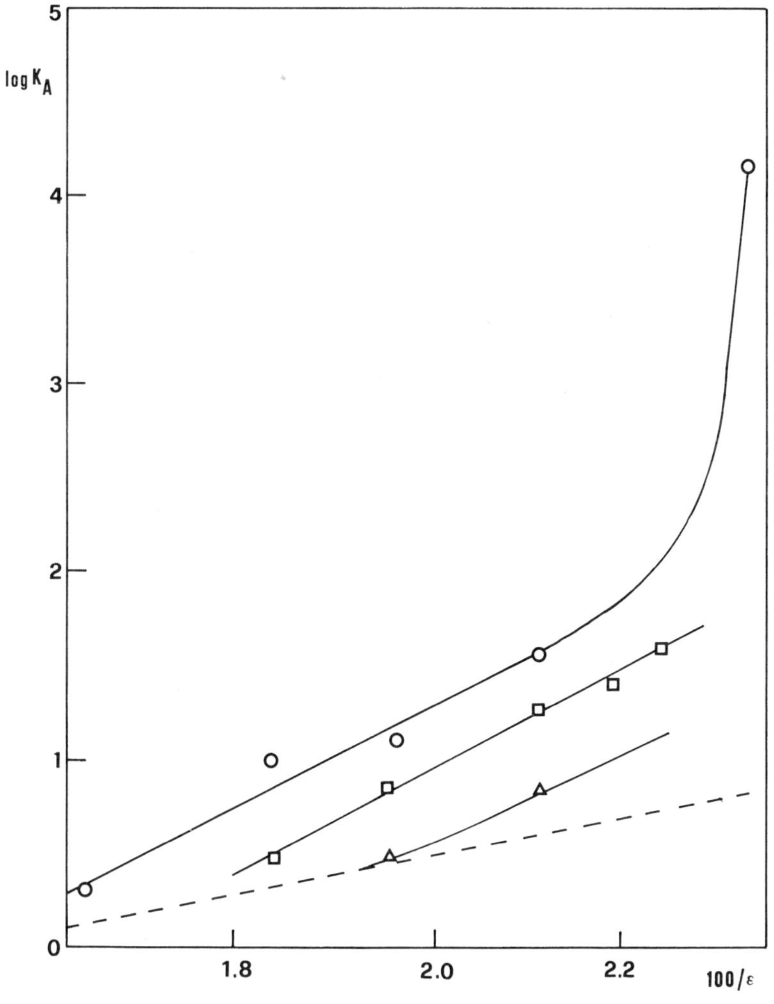

Figure 6. Dependence of association constants on the dielectric constant for LiCl, NaCl, and KCl in water–TMS mixtures at 35°C: (– – –), association constants calculated from the Fuoss equation (a = 2.413 Å); (○), LiCl; (□), NaCl; (△), KCl.

case of alkali chlorides, among which those of Na^+ and K^+ are scarcely soluble in TMS. In the case of LiCl and HCl, as the solvent becomes richer in TMS, log K_A abruptly increases.

Higher association constants than those expected on the basis of Equation 2 have been observed also in several protic solvents by Evans and co-workers (*40*), who explained it by assuming that association between cation M^+ and anion A^- in a protic solvent SH is a two-step process where there is, first, formation of a solvent-separated ion pair and then, following the loss of a solvent molecule by the anion, formation of a contact ion pair.

This happens according to the equations:

$$(SH)_m\, M^+ + A(SH)_n^- \overset{K_1}{\rightleftarrows} M(SH)\ A(SH)_{m+n-1} \qquad (9)$$

$$M(SH)\ A(SH)_{m+n-1} \overset{K_2}{\rightleftarrows} MA(SH)_{m+n-1} + SH \qquad (10)$$

K_1 may be calculated by the Fuoss equation and K_2 becomes larger as the bond between the anion and the solvent molecule weakens. The association constant observed is therefore:

$$K_A = K_1\left(1 + \frac{K_2}{[SH]}\right) \qquad (11)$$

This equation accounts for experimental association constant values higher than theoretical ones because $K_A > K_1$. Moreover, we may deduce that when cations are equal, K_A is smaller for smaller anions, which means that they have a greater charge density and thus are more basic, since K_2 is lower for them. On the contrary, K_A values decreasing with an increase in the size of the anion have been observed in aprotic solvents like TMS (*11*), acetone (*41*), nitrobenzene (*42*), nitromethane (*43*), acetonitrile (*44*), and 1,1,3,3-tetramethylurea (*45*). This order can be understood if one considers that in this class of solvents the anions are scarcely solvated so that association is affected mainly by the strength of cation–anion interaction, which, with the cations being equal, increases with an increase in the anion charge density.

In order to see whether the high K_A values we found in water–TMS mixtures can be explained by the multiple-step mechanism suggested by Evans and co-workers we have studied conductometric behavior of $NaClO_4$ in these mixtures (*19*). The results of these measurements show that $NaClO_4$ is not associated, whereas for NaCl association appears to start from $w_2 \simeq 70$ wt % TMS ($\epsilon \simeq 55$). So, the order observed

$$K_A\ (Cl^-) > K_A\ (ClO_4^-) \qquad (12)$$

served in water–TMS is analogous to that found in aprotic solvents. It therefore excludes the fact that anions forming ion pairs in water–TMS mixtures are preferentially solvated by water, since in that case sodium perchlorate should be more associated than chloride, as suggested by the Evans mechanism. Instead it suggests that ionic association phenomenon in water–TMS mixtures are mainly affected by TMS. In fact we have already seen that the Cl^- ion is scarcely stabilized in TMS and thus tends to form very stable ion pairs. Ionic association in the case of alkali chlorides and hydrochloric acid in water–TMS mixtures may therefore be explained by assuming that the small fraction of chloride ion which interacts with TMS forms very stable ion pairs, despite the high value of the dielectric constant.

Alcohol–TMS Mixtures. Furthermore, convincing evidence concerning the inadequacy of the theoretical model in describing ionic association may be obtained on the basis of results of studies on conductometrical behavior of LiCl in MeOH–TMS and spectrophotometrical behavior of HPi in EtOH–TMS and *tert*-BuOH–TMS mixtures.

Regarding LiCl, Figure 7 shows that even in alcoholic mixtures experimental K_A is higher than the theoretical one in nearly the whole range of the solvent composition. Furthermore, the most interesting fact is the noncoulombic trend of the association constant which decreases even though the dielectric constant decreases on passing from TMS ($\epsilon \simeq 43$) to methanol ($\epsilon \simeq 31$). Given that LiCl is not associated in MeOH, ionic association is ascribable to TMS, which also occurs in its mixtures with water. The trend of log K_A vs. $100/\epsilon$ also can be explained by assuming that ion–solvent interactions strongly influence ionic association, much more than the dielectric constant does. The K_A decrease that occurs when methanol is added to TMS is caused mainly by the greater capacity of methanol as compared with TMS to solvate the ions. This is proved by the strongly negative values of transfer enthalpies from TMS to methanol of both Li^+ (−9.1 K cal/mol) and Cl^- (−4.8 K cal/mol) (*46*).

Noncoulombic variation of ionic association was observed for HPi too, as shown in Figure 8 both for the EtOH–TMS and the *tert*-BuOH–TMS mixtures. Log K_A vs. $100/\epsilon$ is not linear and K_A decreases in these mixtures with a decrease in the dielectric constant, in spite of the theory based on electrostatic control of ion association ($\epsilon_{EtOH} = 23.8$, $\epsilon_{tert\text{-}BuOH} = 11.5$). Like LiCl in MeOH–TMS, the greatest decrease of K_A is observed with small additions of alcohols that are better able to solvate picric acid than TMS with which they establish hydrogen bonds. In fact picric acid is stronger in alcohols ($pK_{EtOH} = 3.7$ and $pK_{tert\text{-}BuOH} = 4.6$) than in TMS ($pK_{HPi} = 7.6$). The trends of K_A values may be explained by preferential solvation of HPi by the alcohols. On this basis an attempt was made to explain more thoroughly the kind of solute–solvent interactions in the case of picric acid. Given the noticeable difference between pK values in TMS and those in EtOH and *tert*-BuOH, we hypothesized that TMS behaves like an inert solvent and picric acid mainly reacts with alcohols according to the reactions (*47*):

$$HPi + nROH \leftrightarrows HPi \cdot nROH$$

$$K_1 = \frac{[HPi \cdot nROH]}{[HPi][ROH]^n} \qquad (13)$$

$$HPi \cdot nROH \rightleftharpoons ROH_2^+ \cdot ROH_{(n-1)}Pi^-$$

$$K_2 = \frac{[ROH_2^+ \cdot ROH_{(n-1)}Pi^-]}{[HPi \cdot nROH]} \qquad (14)$$

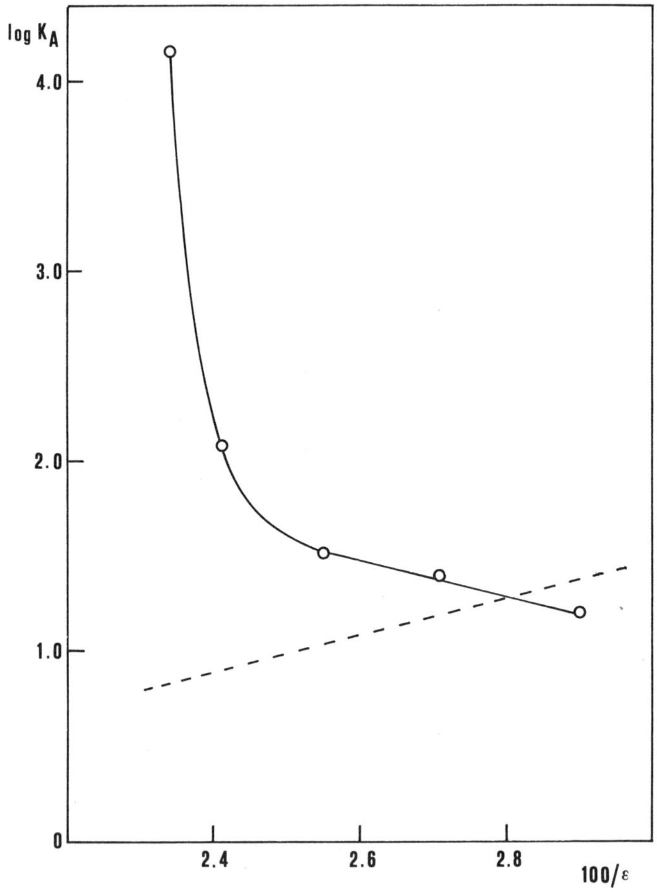

Zeitschrift für Naturforschung

Figure 7. Dependence of association constants on the dielectric constant for LiCl in methanol–TMS mixtures at 35°C: (– – –), association constants calculated from the Fuoss equation (a = 2.413 Å) (20).

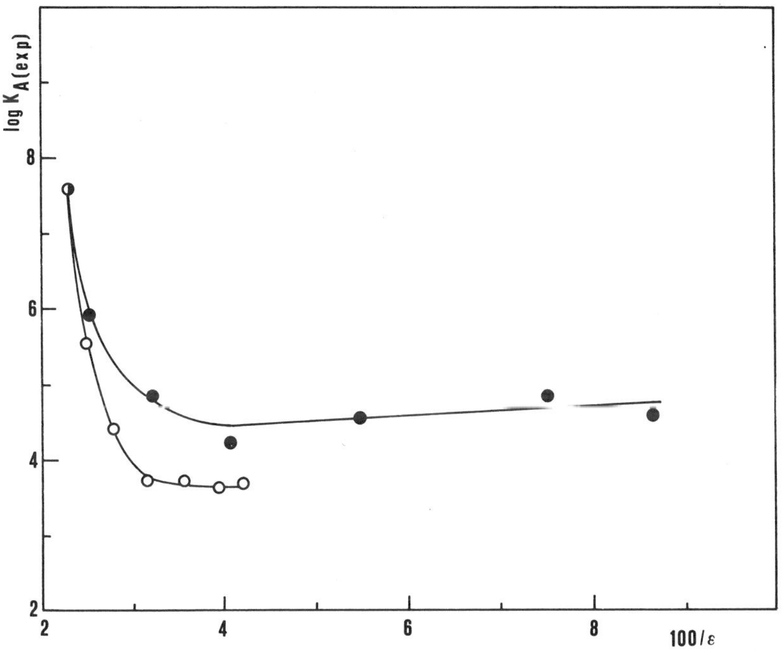

Figure 8. Association constants of picric acid in ethanol–TMS and tert-*butyl alcohol–TMS mixtures: (○), EtOH–TMS; (●),* tert-*BuOH–TMS (21).*

$$ROH_2^+ \cdot ROH_{(n-1)}Pi^- \rightleftharpoons ROH_2^+ + ROH_{(n-1)}Pi^-$$

$$K_3 = \frac{[ROH_2^+][ROH_{(n-1)}Pi^-]}{[ROH_2^+ \cdot ROH_{(n-1)}Pi^-]}. \quad (15)$$

Thus picric acid reacts with n alcohol molecules forming a complex (Equation 13) which rearranges to form an ion pair (Equation 14) which then dissociates (Equation 15). The processes involved in Equations 13 and 14 are conditioned by specific solute–solvent interactions, while the process depicted by Equation 15 is controlled by electrostatics. K_3 is equal to $1/K_A$, where K_A is the theoretical association constant given by Equation 2. It can be easily shown that the experimental association constant is given by the equation:

$$K_{A(exp)} = \frac{K_A}{[ROH]^n K_1 K_2}. \quad (16)$$

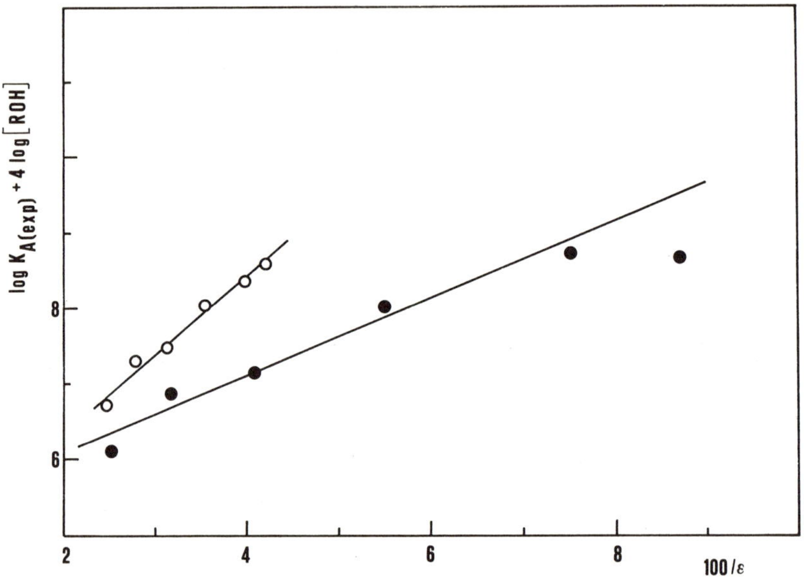

Figure 9. Test of Equation 17 (n = 4) for ionization of picric acid in ethanol–TMS and tert-*butyl alchol–TMS mixtures: (○), EtOH–TMS; (●),* tert-*BuOH–TMS (21).*

Equation 16 may be rewritten as follows:

$$\log K_{A(exp)} + n \log[ROH] = \log \frac{4 \pi N a^3}{3000} - \log K_1K_2 + \frac{0.434 e^2}{100 akT}\left(\frac{100}{\epsilon}\right) \quad (17)$$

It may be assumed that picric acid reacts with four alcohol molecules. One of them neutralizes the acid forming the ion ROH_2^+ and the remaining three molecules solvate the picrate anion by hydrogen bonds with the three nitro groups. D'Aprano and Fuoss have observed a similar mechanism in mixtures of acetonitrile and protic solvents (*48*). If this hypothesis is correct, when substituting the value of four for n in Equation 17, a graph of log $K_{A(exp)}$ + 4 log[ROH] vs. 100/ε should be a straight line. This, in fact, can be seen for both solvent mixtures, as shown in Figure 9.

Literature Cited

1. Lamanna, U., Sciacovelli, O., Jannelli, L., *Gazz. Chim. Ital.* (1966) **96**, 114.
2. Petrella, G., Castagnolo, M., Sacco, A., DeGiglio, A., *J. Solution Chem.* (1976) **5**, 621.

3. Garnsey, R., Prue, J. E., *Trans. Faraday Soc.* (1968) **64**, 1206.
4. Della Monica, M., Jannelli, L., Lamanna, U., *J. Phys. Chem.* (1968) **72**, 1068.
5. Kreshkov, A. P., Aldarova, N. S., Tanganov, B. B., *Zh. Fiz. Khim.* (1970) **44**, 2089; *Chem. Abstr.* (1970) **73**, 124160.
6. Hall, S. K., Robinson, E. A., *Can. J. Chem.* (1964) **42**, 1113.
7. Bordwell, F. G., Imes, R. H., Steiner, E. C., *J. Am. Chem. Soc.* (1967) **89**, 3905.
8. Della Monica, M., Lamanna, U., Senatore, L., *J. Phys. Chem.* (1968) **72**, 2124.
9. Della Monica, M., Lamanna, U., *J. Phys. Chem.* (1968) **72**, 4329.
10. Zipp, A. P., *J. Phys. Chem.* (1973) **5**, 718.
11. Fernández-Prini, R., Prue, J. E., *Trans. Faraday Soc.* (1966) **62**, 1257.
12. Petrella, G., Sacco, A., Castagnolo, M., Della Monica, M., De Giglio, A., *J. Solution Chem.* (1977) **6**, 13.
13. Fuoss, R. M., Onsager, L., Skinner, J. T., *J. Phys. Chem.* (1965) **69**, 2581.
14. Coplan, M. A., Fuoss, R. M., *J. Phys. Chem.* (1964) **68**, 1177.
15. Petrella, G., Castagnolo, M., Sacco, A., Lasalandra, L., *Z. Naturforsch., Teil A* (1972) **27**, 1345.
16. Castagnolo, M., Jannelli, L., Petrella, G., Sacco, A., *Z. Naturforsch., Teil A* (1971) **26**, 755.
17. Sacco, A., Petrella, G., Castagnolo, M., *Z. Naturforsch., Teil A* (1971) **26**, 1306.
18. Castagnolo, M., Petrella, G., *Electrochim. Acta* (1974) **19**, 855.
19. Petrella, G., Castagnolo, M., Sacco, A., *Z. Naturforsch., Teil A* (1975) **30**, 533.
20. Petrella, G., Castagnolo, M., *Z. Naturforsch., Teil A* (1973) **28**, 1149.
21. Castagnolo, M., De Giglio, A., Dell'Atti, A., Petrella, G., *Can. J. Chem.* (1975) **53**, 1651.
22. Kay, R. L., Evans, D. F., *J. Phys. Chem.* (1966) **70**, 2325.
23. Franks, F., "Physico-Chemical Processes in Mixed Aqueous Solvents," F. Franks, Ed., p. 50, Heineman, London, 1967.
24. Broadwater, T. L., Kay, R. L., *J. Phys. Chem.* (1970) **74**, 3802.
25. Kay, R. L., Broadwater, T. L., *J. Solution Chem.* (1976) **5**, 57.
26. Kay, R. L., Broadwater, T. L., *Electrochim. Acta* (1971) **16**, 667.
27. Macdonald, D. D., Smith, M. D., Hyne, J. B., *Can. J. Chem.* (1971) **49**, 2818.
28. Benoit, R. L., Choux, G., *Can. J. Chem.* (1968) **46**, 3215.
29. Fuoss, R. M., *J. Am. Chem. Soc.* (1958) **80**, 5059.
30. Parker, A. J., *Chem. Rev.* (1969) **69**, 1.
31. Sellers, N. G., Eller, P. M. P., Caruso, J. A., *J. Phys. Chem.* (1972) **76**, 3618.
32. Benoit, R. L., Beauchamp, A. L., Domain, R., *Inorg. Nucl. Chem. Lett.* (1971) **7**, 557.
33. Benoit, R. L., Rinfret, M., Domain, R., *Inorg. Chem.* (1972) **11**, 2603.
34. Coetzee, J. F., Bertozzi, R. J., *Anal. Chem.* (1973) **45**, 1604.
35. Ibid. (1971) **43**, 961.
36. Morman, D. H., Harlow, G. A., *Anal. Chem.* (1967) **39**, 1869.
37. Kolthoff, I. M., Bruckenstein, S., Chantooni, M., *J. Am. Chem. Soc.* (1961) **83**, 3927.
38. Eller, P. M. P., Caruso, J. A., *Can. J. Chem.* (1973) **51**, 448.
39. Benoit, R. L., Buisson, C., Choux, G., *Can. J. Chem.* (1970) **48**, 2353.
40. Evans, D. F., Matesich, S. M. A., "The Physical Chemistry of Aqueous Systems," R. L. Kay, Ed., p. 95, Plenum, New York and London, 1973.
41. Evans, D. F., Thomas, J., Nadas, J. A., Matesich, S. M. A., *J. Phys. Chem.* (1971) **75**, 1714.
42. Witschonke, C. R., Kraus, C. A., *J. Am. Chem. Soc.* (1947) **69**, 2472.

43. Wright, C. P., Murray-Rust, D., Hartley, H., *J. Chem. Soc.* (1931) 199.
44. Evans, D. F., Zavoyski, C., Kay, R. L., *J. Phys. Chem.* (1965) **69**, 3878.
45. Barker, B. J., Caruso, J. A., *J. Am. Chem. Soc.* (1971) **93**, 1341.
46. Choux, G., Benoit, R. L., *J. Am. Chem. Soc.* (1969) **91**, 6221.
47. Shedlovsky, T., "Electrolytes," B. Pesce, Ed., p. 146, Pergamon, New York, 1962.
48. D'Aprano, A., Fuoss, R. M., *J. Phys. Chem.* (1969) **73**, 400.
49. Kay, R. L., *J. Am. Chem. Soc.* (1960) **82**, 2099.
50. Goffredi, M., Shedlovsky, T., *J. Phys. Chem.* (1967) **71**, 2176.

RECEIVED February 21, 1978.

7

Enthalpies of Solution of Some Electrolytes with a Large Cation in Mixtures of Water with Aprotic Solvents

C. DE VISSER, W. J. M. HEUVELSLAND, and G. SOMSEN

Department of Chemistry, Free University of Amsterdam,
De Boelelaan 1083, Amsterdam, The Netherlands

*Measured enthalpies of solution of tetra-*n*-butylammonium bromide (Bu_4NBr) in mixtures of water (W) with acetonitrile (ACN) and with ethylene carbonate (EC) are compared with those in mixtures of water with some other aprotic solvents which were reported earlier. The results can be fairly well described by an equation which can be derived either from a cooperative hydration model or from a chemical, pseudo equilibrium model. This equation is tested by varying systematically the nature of the solute and of the cosolvent. In order to describe the experimental results in W–ACN and in W–EC as such, it will be necessary to extend the equation with a term which comprises any specific (nonhydrophobic) interactions of the solute with the (inert) cosolvent.*

The enthalpy of solution of a solute in various solvents is an important datum in studying the behavior of different solutions because it might give direct information on the energetic aspects of both solute–solute and solute–solvent interactions. During the last decade in our laboratory (*see* Ref. *1* and literature cited therein) enthalpies of solution of alkali- and tetraalkylammonium halides have been measured in water and in several nonaqueous solvents including formamide (F), *N*-methylformamide (NMF), *N,N*-dimethylformamide (DMF), *N,N*-dimethylacetamide (DMA), and dimethylsulfoxide (DMSO). The standard enthalpies of transfer ($\Delta H_{tr}°$) (the difference between the standard enthalpies of solution of a certain solute in two solvents) for both alkali bromides and tetraalkyl-

0-8412-0428-4/79/33-177-099$05.00/1

ammonium bromides between two nonaqueous solvents are almost independent of the size of the cation, especially if only aprotic solvents are considered. On the other hand, if water is one of the solvents, ΔH_{tr}° of the tetraalkylammonium bromides deviate as compared with the corresponding alkali bromides since in water these salts show a considerable exothermic shift of the enthalpies of solution. Besides that, enthalpies of solution (ΔH°) of both alkali bromides and tetraalkylammonium bromides are almost proportional to the solvent composition in binary mixtures of two nonaqueous solvents, while in aqueous mixtures the alkali bromides show a similar behavior but the ΔH° of the tetraalkylammonium bromides reach large endothermic maxima in the water-rich region (*see* Ref. *2* and literature cited therein). Generally these phenomena are attributed to the well-known hydrophobic hydration behavior of the tetraalkylammonium salts in water (*3*). Since this special behavior is absent in non-hydrogen-bonded solvents, mixtures of these solvents with water can serve as model systems to study the hydrophobic hydration effect more systematically. It is possible to describe the results for both the enthalpies of dilution (*4*) and for the enthalpies of solution (*2*) in terms of a cooperative hydration model (*2, 5*) although a chemical model (*6*) leads to exactly the same final equation. Both models involve only two parameters: the number of water molecules (n) surrounding a hydrophobic particle and the enthalpic contribution to hydrophobic hydration in pure water, HbW.

In an earlier publication (*6*) we considered the basic assumptions of the models. This chapter briefly reviews this subject and extends the tests by adding new results of the ΔH° of tetrabutylammonium bromide (Bu_4NBr) in mixtures of water with acetonitrile (ACN) and with ethylene carbonate (EC) and preliminary results of some substituted and unsymmetrical tetraalkylammonium salts in mixtures of water with DMF.

Experimental and Results

The enthalpies of solution were measured with a LKB 8700-1 precision calorimetry system. The experimental procedure and test of the instrument have been given before (*6, 7*). EC (Fluka, purissimum) was distilled under reduced pressure and the middle fraction was stored over molecular sieves (4 Å) for at least 48 hr. ACN (Merck, pro analysis) was dried over molecular sieves and used without further purification. The purity of both solvents (determined shortly before use), as deduced from GLC, was always better than 99.8%. The volume fraction of water, determined by K. Fischer titration (*8*) was always less than 3.10^{-4}. The mixed solvents were prepared by weight as shortly as possible before the measurements. ΔH° of Bu_4NBr in W–ACN mixtures have been measured at 25°C while those in W–EC are at 45°C, which is above the melting point of pure EC.

Table I. Standard ΔH° of Bu_4NBr in Mixtures of Water and ACN at 25°C and in Mixtures of Water and EC at 45°C

In W–ACN		*In W–EC*	
X_W	ΔH° *kJ mol⁻¹*	X_W	ΔH° *kJ mol⁻¹*
0	16.80	0	33.08
0.025	11.69	0.030	27.95
0.060	10.83	0.052	25.71
0.109	9.79	0.101	23.59
0.175	9.96	0.151	22.26
0.244	10.79	0.250	21.65
0.329	12.28	0.354	21.86
0.400	13.53	0.499	22.62
0.496	14.87	0.625	23.05
0.578	15.56	0.740	22.87
0.601	15.80	0.850	22.35
0.636	15.95	0.900	21.80
0.720	16.64	0.950	18.48
0.750	17.05	0.970	14.92
0.797	17.82	1.000	5.89
0.850	18.77		
0.901	18.24		
0.920	16.58		
0.953	10.30		
0.976	2.09		
1.000	−8.42		

The concentration of Bu_4NBr in the mixtures varied between 0.003 and 0.01 mol kg^{-1} in which range any concentration dependence appeared to be masked by the experimental error. Therefore, ΔH° was taken to be the average of 2–4 independent measurements agreeing within 150 J mol^{-1}. Final results in ΔH° as a function of the mole fraction of water x_W are given in Table I.

Discussion

Recently (*6*) we found that the ΔH° of Bu_4NBr in W–DMF, W–DMSO, and W–DMA mixtures can be described fairly well in terms of a cooperative hydration model (*2, 5*) leading to the following equation:

$$\Delta H^{\circ}(\mathrm{M}) = (1 - X)\,\Delta H^{\circ}(\mathrm{S}) + X\,\Delta H^{\circ}(\mathrm{W}) + (X^{n} - X)\,\mathrm{HbW} \quad (1)$$

where $\Delta H^{\circ}(\mathrm{M})$ is the standard enthalpy of solution in a mixture of water with a cosolvent S and X is the mole fraction of water. As mentioned before, the same equation is obtained considering the hydration of a tetraalkylammonium ion (R_4N^+) by n cooperatively interacting water molecules as a pseudochemical equilibrium, which can be described by:

$$R_4N^+ + n\,H_2O \overset{K}{\rightleftarrows} R_4N^+ \ldots (H_2O)_n \qquad (2)$$

In this equation $R_4N^+ \ldots (H_2O)_n$ denotes the resulting hydrophobic entity and K is an equilibrium constant. The enthalpic effect of hydrophobic hydration then can be considered as the result of the formation of this hydration complex. In both models the choice of the cosolvent ought to be rather unimportant as long as this solvent does not show specific interactions like hydrogen bonding. As a consequence n and HbW should not vary with the different cosolvents. Besides that, the basic assumption in the concepts is that in the absence of hydrophobic hydration $\Delta H°$ would change proportionally to the solvent composition. In this chapter we will investigate more systematically both aspects.

Influence of Different Solutes. Variation of the solute gives information on the following questions: (1) what happens if hydrophobic hydration cannot or does not occur? In the case of the nonhydrophobic

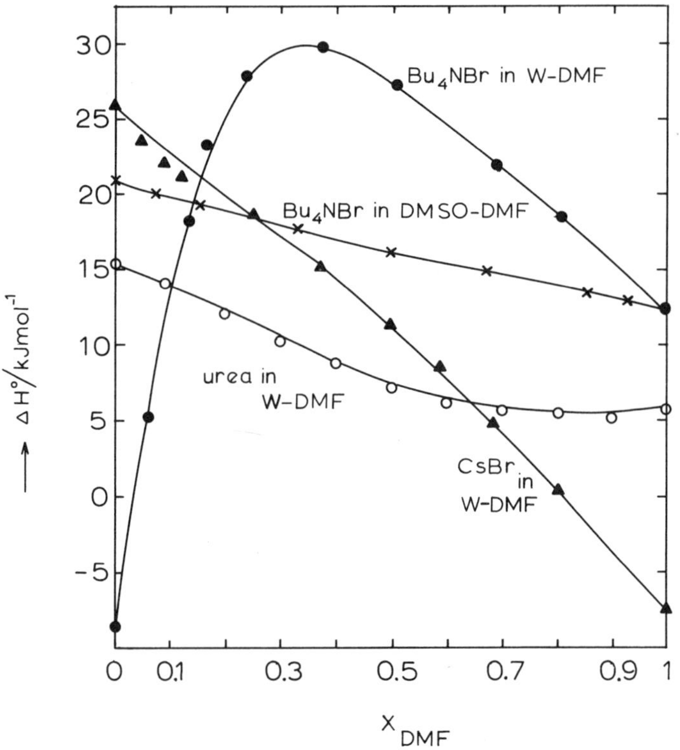

Figure 1. $\Delta H°$ of various solutes in mixtures of DMF with water and with DMSO as a function of the mole fraction of DMF X_{DMF} at 25°C

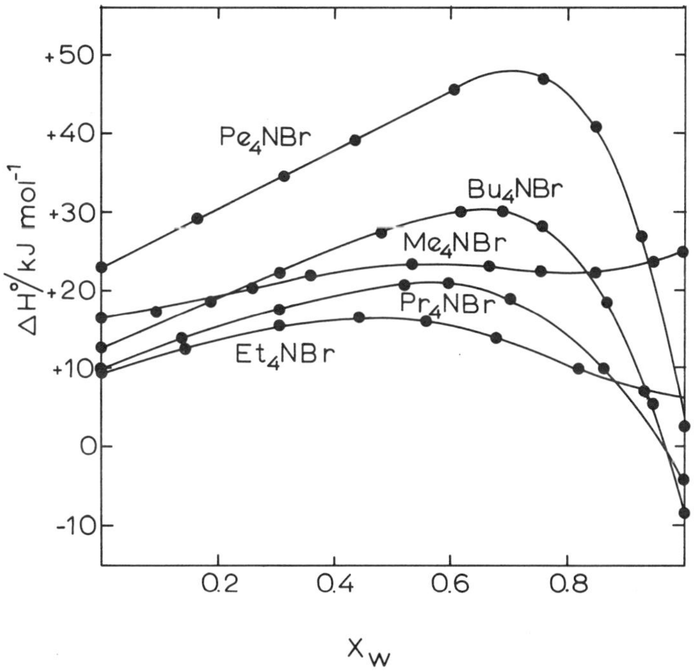

Journal of Physical Chemistry

Figure 2. ΔH° of tetraalkylammonium bromides in mixtures of water with DMF as a function of the mole fraction of water X_w at 25°C (6)

urea we have found that $\Delta H°$ in W–DMF indeed are almost proportional to the solvent composition (9) like for some alkali halides in W–DMSO given in the literature (*10*).

Preliminary results in our laboratory of CsBr in W–DMF display the same behavior (Figure 1). In this context it is also important to measure $\Delta H°$ of a hydrophobic particle in mixtures of two non-hydrogen-bonded solvents. Preliminary results for Bu_4NBr in mixtures of DMSO and DMF yielded a straight line. All of these results are summarized in Figure 2 which clearly demonstrates that only in the case of hydrophobic hydration is a large endothermic maximum in the $\Delta H°$ present. (2) What is the influence of the size of the hydrophobic solute? In Figure 2 $\Delta H°$ in W–DMF are given for Me_4NBr, Et_4NBr, Pr_4NBr, Bu_4NBr, and Pen_4NBr. Details of the measurements have been reported earlier (*6*).

As this figure shows, all five tetraalkylammonium salts display similar profiles although in the case of Me_4NBr the hydrophobic character almost has disappeared. A careful analysis of these results in terms of Equation 1 and after correction of the influence of the Br^- ion yielded a value of the parameter n of about 10, rather independent of the number of C atoms and a value of HbW of about − 4 kJ mol^{-1} per CH_2 group. Similar results have been found by Lindenbaum et al. (*5*) in the case of trialkylphosphates in W–DMF. Therefore, in these cases, Equation 1 provides a good possibility to describe the experimental results. (3) What is the influence of different substitution groups on the hydrophobic character of a tetraalkylammonium salt? In Figure 3 preliminary results of three different solutes are compared with those of Pr_4NBr in W–DMF. When in Pr_4NBr the four terminal CH_3 groups are substituted by four OH groups ($(EtOH)_4NBr$) the $\Delta H°$ again is almost proportional to the solvent composition showing that $(EtOH)_4NBr$ is no longer a hydrophobic solute in water nor in mixtures of water with DMF. This conclusion is in agreement with results from other measurements for this salt in water (*11*). Obviously, introduction of the hydrophilic OH groups converts this salt to a rather normal nonhydrophobic one. When in Pr_4NBr the alkyl

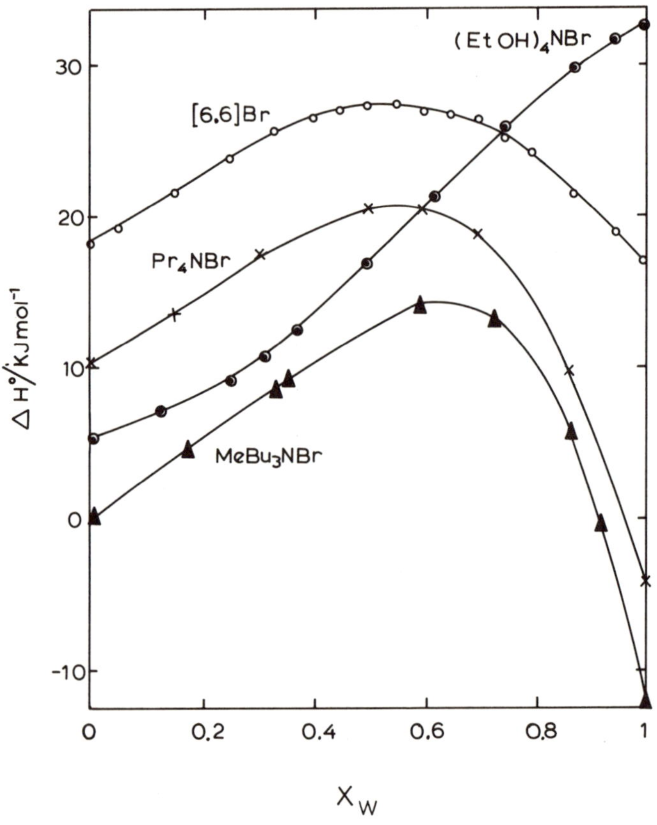

Figure 3. ΔH° of various salts in mixtures of water with DMF as a function of the mole fraction of water X_w at 25°C

chains are two-by-two interconnected (7-azonia [6,6] tridecane bromide), the hydrophobic character is maintained but to a much smaller extent. In view of the hydration model this seems quite reasonable since in this case especially it will be hardly possible for the water molecules to cooperate in forming either subcages or hydration complexes around the alkyl chains owing to the low flexibility of the alkyl groups, the absence of terminal methyl groups, and the wrong shape of the molecule (*12*). In view of other measurements (*12*) this result is not unexpected. Finally, for nonsubstituted but unsymmetrical tetraalkylammonium bromides Equation 1 leads to the same values for HbW and n per C-atom as for the symmetrical tetraalkylammonium salts. For example, in the case of $MeBu_3NBr$, after correction for the Br^- ion, the value of n again is about 10, while HbW per C atom equals about -4 kJ mol^{-1} also. This supports one of the assumptions underlying Equation 1 which states that the magnitude of the hydrophobic effect indeed is additive with the number of CH_2 groups.

Influence of Different Cosolvents. As stated previously the nature of the cosolvent which is added to water to prepare the mixtures ought to be unimportant as long as the role of this solvent is restricted to a dilution of the water, i.e. as long as specific interactions like hydrogen bonding in this solvent are absent. Recently (*6*) we reported ΔH° of Bu_4NBr in mixtures of water and DMF, DMSO, and DMA to verify this idea. However, although the values of HbW and n were quite reasonable in these three cases, the fit of the experimental results to Equation 1 in W–DMSO and W–DMA was less accurate than of those in W–DMF and the values of the calculated parameters HbW and n slightly varied with the different cosolvents. Therefore we extended this test by measuring the ΔH° of Bu_4NBr in W–EC and W–ACN as given in Table I. All results are summarized in Figure 4. The ΔH°–composition profile is quite different in W–EC and W–ACN as compared with that in the other solvent systems. This is especially demonstrated by the endothermic shift of the enthalpies in the nonaqueous region which does not occur in mixtures with the amides or with DMSO. The application of Equation 1 to these results is meaningless, thus our description seems to fail completely for these mixtures. However, in this case it is interesting to consider the behavior of a nonhydrophobic solute in the same solvent mixture. The only reference we could find in the literature is given by Tomkins et al. (*13*) who measured ΔH° of $LiClO_4$ in W–ACN. Also in that case a pronounced minimum in the ACN-rich region occurs. Most probably this minimum is the result of some specific interaction of the relatively small cation or anion with the nonaqueous solvent. Therefore the minimum in the curves in Figure 4 in the W–ACN and W–EC mixtures might be caused by the influence of the anion which is also reflected by the deviation of ΔH° of

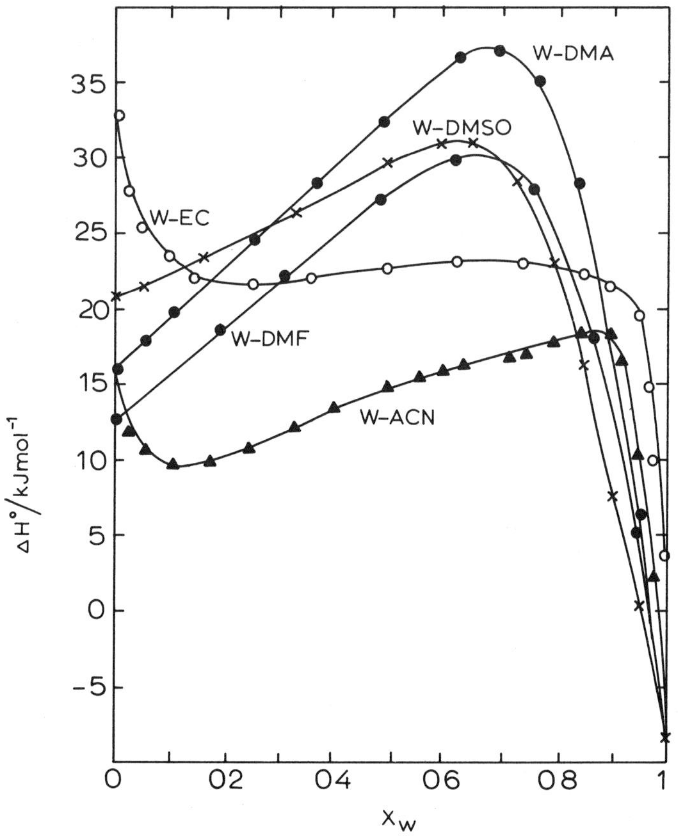

Figure 4. ΔH° of Bu_4NBr in various aqueous mixtures as a function of the mole fraction of water X_w. All results are at 25°C except for those in W–EC which are at 45°C.

$LiClO_4$ in W–ACN from a straight line ("ideal behavior"). Therefore, as a first rough approximation we subtracted this deviation from the $\Delta H°$ values of Bu_4NBr in the same mixture at the corresponding mole fractions. Indeed, a similar profile as in W–DMF, W–DMA, and W–DMSO then occurs. Of course, it would be much better to have data of an alkali bromide since the influence of the relatively large ClO_4^- anion is unknown. A similar reasoning might apply to the results in W–EC although data for other salts in this mixture are not available at present. Of course it is also possible that the minimum in the nonaqueous end is caused by specific interactions of the nonaqueous solvent itself with water. In this respect it should be noted that the partial molar heat capacity of water infinitely diluted in ACN is unexpectedly high (*14*). More experiments in this field are in progress.

Literature Cited

1. de Visser, C., Somsen, G., *J. Chem. Soc., Faraday Trans. 1* (1973) **69,** 1440.
2. de Visser, C., Somsen, G., ADV. CHEM. SER. (1976) **155,** 289.
3. Wen, W. Y., "Water and Aqueous Solutions," R. A. Horne, Ed., p. 613, Wiley Interscience, New York, 1972.
4. Mastroianni, M. J., Pikal, M. J., Lindenbaum, S., *J. Phys. Chem.* (1972) **76,** 3050.
5. Lindenbaum, S., Stevenson, D., Rytting, J. H., *J. Solution Chem.* (1975) **4,** 893.
6. Heuvelsland, W. J. M., de Visser, C., Somsen, G., *J. Phys. Chem.* **(1978) 82,** 29.
7. de Visser, C., van Netten, E., Somsen, G., *Electrochim. Acta* (1976) **21,** 97.
8. Verhoef, J. C., Barendrecht, E., *Anal. Chim. Acta* (1977) **94,** 395.
9. de Visser, C., Grünbauer, H. J. M., Somsen, G., *Z. Phys. Chem. (Frankfurt)* (1975) **97,** 69.
10. Vorob'er, A. F., Monaenkova, A. S., Padúnova, I. D., *Moscow Univ. Chem. Bull. (Engl. Transl.)* (1974) **29,** 14.
11. Wood, R. H., Belkin, F., *J. Chem. Eng. Data* (1973) **18,** 184.
12. LoSurdo, A., Wen, W.-Y., Jolicoeur, C., Fortier, J.-L., *J. Phys. Chem.* (1977) **81,** 1813.
13. Tomkins, R. P. T., Gerhardt, G. M., Lichtenstein, L. M., Turner, P. J., ADV. CHEM. SER. (1976) **155,** 297.
14. de Visser, C., Heuvelsland, W. J. M., Dunn, L. A., Somsen, G., *J. Chem. Soc., Faraday Trans. 1* (1978) **74,** 1159.

RECEIVED February 14, 1978.

8

The Apparent Charge of Ionic Micelles in Aqueous Binary Solvents

CLAUDE TREINER, ANNE LeBESNERAIS, and CHRISTIAN MICHELETTI

Universite Pierre et Marie Curie, Laboratoire d'Electrochimie, 4, Place Jussieu, 75005 Paris, France

The degree of dissociation α of ionic micelles for sodium decylsulfate and trimethyldecylammonium bromide have been determined in water + acetone and water + n-*propanol mixtures from conductance measurements. Using these data and assuming a pseudophase model for the equilibrium between surfactant ions in the bulk and in the micellar phase the variation of the apparent charge of the micelles Z with addition of organic solvents can be calculated. After a small initial decrease, a constant Z value is observed in all cases studied; this result is interpreted as the consequence of two concomitant phenomena: an increase of α caused by the preferential solvation of cations by the organic component and a decrease of the aggregation number of the micelles as the water structure is progressively destroyed by that same component.*

Ionic micelles are charged aggregates formed essentially in water from the cooperative ion–ion interactions of electrolytes constituted by a linear hydrocarbon chain attached to an ionic group and a counterion. The addition of organic substances to ionic micelles may alter the size and charge and ultimately destroy these ionic aggregates (*1*); however, little is known about the behavior of the ionic constituents of the micelles in the mixed solvents and their influence on the micelle itself, although in some cases changes in the property of a micellar solution with addition of organic molecules have been attributed to changes of the ion–solvent interactions (*2, 3, 4, 5*).

0-8412-0428-4/79/33-177-109$05.00/1

Incorporation of that effect was difficult because reliable thermodynamic data were not available for the standard properties of the ionic components of the micelle in the mixed solvents. Thus we have determined recently (*6, 7*) the standard Gibbs free energy of transfer, $\Delta G_t°$, of two ionic surfactants, namely trimethyldecylammonium bromide (*n*DTMABr) and sodium decylsulfate (SDS) from water to water + acetone mixtures at 298.15 K, using a precise vapor pressure technique. We shall show, using the pseudophase model in order to describe the equilibrium between ions in the bulk and in the micellar phase, that the behavior of micelles at the critical micelle concentrations (cmc) can be, in these mixed solvents, essentially deduced from the properties of the ionic constituents at infinite dilution. Furthermore, we have determined the degree of dissociation of the micelles using a conductivity method, and formally treating the electric potential created by the micelle by numerical integration of the Poisson–Boltzmann Equation we have determined the variation of its apparent change with the addition of acetone. The case of water + *n*-propanol mixtures where mixed micelles are formed also will be discussed.

Theory and Experimental

The micellar solution is described using the pseudophase model (*1, 8, 9, 10*), which implies that the cmc corresponds to a maximum concentration of monomers in the bulk solution above which a second phase (the micelle phase) appears. This is a very crude model which has been discussed extensively in the literature. We use it merely because it is the only one tractable in the complex media we are dealing with (ions + micelles + mixed solvent). We may write:
in water,

$$\mu_w = \mu_w° + 2RT \ln (\text{cmc})_w \; (f_\pm)_w = \mu_{\text{mic}}° + RT \ln a_w; \qquad (1)$$

in solvent *s*,

$$\mu_s = \mu_s° + 2RT \ln (\text{cmc})_s \; (f_\pm)_s = \mu_{\text{mic}}'° + RT \ln a_s,$$

where μ_s' is the chemical potential of the monomer in the bulk solution and in the micelle, $\mu_{\text{mic}}°$ the standard chemical potential of the monomer in the micelle, $f_\pm$ the mean activity coefficient of the free ions, and *a* the activity of the monomer in the micelle phase.

If the solvent does not participate to the micelle formation, then the micelle may be considered as a pure phase, then $a_w = a_s = 1$. In that particular case we may also write $\mu_{\text{mic}}° = \mu_{\text{mic}}'°$. Thus we get:

$$\mu_s^\circ - \mu_w^\circ = \Delta G_t^\circ = 2RT \ln \frac{(\mathrm{cmc})_w \, (f_\pm)_w}{(\mathrm{cmc})_s \, (f_\pm)_s} \tag{2}$$

If we consider that the organic solvent participates to the micelle formation, then:

$$\Delta G_t^\circ = 2RT \ln \frac{(\mathrm{cmc})_w \, (f_\pm)_w}{(\mathrm{cmc})_s \, (f_\pm)_s} + RT \ln x\, f + \quad G_t^\circ{}_{(\mathrm{mic})} \tag{3}$$

where X is the mole fraction of monomer surfactant in the mixed micelle formed by the surfactant and the organic molecules (excluding water) and f the activity coefficient of the monomers in the micelle. Finally, the micellar solution may be described using the electrochemical potential formalism with the electrochemical potential of the free ions assumed to be zero (*1*):

$$\Delta G_t^\circ = 2RT \ln \frac{(\mathrm{cmc})_w \, (f_\pm)_w}{(\mathrm{cmc})_s \, (f_\pm)_s} + RT \ln Xf + \mathrm{Ne}\,(\phi_s - \phi_w) + \Delta G_t^{\circ\prime}{}_{\mathrm{mic}} \tag{4}$$

where ϕ_s and ϕ_w are the electric potentials owing to the micelle electric phase in the solvent mixture and in water; they depend on the charge, size, and shape of the micelle and the dielectric constant of the medium. Again, if the solvent does not participate to the micelle formation, $RT \ln Xf = \Delta G_t^\circ{}_{\mathrm{mic}} = 0$ and:

$$\Delta G_t^\circ = 2RT \ln \frac{(\mathrm{cmc})_w \, (f_\pm)_w}{(\mathrm{cmc})_s \, (f_\pm)_s} + \mathrm{Ne}\,(\phi_s - \phi_w) \tag{5}$$

ΔG_t° is the standard Gibbs free energy of transfer of the ionic surfactant between water and the solvent s; $\Delta G_t^\circ{}_{\mathrm{mic}}$ is the same quantity defined between a micelle in water and in solvent s.

The aim of most authors (*3, 11, 12*) in using Equations 2, 3, 4, or 5 was to predict from a given model the variation of the cmc with solvent composition. To our knowledge, however, except in a few cases (*3, 4*), ΔG_t° was ignored and the electric potentials ϕ were calculated with drastic simplifications: e.g., the variation of the apparent charge of the micelle with the addition of the organic component was not taken into account. There is no easy way of determining the composition X of the mixed micelle, so we shall focus our discussion on the simplest case, i.e. that described by Equation 5.

The main goal will be the determination of the apparent charge of the micelle in the binary aqueous solvents. Contradictory findings are found in the literature when the apparent charge is deduced from emf determinations (*12, 13*). We shall discuss these results and compare them with those obtained using a conductivity approach.

Calculation of the Electric Potential of the Micelle

The electric potential ψ created by the micelle was calculated using a numerical integration of the Poisson–Boltzman Equation by the Runge–Kutta method on an electronic computer. We shall outline briefly the method used (*14, 15, 16*). The Poisson–Boltzman Equation can be written (for programming commodities) in the following form:

$$\frac{d^2\phi}{dx^2} = -\frac{Z_M}{2}\left[\exp\left(-\frac{Z_i}{Z_M}\right) - \exp\left(-\frac{Z_j}{Z_M}\right)\right] - \frac{2}{x}\frac{d\phi}{dx} \tag{6}$$

where i and j correspond to the ions of the ionic atmosphere, M to the monomer surfactant. In the present case, $Z_i = -Z_j = Z_M$. $x = \chi R$ where χ is the inverse of the Debye–Hückel distance and R the distance from the micelle at which the potential is calculated. The reduced electric potential is $\phi = (Z_M e\psi/kT)$.

Two initial conditions are necessary in order to solve Equation 6.

(1) At a large distance R from the micelle where $\phi(x_o) << 1$, the Debye–Hückel potential $\psi(R)$ is adopted in order to start the calculation

$$\phi_o = \frac{Z_i e\chi}{kT} A \frac{\exp(-x_o)}{x_o} \quad \text{with } x_o = \chi R \tag{7}$$

and

$$A = \frac{Z_{mic}e}{D}\frac{\exp(\chi a)}{1+\chi a}$$

where a is the minimum distance of approach of a counterion from the micelle center. $Z_{mic} = \alpha n Z_M$ is the apparent charge of the micelle where n is the number of monomer per micelle and α the degree of dissociation of the micelle.

(2) At the distance x_o, the field is given by the following expression:

$$\left(\frac{d\phi}{dx}\right)_{x=x_o} = -\frac{\phi_o(1+x_o)}{x_o}$$

The Runge–Kutta method then can be used in order to calculate at the closest distance $r = a$, ϕ_a and $(d\phi/dx)_{x=x_a}(x_a = \chi a)$. However, at short distances, the Debye–Hückel approximation is no longer valid ($\phi_a > 1$); it then is assumed that at the micelle surface, the field must be coulombian, so that:

$$\left(\frac{d\phi}{dx}\right)_{x=x_a} = -\frac{Z_i Z_{mic}\, e^2\, \phi}{DkT' x^2_a} \qquad (8)$$

The calculation is repeated with a new constant A until the following test is verified.

$$\left(\frac{d\phi}{dx}\right)^{calc.}_{x=x_a} = \left(\frac{d\phi}{dx}\right)^{imposed}_{x=x_a}$$

The calculations have used $R = 40$ Å at long distance; we have checked that for longer distances the results were unchanged. We have also shown that different A values do not change the potential $\phi(R)$ for large distances. Finally, for very small concentrations we obtain asymptotically the coulomb potential for the isolated micelle. We used the same value of a (the contact distance) throughout the calculations ($a = 18.7$ Å) (*17*) for all the solvent mixtures (*see* the Discussion section). The number of monomers per micelle in water are 36 for *n*DTMABr (*18*) and 50 for SDS (*19*). Knowing the degree of dissociation of the ionic micelle from independent measurements, Equation 5 is solved for the apparent charge of the micelle Z_{mic} which is the only unknown parameter. However, α is a function of n through Equation 9 so an iterative procedure is adopted between Equations 6 and 9; the convergence is very rapid.

Determination of the Degree of Dissociation α of the Micelles

We have chosen Evans' conductivity method (*20*); it is shown that α may be obtained from the following equation:

$$(1000\, S_1 - \Lambda_c)\left(\frac{n - m}{n^{4/3}}\right)^2 + \Lambda_c\left(\frac{n - m}{n}\right) - 1000\, S_2 = 0 \qquad (9)$$

where S_1 and S_2 are the slopes of the curve representing the variation of specific conductivity with concentration below and above the cmc; $\alpha = (n - m/n)$ where n is the number of monomers per micelle and m the number of counterions of equivalent conductance Λ_c adsorbed on the micelle. Calculations are made first with the known n values in water and slightly changed after the first result of the apparent charge calculations for the micelles as described in the preceeding section.

Table I. Transference Number of NaBr in a 20 Weight Percent Water + Acetone Mixture at 298.15 K

C (*mol* L^{-1})	t (Na^+)
0.0	0.425
0.005385	0.4210
0.021240	0.4189
0.10665	0.4110

The critical micelle contraction values were obtained using the conductance method from the usual equivalent conductance vs. the square root of concentration plots. The apparatus has been described before (*21*). About 20 experimental points were obtained for each cmc determination. No cmc could be detected above 20 wt % of acetone and 30 wt % of *n*-propanol. Surfactants were studied in water + acetone mixtures at 298.15 K; in water + *n*-propanol mixtures at 308.15 K for comparison purposes with other systems (*12*) (*see* the Discussion section).

In order to obtain Λ_c, the transference number t_c of the counterion has to be known. In the water + *n*-propanol mixtures transference numbers were not available. The following procedure was used: the transference number is known for Na^+ in water + ethanol and water + *tert*-BuOH mixtures (*22, 23*) and its variation with alcohol concentration is small; knowing the necessary conductance data, *t* could be calculated for Na^+ and Br^-. Intermediate values were chosen in the water + *n*-propanol mixtures applying a slight (and known) temperature dependence (*24*). The cmc values are low enough in these media and the transference numbers do not vary significantly with concentration (*22, 23*) so that the value of t_0 corresponding to infinite dilution was adopted. In the water + acetone mixtures the cmc are too high (up to 0.13 mol/L and the preceeding procedure could not be used. An experimental value for $t(Br^-)$ and $t(Na^+)$ in a 20 wt % of acetone mixture was determined using a method based on radiotracer elements. (We are most grateful to Mrs. Perie for performing these experiments for us (*25*).) Table I presents the results obtained and the transference number values used in the calculation of α. For the other water + acetone mixtures, interpolated *t* values were used.

Materials

Acetone and *n*-propanol were Merck reagent-grade chemicals. SDS and sodium dodecylsulfate (SDOS) (Merck 99% pure) were used without purification, the limiting conductance at 298.15 K in water being the purity criterion (*6*); *n*DTMABr (Eastman Kodak) was recrystallized three times from pure acetone and dried 24 hr at 323 K before use.

Results

Tables II and III present the values of Λ, Λ_c, 1000 S_1, 1000 S_2, cmc, and α for *n*DTMABr and SDS in water + acetone and water + *n*-propanol mixtures. In order to calculate Z_{mic} from Equations 2, 3, 4, or 5 we have to use our previous determinations of $\Delta G_t°$ (*6, 7*) (*see* Table V). This could be done for *n*DTMABr and SDS in water + acetone mixtures. Unfortunately, the determination of $\Delta G_t°$ for both surfactants in water + *n*-propanol mixtures by the vapor pressure technique were of insufficient accuracy because of concentration limits; the cmc values being rather low, pressure measurements proved to be difficult in the concentration range required (sufficiently below the cmc value, i.e. $m \leq 0.01$ m/kg) in these particular experiments.

The cmc values of *n*DTMABr and SOS in water are in good agreement with literature values (*26*); our α values are slightly different from Evans' for SDS (0.26 and 0.32 respectively); however, there is a 15 K temperature difference between the two experiments. Furthermore, Evans uses the transference number of the sodium ion at infinite dilution in his calculation. We have also determined α for sodium dodecylsulfate in water at 298.15 K. We found $\alpha = 0.24$. Literature values range from 0.18 to 0.28 according to the physicochemical method used (*27, 28*).

Table II. Calculation of the Degree of Dissociation α of a Micelle at 298.15 K

Wt (%)	*1000 S_1*	*1000 S_2*	Λ	Λ_c	*cmc ($mol\ L^{-1}$)*	α
		n*DTMABr in Water + Acetone Mixtures* $[\Lambda_c = \Lambda(Br^-)]$				
0	72.19	30.21	108.74	66.88	0.0657	0.35
5	64.08	32.00	97.47	59.16	0.0724	0.41
10	56.45	37.75	86.50	51.83	0.0743	0.53
15	51.53	42.98	79.91	47.39	0.0784	0.66
17	47.22	37.84	75.42	44.65	0.1002	0.80
20	—	—	70.46	41.71	0.136	—
		SDS in Water + Acetone Mixtures $[\Lambda_c = \Lambda_c(Na^+)]$				
0	62.10	27.72	116.16	45.07	0.0317	0.26
5	56.60	36.32	106.75	42.27	0.0286	0.35
10	53.26	44.05	97.45	39.36	0.0310	0.49
15	49.35	41.71	88.17	36.41	0.0366	0.51
20	42.77	38.88	78.48	32.73	0.0537	0.56
		SDOS in Water $[\Lambda_c = \Lambda°(Na^+)]^a$				
0	66.17	29.07	—	50.10	0.0079	0.24

[a] The aggregation number is equal to 80 (*29*).

Table III. Calculation of the Degree of Dissociation of a Micelle at 308.15 K: *n*DTMABr in water + *n*-Propanol Mixtures ($\Lambda_c^\circ = \Lambda^\circ(CBr^-)$)

Wt(%)	*1000 S_1*	*1000 S_2*	Λ_c°	*cmc (mol L^{-1})*	α
0	88.37	40.73	80.74	0.0655	0.37
5	80.61	50.72	67.11	0.0443	0.42
10	70.77	58.74	59.41	0.0294	0.56
15	60.76	52.83	52.16	0.0337	0.63
20	52.37	44.85	45.84	0.0523	0.60

Discussion of Thermodynamic Behavior of the Ionic Constituents of Micelles in Mixed Solvents

In order to discuss the behavior of the ionic constituents of the micelles studied here, we find it useful to make use of an extra-thermodynamic approach. Among the many extra-thermodynamic approaches which have been proposed, the tetraphenylboron assumption is the most widely used. However, we have recently shown that it was possible to calculate the standard Gibbs free energy of transfer of a large ion like the tetrabutylammonium ion (in order to minimize the charge effect which cannot be predicted) using the scaled-particle theory (*29*); Desrosiers et al. have made the same calculations for the heat capacity and volume of transfer of that ion from water to various aqueous binary mixtures (*30*). Using our previous determination of ΔG_t° for $n\mathrm{Bu_4NBr}$, $\mathrm{Me_4NBr}$, and $\mathrm{MeOSO_3Na}$ in water + acetone mixtures (*31, 32*), the ΔG_t° value for the bromide ion and hence of all the ionic constituents of the micelles studied could be obtained. We did not have ΔG_t° for NaBr at our disposal, but only for KBr in these mixtures. However, Deligny et al. have shown that ΔG_t° was identical for KBr and RbBr and only 0.3 kcal/mol different (at most) up to 80 wt % from CsBr (*33*). Assuming then that $\Delta G_t^\circ(\mathrm{Na^+}) \approx \Delta G_t^\circ(\mathrm{K^+})$ will not change our conclusions.

The work of cavity formation can be expressed by the following relationship (mole-fraction scale) (*34, 35*):

$$G_{CAV} = G_c + RT \log (RT/V)$$

$$G_c = RT\left[-\log(1-\phi) + \frac{C^2}{1-\phi}3Y + \frac{C}{1-\phi}3X + \frac{\phi}{2}X^2\frac{C^2}{(1-\phi)^2}\right]$$

where

$$\phi = \frac{\pi N}{6V}(Z_1 d^3 + Z_2 b^3)$$

$$X = \frac{\pi N}{6V} (Z_1 d^2 + Z_2 b^2)$$

$$Y = \frac{\pi N}{6V} (Z_1 d + Z_2 b) \text{ and}$$

$$V = M_{12}/d_o$$

$$\text{with } \Delta G_t^\circ \text{ (CAV)} = G_c(\text{CAV})_s - G_c(\text{CAV})_w$$

The *a, b,* and *C* are respectively the diameter (in Å) of each solvent and solute molecules; d_o is the density of the solvent. $M_{12} = M_1Z_1 + M_2Z_2$ where Z_1 is the water-mole fraction; M_1 and M_2 are the molar masses of the solvent.

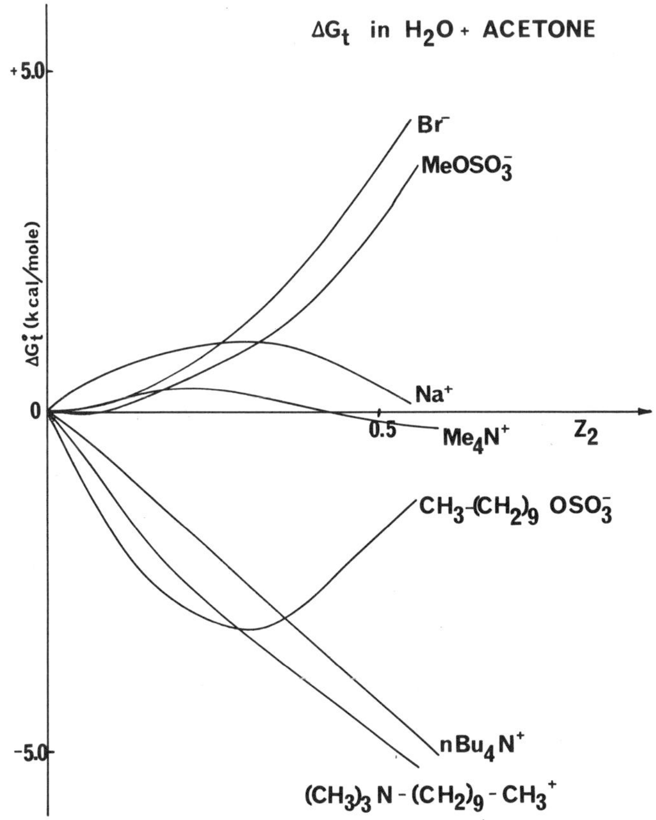

Figure 1. Single-ion standard Gibbs free energy of transfer from water to water + acetone mixtures based on S.P.T. calculations for the tetrabutylammonium ion (mole-fraction scale)

We have adopted our previous estimates for the hard sphere diameters of solute and solvent molecules, i.e. $C = 7.74$ Å ($n\mathrm{Bu_4N^+}$), $d = 2.76$ Å (water), $b = 4.75$ Å (acetone) (*29*).

Figure 1 presents the results of these calculations for all the electrolytes we have studied in water + acetone mixtures. Note that these results for single-ion $\Delta G_t°$ values would be shifted to more negative values by at most 2 kcal/mol (*33*) if we had used Deligny's entirely different extrathermodynamic approach (*33*). $\Delta G_t°$ (−) is positive for the bromide as well as for the methylsulfate ion, a classical behavior for inorganic anions in aqueous binary mixtures, which can be interpreted as a preferential solvation by water molecules or by the Born electric effect. For $\mathrm{Me_4N^+}$, $\Delta G_t°$ (+) is close to zero, an interesting result since it corresponds to Abraham's extrathermodynamic assumption (*36*); $\Delta G_t°$ (+) for the sodium ion is slightly positive in the rich aqueous mixtures and tends towards negative $\Delta G_t°$ values at higher organic compositions.

Following previous investigations (*37, 38*) it may be helpful at this stage to consider $\Delta G_t°$ as made up of several contributions. $\Delta G_t°$ (Exp) = $\Delta G_t°$ (CAV) + $\Delta G_t°$ (SPEC) + $\Delta G_t°$ (EL) + $\Delta G_t°$ (STRC) where subscripts CAV, SPEC, EL, and STRC stand respectively for cavity, specific, electric, and structural effects.

We know from S.P.T. calculations that the cavity term will be negative for the transfer from water to water + acetone mixtures (Table IV). According to the Born Equation $\Delta G_t°$ (EL) is positive. Insofar as the separate consideration of a specific and and an electric term is not arbitrary we may note that in order to have $\Delta G_t°$ (SPEC) < 0 (preferen-

Table IV. Extrathermodynamic Assumption Based on Scaled Particle Theory Calculations for a Tetrabutylammonium Particle in Water + Acetone Mixtures at 298.15 K[a, b]

Wt (%)	d_o (*g cm*$^{-3}$)	$-\Delta G_t°$ (*calc,* n*Bu*$_4$ *N*) (*cal mol*$^{-1}$)	$-\Delta G_t°$ (*exp,* n*Bu*$_4$ *N Br*)[c] (*cal mol*$^{-1}$)
0	0.9971	0	0
5	0.990	135	90
10	0.983	275	195
20	0.969	564	430
30	0.954	945	680
40	0.936	1485	915
50	0.917	2102	1110
60	0.894	2946	1230
80	0.843	4959	835
100	0.790	7165	—

[a] Mole-fraction scale.
[b] $\Delta G_t°$ ($\mathrm{Br^-}$) = $\Delta G_t°$ (exp, $n\mathrm{Bu_4}$ NBr) − $\Delta G_t°$ (calc, $n\mathrm{Bu_4}$ N).
[c] Ref. *32*.

tial interaction of an ion with acetone) we must have: $\Delta G_t°$ (Exp) $<$ $\Delta G_t°$ (CAV)) $+$ $\Delta G_t°$ (EL) which means that $\Delta G_t°$ (EL) $>$ $\Delta G_t°$ (Exp). This is most likely the case for Me_4N^+ and Na^+ for which $\Delta G_t°$ (Exp) is near zero. We conclude that for these two ions, acetone plays a major role in their solvation shells. This result will be useful when we discuss the variation of the degree of dissociation α of the micelles with the addition of organic solvents.

The behavior of the decylsulfate ion is particularly interesting (Figure 1). $\Delta G_t°$ (CAV) is negative for the hydrocarbon chain according to the S.P.T. calculations, but the $MeOSO_3^-$ group being preferentially solvated by water ($\Delta G_t°$ (–) $>$ 0) the opposite behavior of the two different parts of the ion towards the solvent is responsible for the minimum observed; no minimum could be found for the nDTMA$^+$ ion; $\Delta G_t°$ for Me_4N^+ is near zero, as consequently $\Delta G_t°$ should be for $-Me_3N^+$ according to the results of Figure 1, and $\Delta G_t°$ (CAV) being again negative for the nonylhydrocarbon chain.

Degree of Dissociation of Ionic Micelles in Mixed Solvents

Figure 2 compares the variation of α with solvent composition for related surfactants as obtained from emf (*12*) and conductivity data. The emf results (at 308.15 K) have made use of the Botré Equation (*39*), the validity of which has been questioned (*40, 41*). However, the main characteristics of these results are that they qualitatively present the same micelle behavior, i.e. an increase of α with the addition of acetone or propanol to water. The increase is much faster with propanol using the emf method than with the conductivity method and as Miyagishi apparently could not determine an α value above mole fraction 0.02 we suspect some electrode problems in this particular case. In the water + acetone mixtures the two methods yield similar results. Nevertheless, we believe that the conductivity method is more reliable than the emf method. Finally, Mathews et al. (*13*) found no change of α with addition of organic molecules in apparent contradistinction with the present results. However, these authors used scarcely soluble additives (e.g., CCl_4) and Figure 2 shows that α changes slowly with the addition of acetone or propanol.

The n-propanol forms mixed micelles with SDS or nDTMABr but not with acetone or at least not to the same extent (*42*). Consequently, it may be found intriguing that α increases with about the same rate for a cationic, anionic, and/or a mixed micelle. The first effect of adding the organic additives to the aqueous micelles is to solvate the cationic charges, resulting in the prevention of counterions from approaching as near to the micelle surface as they do without the organic additives. In the case of the cationic micelle we have shown that the Me_4N^+ ion and thus the

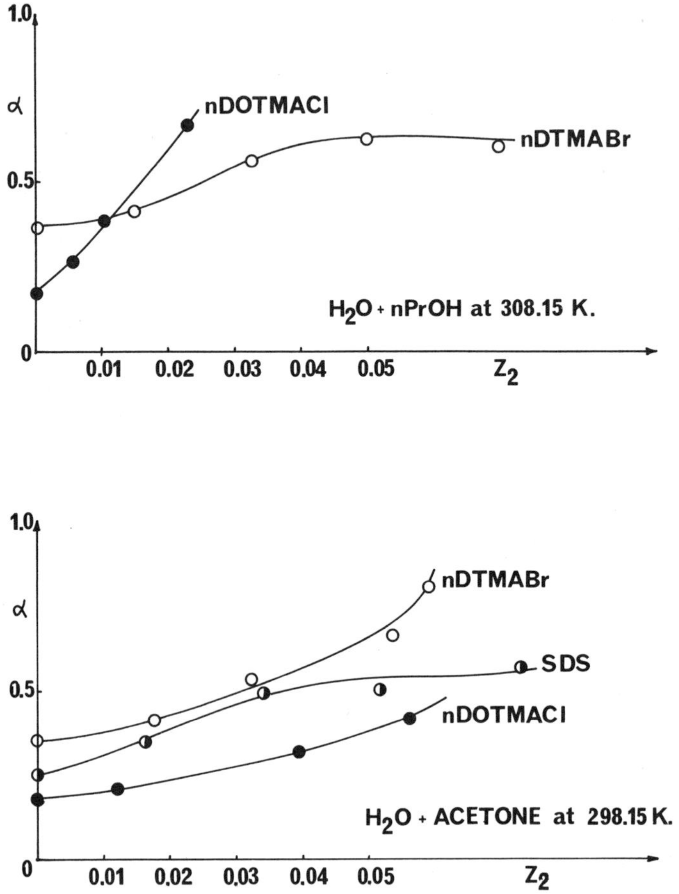

Figure 2. Variation of the degree of dissociation of ionic micelles with solvent composition. (●), Data from emf measurements, (○, ◑), conductance + mobility measurements. D = decyl, DO = dodecyl.

$-Me_3N^+$ group are preferentially solvated by acetone with the bromide ion solvated by water; in the case of the anionic micelle with the negative head-groups solvated by water, it is the counterions (positive) which are solvated by acetone with the same overall result as for the cationic micelle: a larger desolvation energy has to be provided in order for the counterions to approach the micelle surface with the organic molecules solvating the cations preferentially. The faster increase of α in the case of the cationic micelle over that of the anionic micelle in the water + acetone system is in agreement with the results of Figure 1: $-Me_3N^+$ is more strongly solvated than Na^+. The same interpretation holds for the *n*DTMABr + *n*-propanol + water system. Different extrathermodynamic assumptions used for the evaluation of single ions ΔG_t° values in water + ethanol, water + methanol (*43*), and water + *tert*-BuoH (*44*) mixtures have indicated that here also inorganic anions are preferentially solvated by water and cations by the organic solvent. Thus we may apply the same qualitative conclusion for the water + *n*-propanol mixtures with the same consequences as far as the solvation of cationic groups on the micelle by the organic molecules is concerned.

Variation of the Critical Micelle Concentration and Apparent Charge of the Micelles in the Mixed Solvents

We shall first discuss semiquantitatively the behavior of the ionic micelles. Figure 3 presents the variation of the cmc for *n*DTMABr and

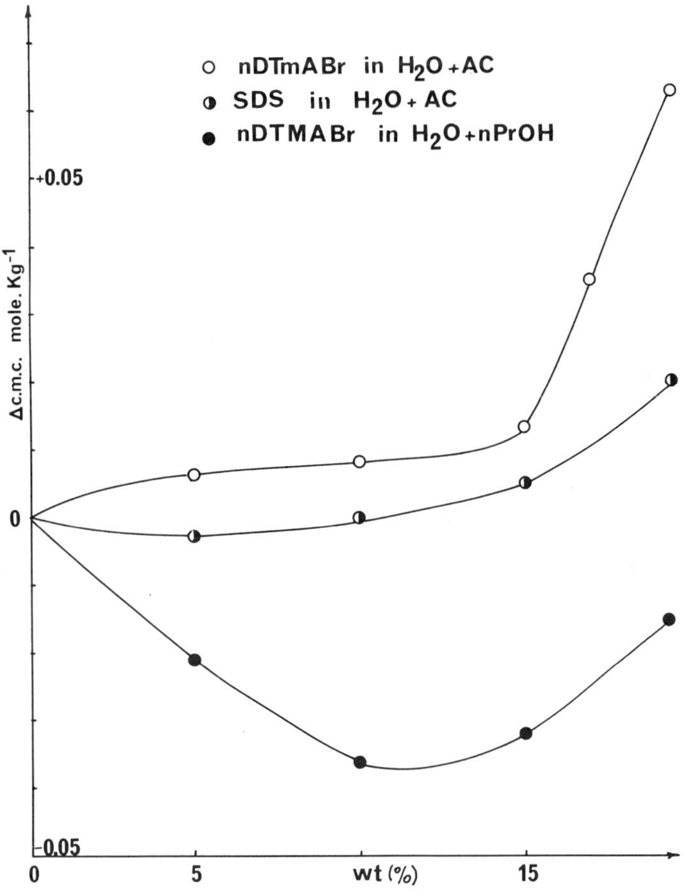

Figure 3. Variation of cmc with solvent composition

SDS in water + acetone and water + n-propanol mixtures. The increase of the cmc with acetone has been observed before in similar cases as well as the minimum for water + propanol mixtures (*1, 42, 43*). It is instructive to discuss the results by first using Relation 3:

$$\Delta G_t^\circ = 2RT \ln \frac{(\text{cmc})_w f_{\pm w}}{\text{cmc}_s f_{\pm s}} + RT \ln Xf + \Delta G_t^\circ \text{ (mic)}$$

The mean activity coefficient of the free ions is calculated using the Debye–Hückel Equation with a distance parameter a = 4.0 Å. If we ignore ΔG_t° (mic), the above relation can be rearranged as:

$$2RT\,\Delta \ln (\text{cmc}) = \Delta G_t^\circ + RT \ln Xf$$

with

$$\Delta \ln \text{cmc} = \ln \text{cmc}_s f_\pm - \ln \text{cmc}_w f_{\pm w} \qquad (15)$$

It is clear from this relation that ΔG_t° is negative (*see* Table V) if there is no mixed-micelle formation ($RT \ln Xf = 0$), Δ ln cmc must be positive: the cmc must increase with the addition of organic solvent (as in the case of acetone). If there is some participation of the organic solvent in the micelle formation (as the case of n-propanol), then $RT \ln Xf$ would be negative, the variation of Δ ln cmc depends on the relative magnitude of ΔG_t° and $RT \ln Xf$. From Figure 3 it can be concluded that $|RT \ln Xf| > |\Delta G_t^\circ|$ below $Z_2 = 0.03$, but that $|RT \ln Xf| < |\Delta G_t^\circ|$ above this propanol composition. The case of SDS in water +

Table V. Aggregation Numbers of Ionic Micelles in Water + Acetone Mixtures at 298.15 K[a]

Surfactant	*Wt (%)*	$-\Delta G_t^\circ$ *(cal mol⁻¹)*	$10^4\,\psi$ *(uescgs)*	n
nDTMABr	0	0	1.6218	36
	5	180	1.4311	30
	10	380	1.2411	19
	15	590	1.1572	14
	17	680	1.2048	13
SDS	0	0	1.8781	50
	5	166	1.4917	27
	10	383	1.4036	18
	15	680	1.2970	16
	20	965	1.1978	15

[a] Mole fraction scale (*6, 7*).

acetone mixtures shows a slight mixed-micelle formation rapidly compensated by the standard free energy term. We shall now focus our attention on the water + acetone system.

nDTMABr and SDS in Water + Acetone Mixtures at 298.15 K. From the above arguments it can be assumed that the terms $RT \ln Xf$ and hence $\Delta G_t°$ (mic) are small in these binary mixtures, so we can consider that $Xf = 1$. We have calculated the apparent charge of the micelle as described above with the same contact distance a for both surfactants, the small difference which should exist being, in our opinion, shadowed by the intrinsic ambiguity of that distance and the crudeness of the model used (*45*).

Table V and Figures 4, 5, and 6 present our essential results. Figures 4 and 5 show the relative magnitude of each term of Equation 5, the electric potential term being represented by the difference between the two curves for $\Delta G_t°$ and $2RT \ln (cmc_w f_{\pm w})/(cmc_s f_{\pm s})$. Figure 6 shows that the number of monomers per micelle n decreases rapidly with the addition of acetone for both surfactants as could be expected although the rate of the change of n could not be predicted; more important is the very small change of the apparent charge of the micelle $Z_{(mic)}$ with the

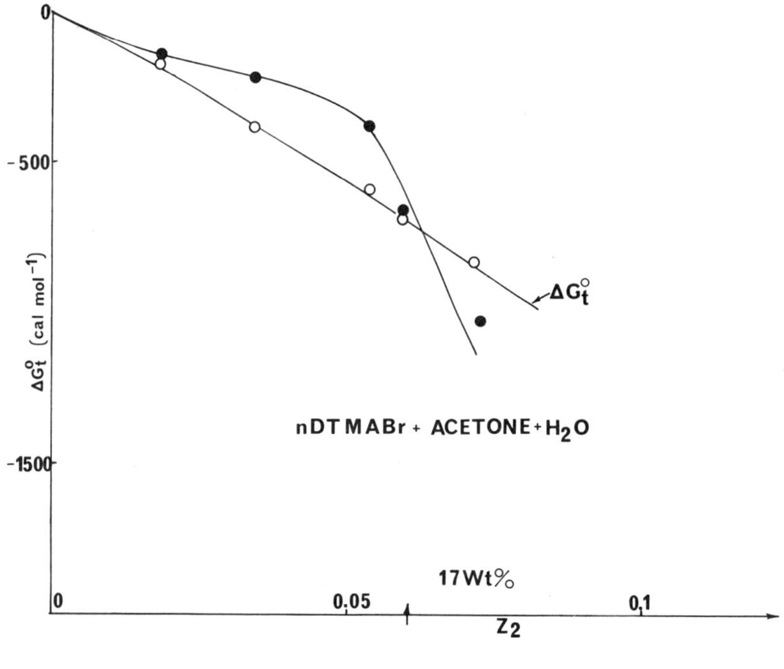

Figure 4. Variation of the terms of Equation 5: ○, $\Delta G_t°$; ●, $2RT \ln (cmc_w f_{\pm w}/cmc_s f_{\pm s})$; *the difference between the two curves represent* $Ne(\psi - \psi_w)$

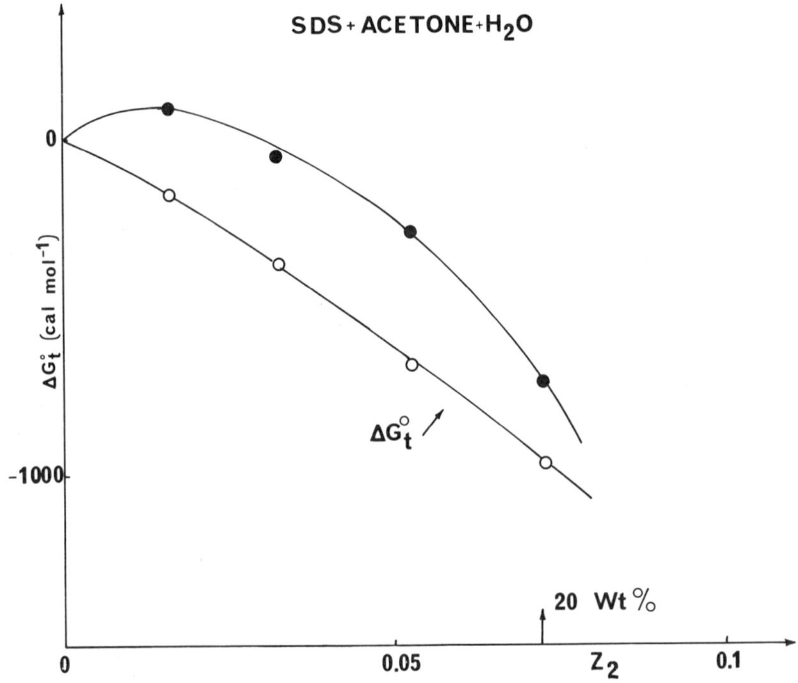

Figure 5. Variation of the terms of Equation 5: ○, $\Delta G_t{}^\circ$; ●, *2RT ln* $(cmc_w f_{\pm w}/cmc_s f_{\pm s})$*, the difference between the two curves represent Ne* $(\psi_s - \psi_w)$

addition of organic solvent; the decrease of *n* being almost entirely compensated for by the increase of α (Figure 7). We may conclude that according to our model the apparent charge of the micelle is apparently not affected by the addition of acetone because of two compensating phenomenon: (a) the number of monomers per micelle decreases as the water structure is destroyed by the addition of acetone but at the same time, (b) the power of acetone to solvate the cations increases the degree of dissociation which compensates the effect of the organic solvent on the structure of the medium as far as the apparent charge is concerned. The difference between the results for *n*DTMABr and SDS essentially may be interpreted by the different variation of α with solvent composition owing to the different effect of acetone on the trimethyammonium group and on the sodium ion. It is interesting to compare our results with those obtained from a more direct method; one of the few studies concerned with the direct determination of aggregation numbers in mixed solvents is the one of Becher (*46*); this author has determined *n* for a nonionic surfactant, POE_{23} (polyoxyethylene lauryl ether) in water + ethanol mixtures using a light-scattering technique associated with micellar weight determinations. The cmc of POE_{23} increases with the addition of ethanol in a manner comparable with that of *n*DTMABr in water + acetone mixtures indicating a similar behavior of both surfactants in these binary mixtures. Figure 6 shows that this is indeed the case even if the shapes of the curves for the ionic and nonionic surfactants are somewhat different; furthermore, the minimum aggregation numbers detected by each method and for each surfactant are practically equal, as they should be, a result which can be looked upon as a confirmation of the validity of our calculations. These observations indicate that our Poisson–Boltzman calculations correspond to a real physical phenomenon and gives some strong support to the simple model used.

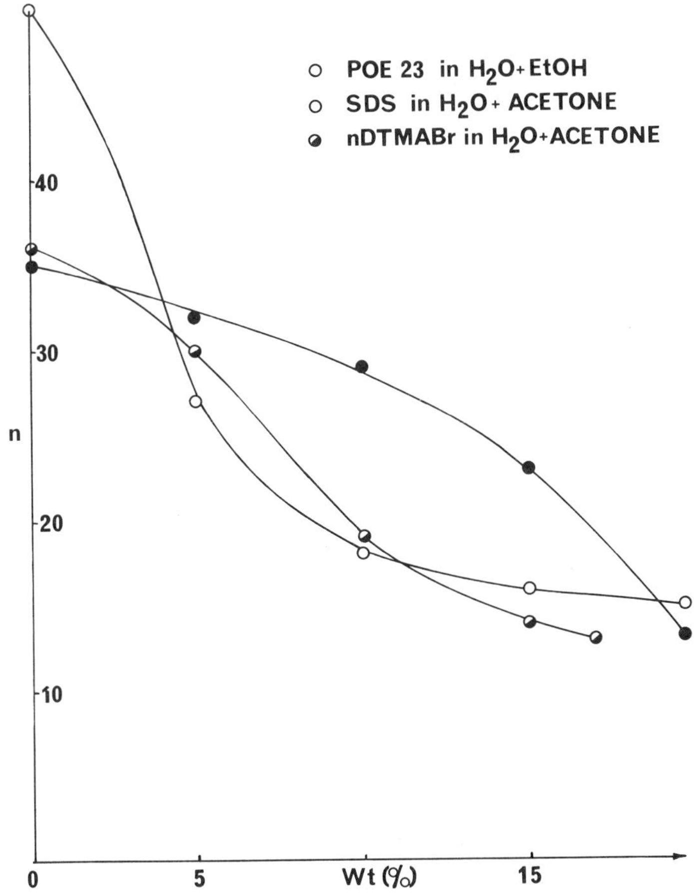

Figure 6. Variation of the apparent charge of the ionic micelles with solvent composition: $Z_{mic} = \alpha n$

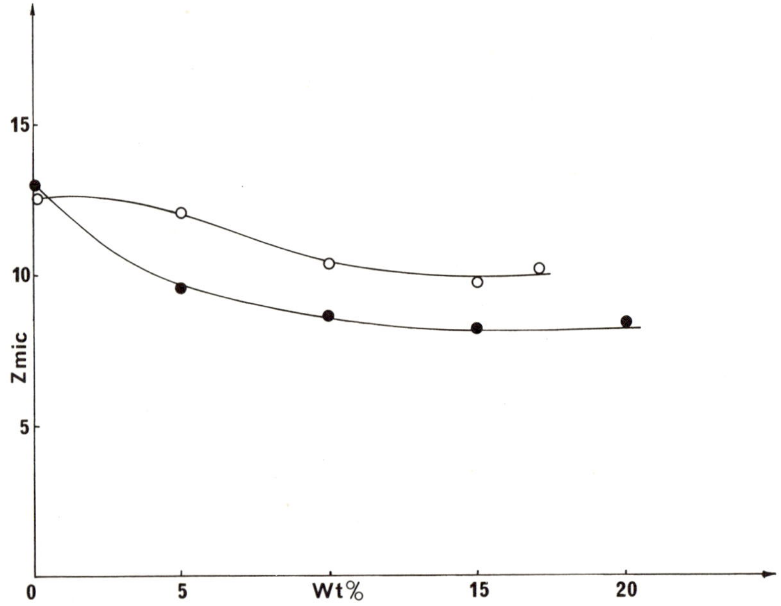

Figure 7. Variations of aggregation numbers for ionic and non-ionic micelles in aqueous binary mixtures at 298.15 K

Note: lower aggregation numbers than those of Figure 6 have been published as deduced from chemical shift spectroscopic determinations (*47*) for analogue systems; discrepancies of that kind are common when thermodynamic and spectroscopic results are compared directly.

Ionic Micelles in Water + *n*-Propanol Mixtures. Our calculations could not be extended to systems with mixed-micelle formation because too many unknown parameters would have to be introduced: standard free energy of transfer of the organic molecule from the bulk to the micellar phase, composition of the mixed micelle, and activity coefficient of the monomer in the mixed micelle. However, if we assume that n is the same for nDTMABr in water + acetone and water + n-propanol, then α can be calculated for this latter system (note that α is not very sensitive to changes of n) (*20*) and thus an apparent charge can be calculated also; it has the same characteristic as the other systems studied; $Z_{(mic)}$ decreases slightly with addition of propanol. However, this result hides the essential features of the mixed-micelle formation: for example, THF which forms mixed micelles as well as n-propanol does with ionic surfactants breaks up these aggregates for very low organic concentrations (*47*). This clearly indicates that factors like structure and shape of the organic molecule play a major role in this phenomenon.

Glossary of Symbols

a_w, a_s = activity of monomer surfactant in a micelle phase
a = critical distance of approach
b = diameter (in Å) of a cosolvent molecule
cmc = critical micelle concentration
C = diameter (in Å) of a solute particle
d = diameter (in Å) of a solvent molecule
d° = density of the solvent
D = dielectric constant of the solvent
e = electronic charge
f = activity coefficient of monomer surfactant in the micelle phase
$f_\pm$ = mean activity coefficient of an ionic surfactant in solution
ΔG_t° = standard Gibbs free energy of transfer of a solute between two solvents
$\Delta G_t^\circ{}_{(\mathrm{mic})}$ = standard Gibbs free energy of transfer of a monomer surfactant between two micelle phases
k = Boltzman constant
M = molar mass of the solvent
m = number of counterions adsorbed on a micelle
n = number of monomer surfactants per micelle
R = gas constant (or distance)
S_1 = slope of conductance vs. square root of concentration below the cmc
S_2 = slope of conductance vs. square root of concentration above the cmc
T = absolute temperature
X = mole fraction of monomer in the micelle phase
x, x_0 = distance
Z_1, Z_2 = mole fraction of binary solvent
Z_{mic} = apparent charge of an ionic micelle
α = degree of dissociation of an ionic micelle
$1/\kappa$ = Radius of the Debye–Hückel atmosphere
ψ = electric potential
ϕ = reduced electric potential
μ_w = the chemical potential of the ionic surfactant in water
μ_s = the chemical potential of the ionic surfactant in solvents
μ° = the standard chemical potential of the ionic surfactant
Λ = equivalent conductance of ionic surfactant
Λ_c = equivalent conductance of counterion

Literature Cited

1. Shinoda, K., Nakagawa, T., Tamamushi, B. I., Isemura, T., "Colloidal Surfactants," Academic, 1963.
2. Benjamin, L., *J. Colloid Interface Sci.* (1966) **22**, 386.
3. Shirahama, K., Matuura, R., *Bull. Chem. Soc. Jpn.* (1965) **38**, 373.
4. Shirahama, K., Hayagishi, M., Matuura, R., *Bull. Chem. Soc. Jpn.* (1969) **42**, 1206, 2123.
5. Gratzer, W. B., Beaven, G. H., *J. Phys. Chem.* (1969) **73**, 2270.
6. Treiner, L., LeBesnerais, A., *J. Chem. Soc., Faraday Trans. 1* (1977) **73**, 44.
7. Bury, R., Treiner, C., *Can. J. Chem.*, in press.
8. Shinoda, K., Hutchinson, E., *J. Phys. Chem.* (1962) **66**, 577.
9. Mukerjee, P., *Adv. Colloid Interface Sci.* (1967) **I**, 241.
10. Tanford, C., "The Hydrophobic Bond," Wiley, New York, 1973.
11. Shirahama, K., Kashiwabara, T., *J. Colloid Interface Sci.* (1971) **36**, 65.
12. Miyagishi, S., *Bull. Chem. Soc. Jpn.* (1974) **42**, 2972.
13. Mathews, W. K., Larsen, J. W., Pikal, M., *Tetrahedron Lett.* (1972) **6**, 513.
14. Hoskin, N. E., *Trans. Faraday Soc.* (1953) **49**, 1971.
15. Kotin, L., Nagasawa, M., *J. Chem. Phys.* (1962) **36**, 873.
16. Emerson, M. F., Holtzer, A., *J. Phys.* (1965) **69**, 3718.
17. Tanford, C., *J. Phys. Chem.* (1972) **76**, 3020.
18. Debye, P., *Ann. N.Y. Acad. Sci.* (1949) **51**, 575.
19. Tartar, H. V., Long, A. L. M., *J. Phys. Chem.* (1955) **59**, 1185.
20. Evans, H. C., *J. Chem. Soc.* (1956) 579.
21. Treiner, C., Justice, J. C., *J. Chim. Phys. France* (1966) 687.
22. Spivey, H. O., Shedlovski, T., *J. Phys. Chem.* (1967) **71**, 2167.
23. Broadwater, T. L., Kay, R. L., *J. Phys. Chem.* (1970) **21**, 3809.
24. Erdey-Gruz, T., Majthenyi, L., *Acad. Sci. Hung.* (1959) **20**, 73.
25. Perie, M., Perie, J., Chemla, M., *Electrochim. Acta* (1974) **19**, 753.
26. Mukerjee, P., complete table of critical micelle concentrations, NBS Washington, 1967.
27. Kresheck, G. C., "Water—a Comprehensive Treatise," F. Franks, Ed., Vol. IV, Plenum, 1977.
28. Anacker, E. W., "Cationic Surfactants," E. Jungermann, Ed., Marcel Dekker, 1970.
29. Treiner, C., Tzias, P., Chemla,, M., Poltoratski, G. N., *J. Chem. Soc., Faraday Trans 1* (1976) **72**, 2007.
30. Desrosiers, N., Desnoyers, J. E., *Can. J. Chem.* (1976) **23**, 3800.
31. Treiner, C., Tzias, P., ADV. CHEM. SER. (1976) **155**, 303.
32. Treiner, C., Lebesnerais, A., *J. Chem. Thermodynamics*, in press.
33. Bax, D., Deligny, C. L., Remijnsee, A. G., *Recl. Trav. Chim. Pays-Bas* (1972) **91**, 1225.
34. Lucas, M., Feillolay, A., *Bull. Soc. Chim. Fr.* (1970) 1267.
35. Pierotti, R. A., *J. Phys. Chem.* (1963) **67**, 1840.
36. Abraham, M. H., *J. Chem. Soc., Faraday Trans. 1* (1973) **69**, 1375.
37. Krishnan, C. V., Friedman, H. L., *J. Phys. Chem.* (1969) **73**, 1572.
38. Jolicoeur, C., Lacroix, G., *Can. J. Chem.* (1973) **51**, 3051.
39. Botre, C., Crescenzi, V. L., Mele, A., *J. Phys. Chem.* (1959) **63**, 648.
40. Shirahama, K., *Bull. Chem. Soc. Jpn.* (1974) **47**, 3165.
41. Miyagishi, S., *Bull. Chem. Soc. Jpn.* (1975) **48**, 2349.
42. Emerson, M. F., Holtzer, A., *J. Phys. Chem.* (1967) **71**, 3320.
43. Popovych, O., *Crit. Rev. Anal. Chem.* (1970) **73**, 1.
44. Wells, C. F., *J. Chem. Soc. Faraday Trans. 1* (1976) **72**, 601.
45. Mukerjee, P., *Phys. Chem.* (1969) **73**, 2054.
46. Becher, *J. Colloid, Sci.* (1965) **20**, 728.
47. Muller, N., Johnson, T. W., *J. Phys. Chem.* (1969) **73**, 2042.

RECEIVED March 2, 1978.

9

The Solvation of Chromium(III) Ion in Mixed Solvents

EDWARD L. KING

Department of Chemistry, University of Colorado, Boulder, CO 80309

The solvation of chromium(III) ion in certain mixed-solvent systems has been studied in experiments which are relatively free of ambiguity. The exchange of solvent molecules between the mixed solvent and the solvated species $Cr(OH_2)_w(So)_n^{3+}$ *(So = organic solvent component) is a very slow process. The species with solvation shells having different compositions can be separated from one another by column ion-exchange procedures. Analytical procedures based upon such separations allow evaluation of equilibrium constants for reactions involving replacement of coordinated water by the polar organic component. These equilibrium constants are reviewed in this chapter with attention focused upon the dependence of the equilibrium constants upon solvent composition, and the relationship of relative values of the equilibrium constants to the statistically expected values.*

The very exothermic solvation of gaseous metal ions in liquid water (*1*) reflects both the formation of strong coordinate bonds between the

$$M^{n+}(g) + w\,H_2O(l) = M(OH_2)_w^{n+}(aq)$$

$$-\Delta U \simeq 200\text{–}500 \text{ kJ mol}^{-1} \qquad n = 1$$

$$-\Delta U \simeq 1000\text{–}2000 \text{ kJ mol}^{-1} \qquad n = 2$$

$$-\Delta U \simeq 3000\text{–}5000 \text{ kJ mol}^{-1} \qquad n = 3$$

metal ion and the oxygen of the water molecule and hydrogen bonding interactions of the bulk solvent with the peripheral hydrogen atoms of the hydrated cation. In the hydration of gaseous chromium(III) ion, approximately 70% of the energy change is caused by the formation of the coordinate bonds (*1, 2*):

0-8412-0428-4/79/33-177-129$05.00/1

$$Cr^{3+}(g) + 6\,H_2O(g) = Cr(OH_2)_6{}^{3+}(g)$$
$$\Delta U \simeq -3.1 \times 10^6 \text{ J mol}^{-1}$$
$$Cr(OH_2)_6{}^{3+}(g) = Cr(OH_2)_6{}^{3+}(aq)$$
$$\Delta U \simeq -1.1 \times 10^6 \text{ J mol}^{-1}$$

The kinetic inertness of chromium(III) has allowed establishment of the primary hydration of this ion by six water molecules (*3*), as represented in these equations. The exchange of oxygen-18 between hexaaquachromium(III) ion and water in aqueous solution (*3, 4*) and the substitution of coordinated water in hexaaquachromium(III) ion by another ligand (e.g., methyl alcohol (*5*)) are very slow:

$$Cr(OH_2)_6{}^{3+}(aq) + H_2\overset{\times}{O} \rightleftarrows Cr(OH_2)_5(\overset{\times}{O}H_2)^{3+} + H_2O$$
$$t_{1/2} \simeq 2 \times 10^5 \text{ sec } (t = 27°C)$$
$$Cr(OH_2)_6{}^{3+}(aq) + CH_3OH \rightleftarrows Cr(OH_2)_5(OHCH_3)^{3+} + H_2O$$
$$t_{1/2} = 9.0 \times 10^4 \text{ sec } (t = 30.0°C, Z = 0.306)$$

This inertness makes it possible to establish unambiguously the composition of the first coordination shell of chromium(III) in mixed solvents of water with a polar organic component. In the studies to be reviewed here, column ion-exchange procedures were used to accomplish this goal. After allowing chromium(III) to come to equilibrium in an acidified mixed solvent at an elevated temperature (60°–70°C) where the rate is conveniently high, the chromium(III) ion with its first coordination shell intact is taken into a cation-exchange resin phase at a low temperature (0–3°C). All of the free organic solvent component then is rinsed away with dilute aqueous acid. More concentrated aqueous acid then is used to elute the mixture of differently solvated chromium(III) species. Direct analysis of the eluent allows determination of the binding of organic solvent by chromium(III) ion. Although acquisition of data is straight forward, the interpretation of data has several aspects which must be explored in some detail.

Values of $\overline{n}$ for H_2O–CH_3OH, H_2O–C_2H_5OH, and H_2O–$(CH_3)_2SO$ Systems as a Function of Solvent Composition

Use of short ion-exchange columns (3–6 cm) and concentrated acid (about 3*M* H_2SO_4) as the eluting agent allows recovery of essentially all of the chromium(III) in a single elution peak. From analysis of this eluent the average amount of organic solvent coordinated by chromium(III) is obtained. The results of such studies for the systems H_2O–CH_3OH (*6*), H_2O–C_2H_5OH (*7*), and H_2O–$(CH_3)_2SO$ (*8*) are given in Figure 1, which

$$\bar{n} = \frac{\text{moles of coordinated organic solvent}}{\text{moles of chromium(III)}}$$

presents values of $\bar{n}$ as a function of the ratio of amounts of the two solvent components, $Z/(1 - Z)$, where Z is the mole fraction of organic component. (In the calculation of Z, only the solvent components are taken into account.) We see in this figure that chromium(III) discriminates in its coordination shell in favor of dimethyl sulfoxide over water but in favor of water over methyl alcohol or ethyl alcohol. (Although direct analysis for coordinated water was not made, nothing either in these data or in other data (e.g., spectra of differently solvated chromium(III) ions) suggests that the coordination number of chromium(III) changes as coordinated water molecules are replaced by other polar molecules.) One simple measure of this discrimination is the quotient κ,

$$\kappa = \frac{\bar{n}/(6 - \bar{n})}{Z/(1 - Z)}$$

values of which are presented as a function of $\bar{n}$ in Figure 2. If the relative solvation in the first coordination shell of these species is strictly random and if the appropriate solvent composition parameter for use in the correlation of $\bar{n}$ data is Z, the value of this quotient would be unity. If the relative solvation is not random and if the appropriate solvent composition parameter is Z, the value of κ would be constant. This is approximately the case for the H_2O–$(CH_3)_2SO$ system ($\kappa \simeq 4.9$; $\log \kappa = 0.69$).

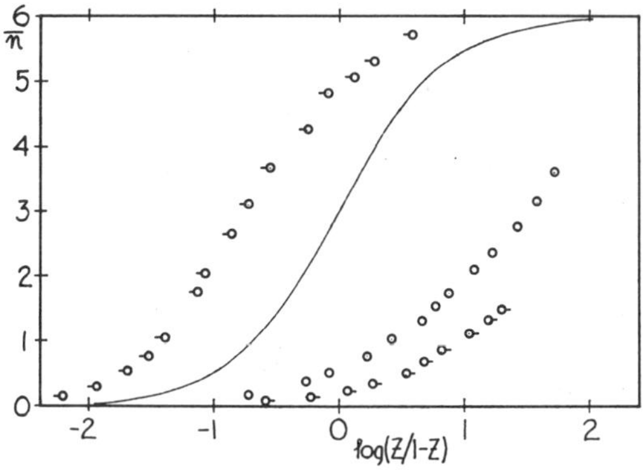

Figure 1. The average solvation of chromium(III) ion in mixed solvents. (–O) H_2O–$(CH_3)_2SO$, t = 60°C; (O) H_2O–CH_3OH, t = 60°C; (O–) H_2O–C_2H_5OH, t = 50°C. The solid line is for random solvation, $\bar{n}/(6 - \bar{n}) = Z/(1 - Z)$.

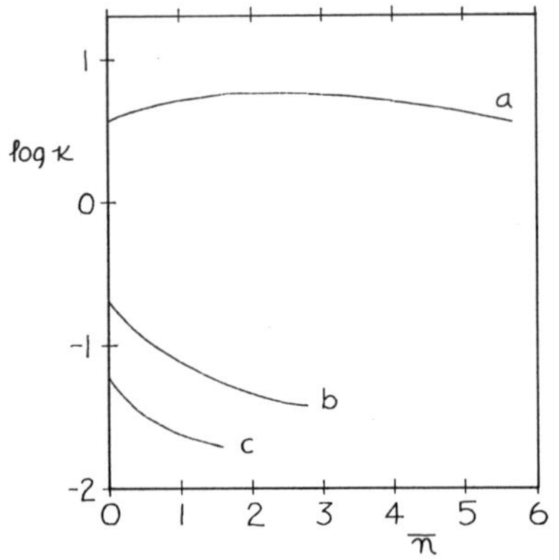

Figure 2. The discrimination factor κ as a function of $\bar{n}$.

$$\kappa = \frac{\bar{n}/(6-\bar{n})}{Z/(1-Z)}$$

(a) H_2O–$(CH_3)_2SO$; range of κ = 3.5–5.9. (b) H_2O–CH_3OH; range of κ = 0.16–0.04. (c) H_2O–C_2H_5OH; range of κ = 0.05–0.02.

The values of κ confirm the already stated qualitative conclusions which were drawn from Figure 1. However, there are more informative data and more detailed methods of data treatment, which now will be presented.

Concentrations of Individual Differently Solvated Species

Use of longer ion-exchange columns (100–150 cm) and lower, variable concentrations of eluting agent has allowed separation from one another of species

$$Cr(OH_2)_{6-n}(So)_n^{3+}$$

in the systems for which $\bar{n}$ data have already been presented (So = organic solvent component), So = CH_3OH (*9*), C_2H_5OH (*7, 9*), and $(CH_3)_2SO$ (*8*). Typical elution profiles for each of these systems presented in Figure 3 show the separation of geometrical isomers for the compositions with n = 2, 3, and 4 in the system $Cr(OH_2)_{6-n}(OS(CH_3)_2)_n^{3+}$ but no isomer separation for the water–alcohol systems. Interpretation of data for the H_2O–$(CH_3)_2SO$ system will involve identification of the isomers. One possible interpretation for the water–alcohol systems involves indirect assessment of the relative stabilities of the isomers for n = 2, 3, and 4, despite the failure to separate them.

For the system $Cr(OH_2)_wG_g^{3+}$ (G = 1,3-propanediol) (*10*), the elution profile given in Figure 4 shows separation of two different species with g = 1, a composition for which there are not geometrical isomers; but with this organic ligand, there is the possibility of species with two different compositions

$Cr(OH_2)_5G^{3+}$ (with G as a monodentate ligand)

$Cr(OH_2)_4G^{3+}$ (with G as a chelated ligand).

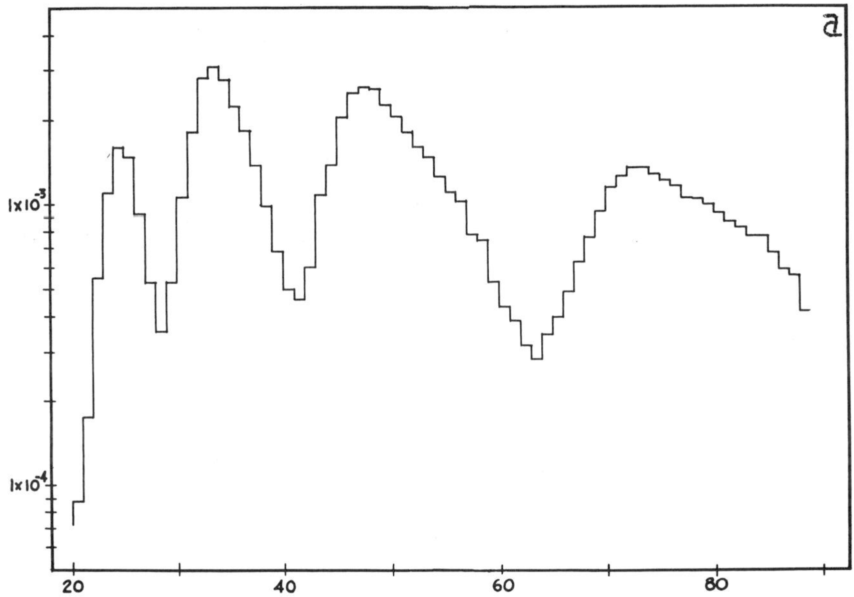

Figure 3a. Elution profile. H_2O–CH_3OH system, 60°C; Z = 0.996. Eluting agent: 1.8M H_2SO_4; in order of elution, species contain 3, 4, 5, and 6 molecules of CH_3OH.

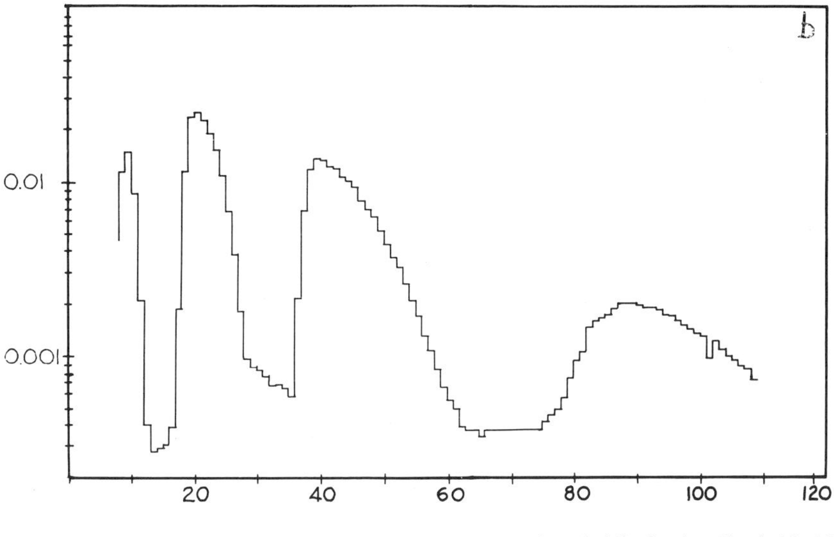

Journal of the American Chemical Society

Figure 3b. Elution profile. H_2O–C_2H_5OH system, 75°C; Z = 0.949. Eluting agent: 3.0M H_2SO_4; in order of elution, species contain 0, 1, 2, and 3 molecules of C_2H_5OH (7).

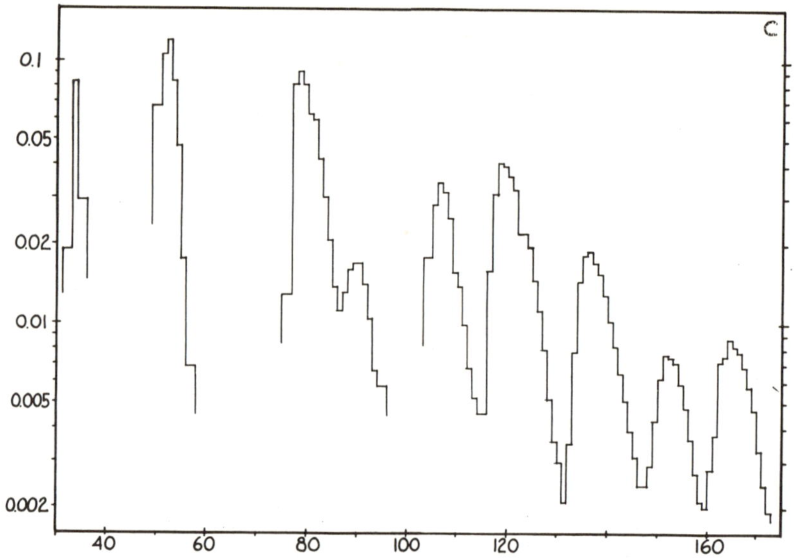

Journal of the American Chemical Society

Figure 3c. H_2O–$(CH_3)_2SO$ system, 60°C; Z = 0.095. Eluting agent 1.5–6.0M H_2SO_4 (gradient elution); in order of elution, species contain 0, 1, 2(cis), 2(trans), 3(fac), 3(mer), 4(cis), 5, 4(trans) molecules of $(CH_3)_2SO$ (8).

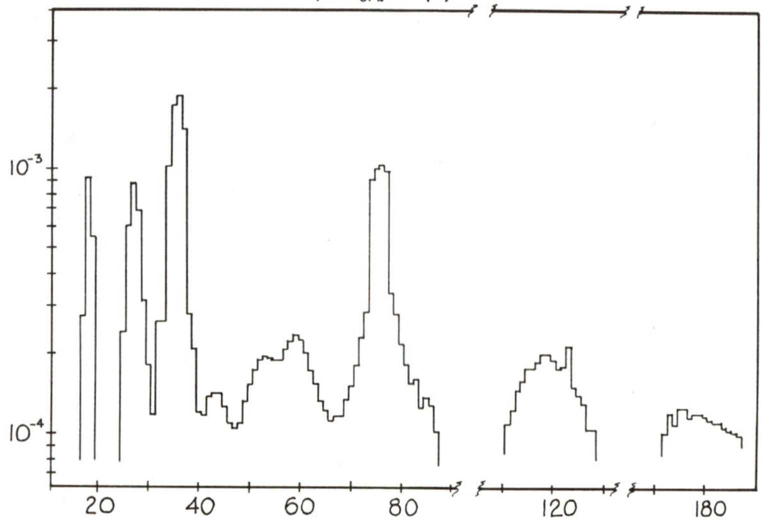

Journal of the American Chemical Society

*Figure 4. Elution profile for species of chromium(III) equilibrated in 1,3 propanediol(G)–H_2O system. (Z = 0.821, t = 60°C). Eluting agent: 3.0M H_2SO_4; in order of elution, Peaks **A–I**, species are: **A**, $Cr(OH_2)_6^{3+}$; **B**, $Cr(OH_2)_5G^{3+}$; **C**, $Cr(OH_2)_4G^{3+}$; **D**, $Cr(OH_2)_4G_2^{3+}$ (mixture of isomers); **E**, $Cr(OH_2)_3G_2^{3+}$; **F**, $Cr(OH_2)_3G_2^{3+}$; **G**, $Cr(OH_2)_2G_2^{3+}$ (mixture of isomers); **H**, $Cr(OH_2)G_3^{3+}$; **I**, CrG_3^{3+} (10).*

Interpretation of data for this system $Cr(OH_2)_wG_g^{3+}$ involves indirect evaluation of w, the number of coordinated water molecules.

With the concentrations of individual species determined, it is possible to evaluate the equilibrium constants for the formation of each differently solvated species, and this can be done as a function of the solvent composition.

Identification of Isomers in the $Cr(OH_2)_{6-n}(OS(CH_3)_2)_n^{3+}$ System

Guidance in the assignment of structure to isomeric species $Cr(OH_2)_{6-n}(OS(CH_3)_2)_n^{3+}$ for $n = 2$, 3, and 4 is obtained by comparing the relative stabilities of the two species having each of these compositions with the statistically expected values. The observed relative concentrations of the more easily eluted **A** isomers and the less easily eluted **B** isomers leads to the conclusion that the more easily eluted isomer is the cis isomer for $n = 2$ and 4 and the fac isomer for $n = 3$. This is summarized in Table I. Although it cannot be asserted that the isomeric species in this system should have exactly the statistically expected relative concentrations, the assignment proposed on this basis is more rational than the opposite one in which the observed relative concentration would differ from the statistically expected ones by factors of 22, 2.8, and 10, respectively, for $n = 2$, 3, and 4. These isomer identifications for $n = 2$ and 3 are supported by observations of the amounts of the isomeric species with $n = 3$ formed from each of the species with $n = 2$ (*11*),

$$Cr(OH_2)_4(OS(CH_3)_2)_2^{3+} + (CH_3)_2SO \longrightarrow Cr(OH_2)_3(OS(CH_3)_2)_3^{3+} + H_2O$$

The results of these experiments with short reaction periods, used to minimize the occurrence of other reactions, are presented in Table II.

This order of elution for isomeric species with $n = 2$, 3, and 4 is observed also for chromium(III) species coordinated by water and pyridine *N*-oxide (*12*).

Table I. Relative Stabilities of Isomeric Species $Cr(OH_2)_{6-n}(OS(CH_3)_2)_n^{3+}$

		([A]/[B]) (statistically expected)	
n	$([A]/[B])_{exp}$ [a]	A = cis or fac	B = cis or fac
2	5.5	4	0.25
3	0.54	0.67	1.5
4	2.5	4	0.25

[a] Experimental values extrapolated to $Z = 0$; **A** and **B** are more easily and less easily eluted isomers, respectively.

Table II. Yields of Isomeric $Cr(OH_2)_3(OS(CH_3)_2)_3^{3+}$ from Each Isomeric $Cr(OH_2)_4(OS(CH_3)_2)_2^{3+}$

	Ratio[a] [A–$Cr(OS(CH_3)_2)_3^{3+}$]/[B–$Cr(OS(CH_3)_2)_3^{3+}$]				
		Expected on Statistical Basis[c]			
Reactant[a]	*Observed*[b]	*I*	*II*	*III*	*IV*
A–$Cr(OS(CH_3)_2)_2^{3+}$	0.73	1	1	0	∞
B–$Cr(OS(CH_3)_2)_2^{3+}$	0.026	0	∞	1	1

[a] A isomer is more easily eluted isomer; B isomer is less easily eluted isomer. Coordinated H_2O not shown.

[b] Each isomeric $Cr(OH_2)_4(OS(CH_3)_2)_2^{3+}$ was placed in a mixed solvent with $Z = 0.705$, $[H^+] = 0.04$ mol L^{-1} for a 13.5-hr reaction period at 27°C. During this period, ~ 10% conversion to $Cr(OH_2)_3(OS(CH_3)_2)_3^{3+}$ occurred.

[c] (I) A–$Cr(OS(CH_3)_2)_2^{3+}$ = cis; A–$Cr(OS(CH_3)_2)_3^{3+}$ = cis. (II) A–$Cr(OS(CH_3)_2)_2^{3+}$ = cis; A–$Cr(OS(CH_3)_2)_3^{3+}$ = trans. (III) A–$Cr(OS(CH_3)_2)_2^{3+}$ = trans; A–$Cr(OS(CH_3)_2)_3^{3+}$ = cis. (IV) A–$Cr(OS(CH_3)_2)_2^{3+}$ = trans; A–$Cr(OS(CH_3)_2)_3^{3+}$ = trans.

Equilibrium Constants for the Formation of Individual Species

For the purposes of comparing data for the several systems of monodentate ligands, the reactions in which one water molecule is replaced,

$$\frac{1}{n}\,Cr(OH_2)_6^{3+} + So \rightleftarrows \frac{1}{n}\,Cr(OH_2)_{6-n}(So)_n^{3+} + H_2O$$

will be considered. Values of equilibrium constants

$$K_n = \left\{ \frac{[Cr(OH_2)_{6-n}(So)_n^{3+}]}{[Cr(OH_2)_6^{3+}]} \right\}^{1/n} \times \frac{a_{H_2O}}{a_{So}}$$

are related to the complete thermodynamic equilibrium constants, K_n°, for these reactions by the equation,

$$K_n = K_n^\circ \times \left\{ \frac{\gamma(Cr(OH_2)_6^{3+})}{\gamma(Cr(OH_2)_{6-n}(So)_n^{3+})} \right\}^{1/n}$$

in which $\gamma(Cr(OH_2)_6^{3+})$ and $\gamma(Cr(OH_2)_{6-n}(So)_n^{3+})$ are medium activity coefficients for the indicated species. (That is, the ratio of activity coefficients is defined as unity in some particular solvent.) Thus, a dependence of the value of K_n upon solvent composition is caused by nonconstancy of the ratio of activity coefficients of the differently solvated metal ions as the solvent composition changes. For the H_2O–CH_3OH

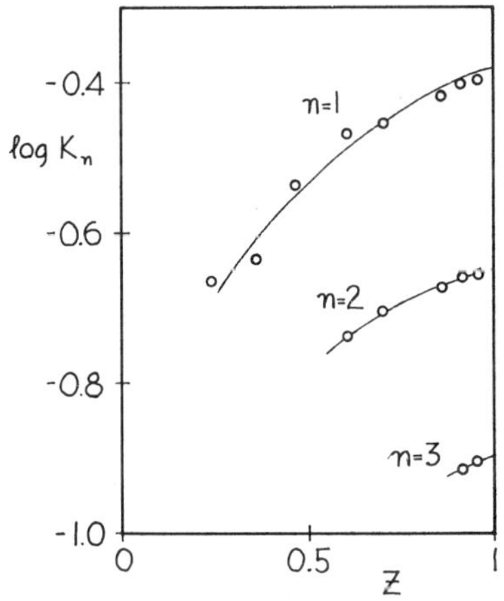

Figure 5. The dependences of K_n (n = 1, 2, and 3) for water–ethanol system at 75°C upon solvent composition

system, the values of K_1, K_2, and K_3 do not depend appreciably upon solvent composition. For the H_2O–C_2H_5OH system and the H_2O–$(CH_3)_2$-SO systems, the dependences are shown in Figures 5 and 6. The values of K_n for the respective systems increase with increasing ethanol content of the solvent and decrease with increasing dimethyl sulfoxide content of the solvent.

For the water–propanediol system, the number of water molecules coordinated to chromium(III) in each species cannot be taken as six minus the number of coordinated molecules of glycol. This quantity has been determined by considering the dependence of the species concentration upon the solvent composition. The extents of formation of the two species containing one glycol molecule can be expected to show

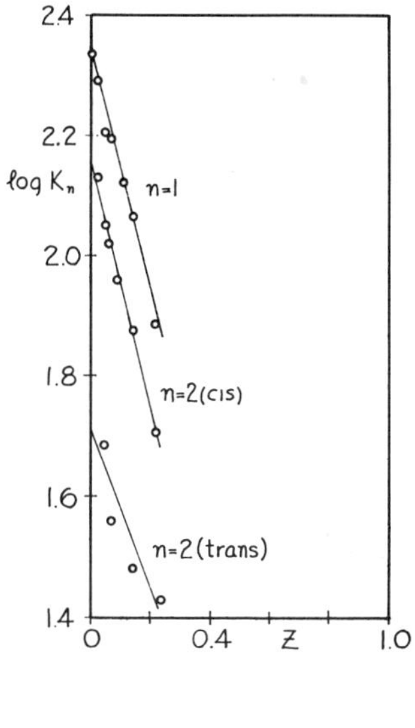

Figure 6. The dependences of K_n (n = 1, 2(cis) and 2(trans)) for H_2O–$(CH_3)_2SO$ system at 60°C upon solvent composition

$$Cr(OH_2)_6^{3+} + G \rightleftarrows Cr(OH_2)_5G^{3+} + H_2O$$
$$Cr(OH_2)_6^{3+} + G \rightleftarrows Cr(OH_2)_4G^{3+} + 2H_2O$$

different dependences upon solvent composition because different numbers of water molecules appear in the two balanced chemical equations. Table III summarizes the calculations which lead to the conclusion that the more easily eluted monoglycol species is $Cr(OH_2)_5G^{3+}$. Under these indirectly derived assignments of composition (i.e., the values of w in $Cr(OH_2)_wG^{3+}$), the values of K_{15} and K_{14} decrease slightly with increasing Z, the ratios $K(Z = 0.821)/K(Z = 0.356)$ being 0.70 and 0.65, respectively. Under the opposite assignment of composition, the values of Q_{15} increase with increasing Z, the ratios $Q(Z = 0.821)/Q(Z = 0.356)$ being 2.4 and the values of Q_{14} decrease, the ratio $Q(Z = 0.821)/Q(Z = 0.356)$ being 0.20. The smaller medium dependence of the values of Q_{15} and Q_{14} under the first of the assignments of composition leads to the conclusion that this assignment is correct. This assignment also gives an elution-order dependence upon the number of coordinated water molecules which is the same as that which is proposed for species in this system with $g = 2$ and 3 (to be discussed). (In this correlation, the solvent composition parameters being used are mole fractions, not activities. The ethylene glycol–water system is close to ideal, and it is assumed that this is true also for the 1,3-propanediol system.)

Correlation of Stabilities of Differently Solvated Species

The values of K_n for the systems involving monodentate ligands, summarized in Table IV, can be compared with one another after an appropriate correction for the statistical factor in each reaction. This correction is made by considering the symmetry numbers of reactant (σ_R) and product species (σ_P) (*14*). The values of K_n, corrected in this manner, are given in parentheses in Table IV.

$$K_n(\text{corr}) = K_n \text{ x } \left(\frac{\sigma_P}{\sigma_R}\right)^{1/n}$$

For the water–dimethyl sulfoxide system, values of the eight corrected equilibrium constants are all encompassed by a factor of 1.6. The one-parameter equation

$$K_n = 45 \text{ x } \left(\frac{\sigma_R}{\sigma_P}\right)^{1/n}$$

generates values of K_n ($n = 1$, 270; $n = 2$(cis), 156; $n = 2$(trans), 78; $n = 3$(fac), 90; $n = 3$(mer), 103; $n = 4$(cis), 84; $n = 4$(trans), 59; and

Table III. Solvent Composition Dependence of Q_{15} and Q_{14} Under Alternate Assignments of Values of w for $Cr(OH_2)_wG^{3+}$ [a]

	$A = Cr(OH_2)_5G^{3+}$		$A = Cr(OH_2)_4G^{3+}$	
Z	Q_{15}	Q_{14}	Q_{15}	Q_{14}
0.356	0.50	0.22	0.34	0.32
0.584	0.42	0.23	0.55	0.17
0.821	0.35	0.143	0.80	0.063

[a] $Q_{15} = \frac{[Cr(OH_5G^{3+}](1 - Z)}{[Cr(OH_2)_6{}^{3+}]Z}$; $Q_{14} = \frac{[Cr(OH_2)_4G^{3+}](1 - Z)^2}{[Cr(OH_2)_6{}^{3+}]Z}$.

$n = 5, 64$), which agree with the experimental values with an average difference of about 14%. Since there is a mild trend in the statistically corrected values of K_n, with K_n(corr) increasing as n increases, it can be anticipated that correlations of the values of K_n with two-parameter equations would give an improved fit. Attempts at such correlations probably are not warranted by the quality of the experimental values. Uncertainties both in the analytical data and in the extrapolations to a common solvent composition make it likely that each value of K_n is uncertain to $\pm$ 10%.

The statistical corrections of K_n values for the water–alcohol systems $Cr(OH_2)_{6-n}(OHR)_n{}^{3+}$ have been made under the assumption that isomeric species for $n = 2$, 3, and 4 are present in the statistically expected relative

Table IV. Values of K_n for Solvent Exchange Reactions

$$K_n = \left\{ \frac{[Cr(OH_2)_{6-n}(So)_n{}^{3+}]}{[Cr(OH_2)_6{}^{3+}]} \right\}^{1/n} \times \frac{a_{H_2O}}{a_{So}}$$

	$H_2O–(CH_3)_2SO$ [a]	$H_2O–CH_3OH$ [b]	$H_2O–C_2H_5OH$ [c]
K_1	220 (37) [d]	0.65 (0.11) [d]	0.42 (0.070) [d]
K_2(cis)	140 (40) [d]	0.37 (0.10) [e]	0.23 (0.060) [e]
K_2(trans)	61 (35) [d]		
K_3(fac)	92 (46) [d]	0.21 (0.079) [e]	0.13 (0.047) [e]
K_3(mer)	110 (48) [d]		
K_4(cis)	92 (49) [d]	0.13 (0.065) [e]	0.073 (0.037) [e]
K_4(trans)	73 (55) [d]		
K_5	75 (52) [d]	0.080 (0.056) [d]	
K_6		0.044 (0.044) [d]	

[a] From Ref. *8;* values have been extrapolated to $Z = 0$ (i.e., pure water) ($t = 60°C$).
[b] From Ref. *9* ($t = 60°C$).
[c] From Ref. *9* ($t = 75°C$); values for K_1 and K_2 have been extrapolated to $Z = 1$ (i.e., pure C_2H_5OH); values of K_3 and K_4 are experimental values at $Z = 0.96 - 1.0$.
[d] Statistically corrected value.
[e] Statistically corrected under the assumption that isomers are present in the statistically expected relative amounts.

amounts (*14*). These corrected values show the trend with increasing n, which is the opposite of the mild trend displayed by the chromium(III) species of the water–dimethyl sulfoxide system. Two approaches have been proposed for correlating the trends of K_n(corr) values in the water–alcohol systems (*9*). In one approach, the variation is attributed to a destabilization of species with alcohol molecules in cis positions relative to one another. In the other approach, it is postulated that a linear dependence of log K_n(corr) upon n reflects a dependence of the chromium(III)–ligand bond strength upon the composition of the coordination shell. The relative merits of these two approaches are not distinguished by the quality of the associated quantitative correlations. The steric destabilization of species in which alcohol molecules are cis to one another is an attractive postulate, but it cannot be proved by the data for this system. No such effect is observed for species in the water–dimethyl sulfoxide system.

For chromium(III) species in the water–1,3 propanediol system, the statistical correlation of stabilities of species with different compositions involves two adjustable parameters, q_1, an intrinsic equilibrium constant for replacement of one water molecule by a glycol molecule (to give a monodentate species) and q_2, an intrinsic equilibrium constant for replacement of two water molecules by a glycol molecule (to give a chelate species). Table V presents a correlation of values of Q_{gw} for g =

$$Q_{gw} = \frac{[Cr(OH_2)_w G_g^{3+}](1 - Z)^{6-w}}{[Cr(OH_2)_6^{3+}]\, Z^g}$$

1 and 2 with the two parameters q_1 and q_2. That five experimental quantities can be reproduced by two parameters with an average difference of about 18% supports the assignments of composition (i.e., the value of w in $Cr(OH_2)_w G_g^{3+}$) upon which the calculation was based. As in the systems involving monodentate ligands, the statistical factors are dominant in establishing relative values of the equilibrium constants.

Table V. Comparison of Experimental Values of Q_{gw} (g = 1 and 2) at Z = 0.5 with Values Calculated Using q_1 = 0.032 and q_2 = 0.011

	Q_{15}	Q_{14}	Q_{24}	Q_{23}	Q_{22}
Exp. Value[a]	0.45	0.22	0.054	0.059	0.020
Calcd. Value[b]	0.38	0.26	0.061	0.068	0.015

[a] Interpolated to $Z = 0.50$.
[b] $Q_{15} = 12\, q_1$; $Q_{14} = 24\, q_2$; $Q_{24} = 48\, q_1^2 + 12\, q_1^2 = 60\, q_1^2$; $Q_{23} = 96\, q_1q_2 + 96\, q_1q_2 = 192\, q_1q_2$; $Q_{22} = 96\, q_2^2 + 24\, q_2^2 = 120\, q_2^2$. (The values of Q_{24}, Q_{23}, and Q_{22} are composite values for the two geometrical isomers with each composition.)

(Studies of the interaction of chromium(III) with 1,2 diols, ethylene glycol (*15*), and 1,2 propanediol (*10*) reveal kinetic anomalies which remove these systems from the scope of the present paper.)

Discussion

The inertness of chromium(III) ion has allowed a complete characterization of the solvation of this ion in several mixed-solvent systems. For this inert transition metal ion of charge 3^+, concern with the first shell coordination of solvent molecules can be viewed as a subdivision of coordination chemistry, a point of view less easily applicable to labile systems of metal ions with lower charge. Thus the approach used here is different from that used by A. K. Covington and co-workers (*16*) in their NMR studies of labile systems (e.g., sodium and cesium chlorides in water–methanol solutions (*17*)).

With concentrations of individual species determined directly for each solution studied, an equilibrium constant can be calculated as a function of solvent composition. The dependence of the equilibrium constants, as defined here with the solvent activity coefficients included, reveals selective outer-sphere interaction. (Equilibrium constants which involve the ratio $(1 - Z)/Z$ instead of a_{H_2O}/a_{So} may show less variation than those being considered here and this approximate constancy may be fortuitous.) Covington (*16*) has presented such single medium-independent parameters with which he correlated data for the systems being considered here. These values, which can be compared with values of κ presented in Figure 2, are $K = 0.03$ (C_2H_5OH), $K = 0.07$ (CH_3OH), and $K = 5.0$ ($(CH_3)_2SO$). (This value for the dimethyl sulfoxide system was given (*16*) as 0.2, but this is the reciprocal of the appropriate value.) It is clear from Figure 2 that these values are approximately average values of κ for the range of solvent composition studied, but it also is clear that independence of solvent composition is not achieved in the alcohol systems by using equilibrium constants involving $(1 - Z/Z)$.

To rationalize the observed trends in the values of κ, as defined here, it is necessary to propose that the dominant outer-sphere interactions are: the $Cr(OH_2)_{6-n}(OS(CH_3)_2)_n{}^{3+}$ system, a hydrogen bonding interaction between solvent dimethyl sulfoxide and coordinated water

$$\begin{array}{l} \quad\;\; H \\ \quad\;\; | \\ Cr\text{–}OH \text{ - - - - } OS(CH_3)_2, \end{array}$$

which causes increased stability of hexaaquachromium(III) ion relative to species with fewer coordinated water molecules as the dimethyl sulfoxide content of the solvent increases; and the $Cr(OH_2)_{6-n}(OCH_2$

$H_5)_n{}^{3+}$ system, a hydrogen bonding interaction between solvent water and coordinated water

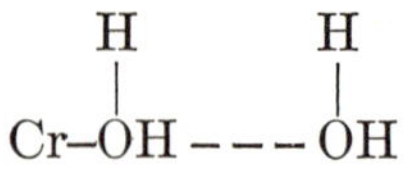

which causes increased stability of hexaaquachromium(III) ion relative to species with fewer coordinated water molecules as the water content of the solvent increases. (For the water–glycol system, the decrease in Q_{gw} with increasing Z is not simply rationalized in terms of interactions such as those just used for the other systems. The trend is the same as observed for the water–dimethyl sulfoxide system, but it seems unlikely that a dominant outer-sphere interaction would involve coordinated water and solvent glycol.)

Correlation of data such as those considered here under the assumption of solution ideality (i.e., by replacing a_{H_2O}/a_{So} with $(1 - Z)/Z$ in the equations for K_n) gives very different dependences of the equilibrium constants upon solvent composition than those presented in Figures 5 and 6. For the H_2O–C_2H_5OH system, the value of K_1 involving $(1 - Z)/Z$ decreases with increasing Z, $K_1(Z = 0.95)/K_1(Z = 0.24) \simeq 0.5$, and for the H_2O–$(CH_3)_2SO$ system the value of K_1 involving $(1 - Z)/Z$ increases with increasing Z, $K_1(Z = 0.14)/K_1(Z = 0.04) \simeq 1.5$. The values of K_1 calculated in this way are not independent of medium, and there seems to be little reason to prefer this correlation of data. Certainly the possibility of rationalizing the dependences of K_n values upon Z in terms of preferential outer-sphere interactions is absent if the equation for $(K_n/K_n{}^\circ)$ involves two quotients of activity coefficients:

$$\frac{K_n}{K_n{}^\circ} = \frac{\gamma(So)}{\gamma(H_2O)} \times \left\{ \frac{\gamma(Cr(OH_2)_6{}^{3+})}{\gamma(Cr(OH_2)_{6-n}So_n{}^{3+})} \right\}^{1/n}$$

Although a statistical factor contributes to the equilibrium constants for many types of reactions, it is in reactions of the type being considered here—the replacement of one neutral ligand by another neutral ligand—that this factor may be the principal factor in causing variation of K_n with n. Other series of reactions of this type are the formation of ammonia complexes in aqueous solution. The variation of K_n with n observed for the chromium(III)/water–methyl alcohol system is slightly smaller than observed for ammonia complexes of cobalt(II) (*18*) and nickel(II) (*19*):

	$\Delta \log K_n(\mathrm{corr})/\Delta n$
$Cr^{III}/H_2O–CH_3OH$	−0.082
$Co^{II}/H_2O–NH_3$	−0.093
$Ni^{II}/H_2O–NH_3$	−0.112

Thorough studies on other systems will be needed before the root of these trends can be discerned.

Glossary of Symbols

So = organic solvent component
w = number of water molecules coordinated to metal
n — number of organic solvent molecules coordinated to metal
(g) = gas
n = change on ion
(l) = liquid
M = a metal
ΔU = change in internal energy
$t_{\frac{1}{2}}$ = half-time for reaction
t = temperature °C
Z = mole fraction of organic solvent component in mixed solvent
$\bar{n}$ = average number of molecules of organic solvent component coordinated to metal
κ = a discrimination factor; defined as $K = [\bar{n}/(6 - \bar{n})]/[Z/(1 - Z)]$
G = 1,3-propanedial
a_{H_2O} = activity of water
a_{So} = activity of organic solvent component
$\gamma(Cr(OH_2)_6{}^{3+})$ = medium activity coefficient for $Cr(OH_2)_6{}^{3+}$
$\gamma(Cr(OH_2)_{6-n}So_n{}^{3+})$ = medium activity coefficient for $Cr(OH_2)_{6-n}So^{3+}$
K_n = equilibrium constant:

$$K_n = \left\{\frac{[Cr(OH_2)_{6-n}(So)_n{}^{3+}]}{[Cr(OH_2)_6{}^{3+}]}\right\}^{1/n} \times \frac{a_{H_2O}}{a_{So}}$$

$K_n°$ = equilibrium constant:

$$K_n° = \left\{\frac{[Cr(OH_2)_{6-n}(So)_n{}^{3+}]\ \gamma(Cr(OH_2)_{6-n}(So)_n{}^{3+})}{[Cr(OH_2)_6{}^{3+}]\ \gamma(Cr(OH_2)_6{}^{3+})}\right\}^{1/n} \times \frac{a_{H_2O}}{a_{So}}$$

$[Cr(OH_2)_6{}^{3+}]$ = concentration of $Cr(OH_2)_6{}^{3+}$
σ_P = symmetry number of product
σ_R = symmetry number of reactant

Q_{wg} = equilbrium constant:

$$Q_{gw} = \frac{[Cr(OH_2)_wG_g^{3+}](1 - Z)^{6-w}}{[Cr(OH_2)_6^{3+}]Z^g}$$

q_1 = an intrinsic equilibrium constant for one molecule of glycol replacing one molecule of water

q_2 = an intrinsic equilibrium constant for one molecule of glycol replacing two molecules of water

$\gamma(S_o)$ = activity coefficient of organic solvent component

$\gamma(H_2O)$ = activity coefficient of water

Literature Cited

1. Hunt, J. P., "Metal Ions in Aqueous Solution," p. 16, W. A. Benjamin, New York, 1963.
2. Basolo, F., Pearson, R., "Mechanisms of Inorganic Reactions," 2nd ed., p. 63, Wiley, New York, 1967.
3. Hunt, J. P., Taube, H., *J. Chem. Phys.* (1951) **19,** 602.
4. Hunt, J. P., Plane, R. A., *J. Am. Chem. Soc.* (1954) **76,** 5960.
5. Baltisberger, R. J., King, E. L., *J. Am. Chem. Soc.* (1954) **86,** 795.
6. Jayne, J. C., King, E. L., *J. Am. Chem. Soc.* (1964) **86,** 3989.
7. Kemp, D. W., King, E. L., *J. Am. Chem. Soc.* (1967) **89,** 3433.
8. Scott, L. P., Weeks, T. J., Bracken, D. E., King, E. L., *J. Am. Chem. Soc.* (1969) **91,** 5219.
9. Mills, C. C., King, E. L., *J. Am. Chem. Soc.* (1970) **92,** 3017.
10. Showalter, K. C., King, E. L., *J. Am. Chem. Soc.* (1976) **98,** 8087.
11. Vanderheiden, D. B., King, E. L., *J. Am. Chem. Soc.* (1973) **95,** 3860.
12. Weeks, T. J., King, E. L., *J. Am. Chem. Soc.* (1968) **90,** 2545.
13. Benson, S. W., *J. Am. Chem. Soc.* (1958) **80,** 5151.
14. King, E. L., *J. Chem. Educ.* (1966) **43,** 478.
15. Klonis, H. B., King, E. L., *Inorg. Chem.* (1972) **11,** 2933.
16. Covington, A. K., Newman, K. E., "Thermodynamic Behavior of Electrolytes in Mixed Solvents," ADV. CHEM. SER. (1976) **155,** 153.
17. Covington, A. K., Newman, K. E., Lilley, T. H., *J. Chem. Soc., Faraday Trans 1* (1973) **69,** 973.
18. Bjerrum, J., "Metal Ammine Formation in Aqueous Solution," p. 187, Copenhagen, Haase, 1941.
19. Rydberg, J., *Acta Chem. Scand.* (1961) **15,** 1723.

RECEIVED February 13, 1978.

10

Preferential Solvation of Some Electrolytes by Water and Diethyl Ether in Sulfolane

E. MILANOVA, S. Y. LAM, B. DESJARDINS, and R. L. BENOIT

Département de Chimie, Université de Montréal, Montréal, Québec, Canada

Comparisons are made of the effects of water and diethyl ether—two oxygen bases (R_2O)—on the thermodynamic properties of acids in sulfolane. The acids (HA) include $HSbCl_6$, HCF_3SO_3, HCH_3SO_3, HCF_3CO_2, and HCH_3CO_2. Vapor pressure and conductivity measurements as well as calorimetric determinations were made on 0.1–0.5M acid solutions as a function of R_2O concentration up to 1M ($X_{R_2O} < 0.1$). Equilibrium constants and enthalpy changes are obtained for the formation of both ionic species $H^+(R_2O)_n$ and uncharged species $R_2O \cdot \cdot HA$. The proton solvates $H^+(R_2O)_n$ have n = 1, 2, 3, 4 . . . for H_2O but only n = 1 for Et_2O. The species $H_2O \cdot \cdot HA$ are more stable than $Et_2O \cdot \cdot HA$. The formation of H_3O^+ is more exothermic than that of Et_2OH^+. These results are discussed in relation to gas phase data. The very large solvation enthalpy of gaseous H_3O^+ is noteworthy.

The study of the behavior of electrolytes in mixed solvents is currently arousing considerable interest because of its practical and fundamental implications (*1*). Among the simpler binary solvent mixtures, those where water is one component are obviously of primary importance. We have recently compared the effects of small quantities of water on the thermodynamic properties of selected 1:1 electrolytes in sulfolane, acetonitrile, propylene carbonate, and dimethylsulfoxide (DMSO). These four compounds belong to the dipolar aprotic (DPA) class of solvents that has received a great deal of attention (*2*) because of their wide use as media for physical separations and chemical and electrochemical reactions. We interpreted our vapor pressure, calorimetry, and NMR results in terms of preferential solvation of small cations and anions by water and obtained

0-8412-0428-4/79/33-177-145$05.50/1

free energy and enthalpy values for the binding of the first water molecule. The basicity of the solvent was found to be the main factor in determining the energetics of the bond formation (*3*).

The work with which we are chiefly concerned here is an extension of these investigations of the effects of water on the thermodynamic properties of electrolytes in DPA solvents. The electrolytes considered are acids (HA), whose importance as a class of electrolytes derives from their involvement in many chemical reactions, either as reactants or as catalysts. In conjunction with these investigations, a parallel study was carried out; water was replaced by diethyl ether (Et_2O) to determine the extent to which the hydrogen bond donor properties of the water molecule affect the interactions between HA, H_2O, and the solvent. For comparison, some additional experiments were included that used as electrolytes a lithium salt and a chloride salt and H_2S instead of H_2O.

Sulfolane (tetramethylenesulfone) was selected as the most suitable DPA solvent for this work because (1) its very low volatility makes our total vapor pressure measurements easier to interpret, and (2) its weak basicity favors both a higher activity of water and stronger interactions between water and acids. The acids studied range from strong to weak (in sulfolane) and include hexachloroantimonic acid ($HSbCl_6$), trifluoromethanesulfonic acid (HCF_3SO_3), methanesulfonic acid (HCH_3SO_3), trifluoroacetic acid (HCF_3CO_2), and acetic acid (HCH_3CO_2). (For simplicity, the formula $HSbCl_6$ is used to represent the hypothetical acid formed when antimony pentachloride is added to HCl solutions. The hexachloroantimonate anion, $SbCl_6^-$, is well characterized, and salts such as $Et_2OH^+SbCl_6^-$ are known (*4*).) The lithium and chloride salts were lithium perchlorate ($LiClO_4$) and tetraethylammonium chloride (NEt_4Cl). The experiments include measurements of total vapor pressure and electrical conductivity as well as calorimetric determinations on 0.1–0.5*M* electrolyte solutions as a function of added water and ether concentrations up to 1*M* (X ~ 0.1).

Experimental Section

Materials. Sulfolane (99% purity) (Aldrich) was treated with calcium hydride and distilled under reduced pressure. The freshly prepared solvent had a specific conductivity of $1.0 \times 10^{-7}\ \Omega^{-1}\ cm^{-1}$ and a residual water content of $8 \times 10^{-3}M$ as determined by Karl Fisher titration. Conductivity water and reagent grade ether (Baker) were used. Glacial acetic acid (CIL), trifluoroacetic acid (Baker), and trifluoromethanesulfonic acid (3M) were used as received. All these acids had a minimum purity of 99.5% as determined by titration with standard sodium hydroxide. Methanesulfonic acid (Eastman), distilled under reduced pressure, had a purity of 99.6%. Sulfolane solutions of these acids were prepared by weight, and the acid concentrations were checked by acidimetry after the samples were flooded with water. The solutions of $HSbCl_6$ in sulfolane were prepared by adding known amounts of $SbCl_5$ (99.9%) (Baker) to HCl solutions of known concentration, which, in turn, were obtained by bubbling anhydrous HCl (Liquid Air) in the solvent. Tetraethylammonium trifluoromethanesulfonate was prepared by neutralizing a 25% aqueous solution of NEt_4OH (BDH) with HCF_3SO_3. $NEt_4CF_3SO_3$ is very soluble in water ($\sim$ 22 mol/L of water at 25°C), in dichloromethane ($\sim$ 6.7 mol/L of solvent at 25°C), and in chloroform ($\sim$ 8.1 mol/L of solvent at 25°C) but almost insoluble in CCl_4. Consequently, $NEt_4CF_3SO_3$ was extracted from its aqueous solution by two portions of dichloromethane. The resulting extracts were combined and filtered, and the solvent was evaporated. The salt was redissolved in chloroform and reprecipitated by addition of CCl_4. The dried salt had a melting point of 164.5°C. Tetraethylammonium chloride (Eastman) and lithium perchlorate (Smith) were dried in vacuum desiccators over phosphorus pentoxide.

Calorimetry. Heats of solution were measured at 30°C using an LKB model 8725-2 isoperibol calorimeter. Details of the procedure have been given (*5*). The heats of mixing known weights of water or ether with 25 ml of 0.1–0.5*M* acid solutions in sulfolane were determined. Some of the acid solutions contained some known concentrations of water as well. In addition, the heats of solution of $NEt_4CF_3SO_3$ in sulfolane and in a 0.211*M* $HSbCl_6$ solution in sulfolane were determined at three concentrations between 0.02 and 0.04*M*.

Vapor Pressure Measurements. Total vapor pressures were measured at 30°C with a Texas Instruments quartz spiral gauge. The procedure used was similar to that given previously (*5*). The concentration of water after the experiment was checked by Karl Fischer titration, while that of ether was found by weighing the cell before and after the vapor pressure determination; the loss of weight was that of ether. Since Sb(V) interferes with the Karl Fischer titration of water, the water concentration in the $HSbCl_6$ solutions was also obtained from the loss of weight of the cell.

Conductivity. The conductivity bridge has been described (*6*). A 0.01000*M* standard KCl solution was used to calibrate the Beckman conductivity cell. A value of 0.488 cm^{-1} was obtained for the cell constant. Measurements were made at 30°C on 50 ml of 0.1–0.5*M* acid solutions in sulfolane with successive additions of water or ether from a 2-ml Teflon glass syringe. All acid solutions were processed in a glove box.

Results

The results of some ancillary experiments to establish the nature of the solutions of HCF_3SO_3 and $HSbCl_6$ in sulfolane are considered first. A value of the ionization constant $K_d°$ of HCF_3SO_3 was estimated from the specific conductance L of 0.1–0.2*M* HCF_3SO_3 solutions, extrapolated to zero water concentration, and from values of the equivalent conductance Λ of $H^+CF_3SO_3^-$, obtained, in turn, from the equivalent conductances of $H^+SbCl_6^-$, a strong electrolyte in sulfolane (*7*), $NEt_4^+SbCl_6^-$, and $NEt_4^+CF_3SO_3^-$. By extrapolation it is not possible to obtain the equivalent conductance of HCF_3SO_3 at infinite dilution because of the strong influence of residual water on the dissociation of the acid. Using the reduced Debye-Hückel Equation for the activity coefficients of H^+ and $CF_3SO_3^-$, the activity constant $K_d°$ for HCF_3SO_3 was calculated to be 2.5×10^{-4}, so that HCF_3SO_3 is nearly as strong an acid as $HClO_4$ in sulfolane (*7, 8*). Trifluoromethanesulfonic acid is therefore preferred to perchloric acid as a reagent for acid–base titrations in sulfolane (*9*). The risk of explosions when handling solution of $HClO_4$ in sulfolane (*7*) must again be forcefully emphasized. The results of our vapor pressure measurements on HCF_3SO_3 solutions in sulfolane indicate that the pressure, P (mm Hg), is proportional to the acid concentration, C_{HA} (M), up to 0.2M. The pressures, P, are: 0.16 mm (0.10M); 0.33 mm (0.20M); 0.57 mm (0.50M); and 0.71 mm (1.00M). The vapor pressure of pure HCF_3SO_3 at 30°C was measured as 3.01 mm, so the observed negative deviation from Henry's Law at increasing concentrations of HCF_3SO_3 is expected. (This value is in good agreement on a log P vs. $1/T$ plot with vapor pressures obtained at three higher temperatures (*11*) and therefore invalidates the value of 1.0 mm Hg at 42°C (*11*).) Formation of a dimer $(HCF_3SO_3)_2$ at high concentrations might explain the deviation. Dimerization of HCF_3SO_3 in acetonitrile solutions has been postulated by Kolthoff (*10*) on the basis of conductivity measurements although the influence of residual water was apparently not taken into account in the interpretation. The heats of solution of solid $NEt_4CF_3SO_3$ were +2.17 kcal mol^{-1} in pure sulfolane and +0.36 kcal mol^{-1} in a 0.212M $HSbCl_6$ solution in sulfolane; thus, the heat of ionization $\Delta H_d°$ of HCF_3SO_3 is +1.81 kcal mol^{-1}.

Concerning the nature of $HSbCl_6$ solutions in sulfolane, we have reported (*7*) that adding $SbCl_5$ (1) to weakly conducting solutions of HCl leads to an initial linear increase of L with $C^{\frac{1}{2}}_{SbCl_5}$ and an extrapolated value of $\Lambda°$ near 11.5 Ω^{-1} cm^2 equiv^{-1}. We took this to indicate that the equilibrium constant K_f for Reaction 1 is large

$$HCl + SbCl_5 = HSbCl_6 \qquad (1)$$

and that $HSbCl_6$ is a strong electrolyte. Additional conductivity determinations, in which we varied the initial HCl concentration and added higher concentrations of $SbCl_5$, led us to estimate for K_f a value of the order of $10^{2.5}$. Further vapor pressure measurements of solutions containing HCl and $SbCl_5$, with molar ratios C_{SbCl_5}/C_{HCl} slightly above 1, showed a residual HCl vapor pressure that was used to estimate the HCl concentration from Henry's Law constant for HCl (*5*). The value of K_f thus calculated, $10^{2.7 \pm 0.2}$, was retained.

We now examine the vapor pressure, conductivity, and calorimetry results obtained when water and ether were added to the acid solutions. The vapor pressure data in Figure 1 give the measured total pressure,

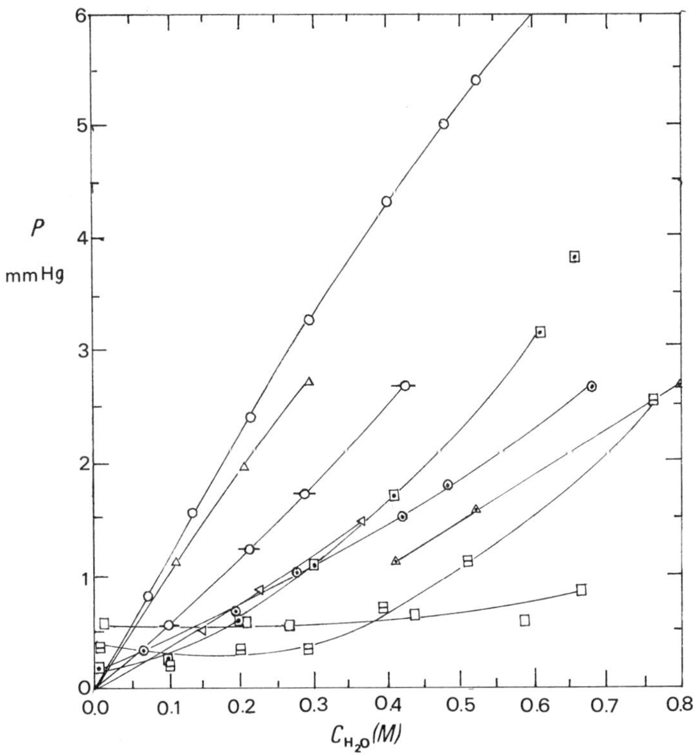

Figure 1. Vapor pressure of sulfolane solutions of electrolytes as a function of water concentration at 30°C: (○), sulfolane; (△), 0.50M HCH_3CO_2; (–○–), 0.50M HCF_3CO_2; (◬), 0.53M HCH_3SO_3; (⊡), 0.10M HCF_3SO_3; (⊟), 0.20M HCF_3SO_3; (□), 0.50M HCF_3SO_3; (◬), 0.50M NEt_4Cl; (⊙), 0.51M $LiClO_4$

P (mm Hg), as a function of the molar concentration of added water, C_{H_2O}, for 0.1–0.5*M* acid solutions in sulfolane. The corresponding data obtained for added ether are plotted in Figure 2. The vapor pressure of pure sulfolane at 30°C is 0.02 mm Hg, so the total pressure, *P*, can be taken as the sum of the vapor pressures of unbound water or ether (p_{H_2O} or p_{Et_2O}) and acid (p_{HA}). Although the contribution of p_{HA} to *P* is small because the acids HA studied have a low volatility, it is the decrease of p_{HA} that explains the initial small variations of *P* with water concentration when $0 < C_{H_2O}/C_{HA} < 1$ for HCF_3SO_3 and $HSbCl_6$. The curves giving the vapor pressures p_{H_2O} and p_{Et_2O} of sulfolane solutions of water and ether, in the absence of electrolytes, are also plotted in Figures 1 and 2, respectively, against the molar concentrations of water and ether. It is clear that the presence of HA lowers the vapor pressure of water and ether, the decrease being larger for water. Further, for a given water

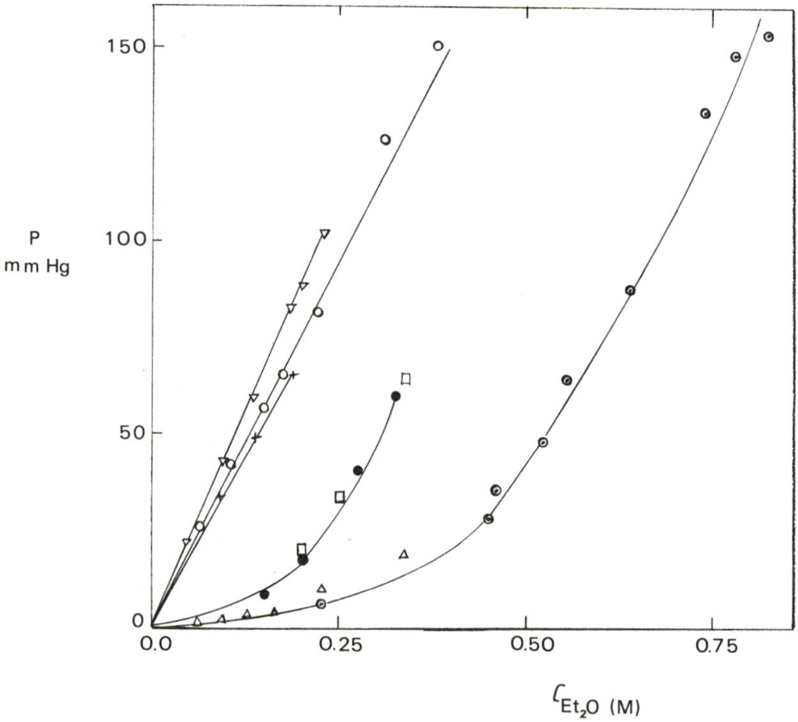

Figure 2. Vapor pressure of sulfolane solutions of acids as a function of diethyl ether concentration at 30°C: (▽), sulfolane; (+), 0.50M HCF_3SO_2; (○), 0.48M HCH_3SO_3; (⊙), 0.51M HCF_3SO_3; (△), 0.44M HCF_3SO_3; (□), 0.25M HCF_3SO_3; (●), 0.20M $HSbCl_6$ ($C_{SbCl_5}/C_{HCl} = 1.00$).

concentration the values of p_{H_2O} decrease in the order HCH_3CO_2 < HCF_3CO_2 < HCH_3SO_3 < HCF_3SO_3.

The specific conductances, L (Ω^{-1} cm^{-1}), of 0.1–0.5M solutions of the acids HA are plotted in Figures 3 and 4 against the concentrations of added water and ether, respectively. L increases rapidly at first with R_2O concentration and more so with water than with ether. However, the addition of Et_2O to $HSbCl_6$ causes L to decrease linearly, with the molar ratio of C_{Et_2O}/C_{HA}, until a ratio of 1.00 is reached; then L remains constant. For a given R_2O concentration, the values of L for the different acids (HA) increase in the order HCH_3CO_2 < HCF_3CO_2 < HCH_3SO_3 < HCF_3SO_3.

The calorimetric data obtained for the addition of water and ether to acid solutions are summarized in Tables I and II, respectively. The addition of water or ether (R_2O) to the solutions of acid (HA) in sulfolane gives rise to a series of reactions. We have considered the following equilibria, with the corresponding equilibrium constants and enthalpy changes:

$$HA + R_2O \rightleftharpoons R_2O \cdot \cdot HA \qquad K, \Delta H^\circ \qquad (2)$$

$$H^+ + R_2O \rightleftharpoons R_2OH^+ \qquad K_1, \Delta H_1^\circ \qquad (3)$$

$$(R_2O)_{n-1}H^+ + R_2O \rightleftharpoons (R_2O)_nH^+ \qquad K_n, \Delta H_n^\circ \qquad (4)$$

in addition to

$$HA \rightleftharpoons H^+ + A^- \qquad K_d, \Delta H_d^\circ \qquad (5)$$

The interpretation of our results in terms of the previous reactions is derived mainly from the treatment of the vapor pressure data. The conductivity results for the Et_2O–$HSbCl_6$ system point conclusively to the formation of a single protonated species, Et_2OH^+ (Figure 4) and also indicate that Λ (Et_2OH^+) $<$ Λ (H^+). Other useful conductivity data are those for the H_2O–HCF_3SO_3 system, which facilitate the computation of the protonation constants K_1 and K_2 from the vapor pressure curves. The difficulty in fully exploiting the conductivity results stems from the fact that the calculation of ionic concentrations of H_3O^+, $(H_2O)_2H^+$, . . . from the specific conductance, L requires the values of the equivalent conductance, Λ, of $H_3O^+A^-$, $(H_2O)_2H^+A^-$, and so forth; these values, which are also affected by the decrease in the viscosity that occurs when water is added, can be known only approximately. Once the equilibrium constants K, K_1, K_n have been determined, the calorimetric data are used to calculate the enthalpy changes ΔH for some of the reactions.

We consider first the vapor pressure data of Figure 1 that were obtained when water was added to 0.5M sulfolane solutions of the weaker acids, HCH_3CO_2, HCF_3CO_2, and HCH_3SO_3. For a given water concentration, C_{H_2O}, the corresponding water vapor pressure, $p_{H_2O} = P$, was read from the graph; the contribution of p_{HA} to P was neglected (HCH_3CO_2 (0.36 mm Hg), HCF_3CO_2 (0.12 mm Hg), HCH_3SO_3 (about 0.10 mm Hg)). Next, the unbound water concentration (H_2O) was read from the curve, giving p_{H_2O} for sulfolane solutions of water. The value of the association constant, K, for Reaction 2 was then calculated using a treatment described in the text of Rossotti (*12*). The conductivity data plotted in Figure 3 show that the concentration of ionic species increases with C_{H_2O}, as would be expected from Reactions 3 and 5. Taking a lower limit of 12 Ω^{-1} cm^2 equiv^{-1} for the equivalent conductance of the ionic species leads to concentrations of ionic species that are less than 4% of the concentration of added water (C_{H_2O}), for the HCH_3SO_3 solution. For the HCF_3CO_2 solution the concentration of ionic species is

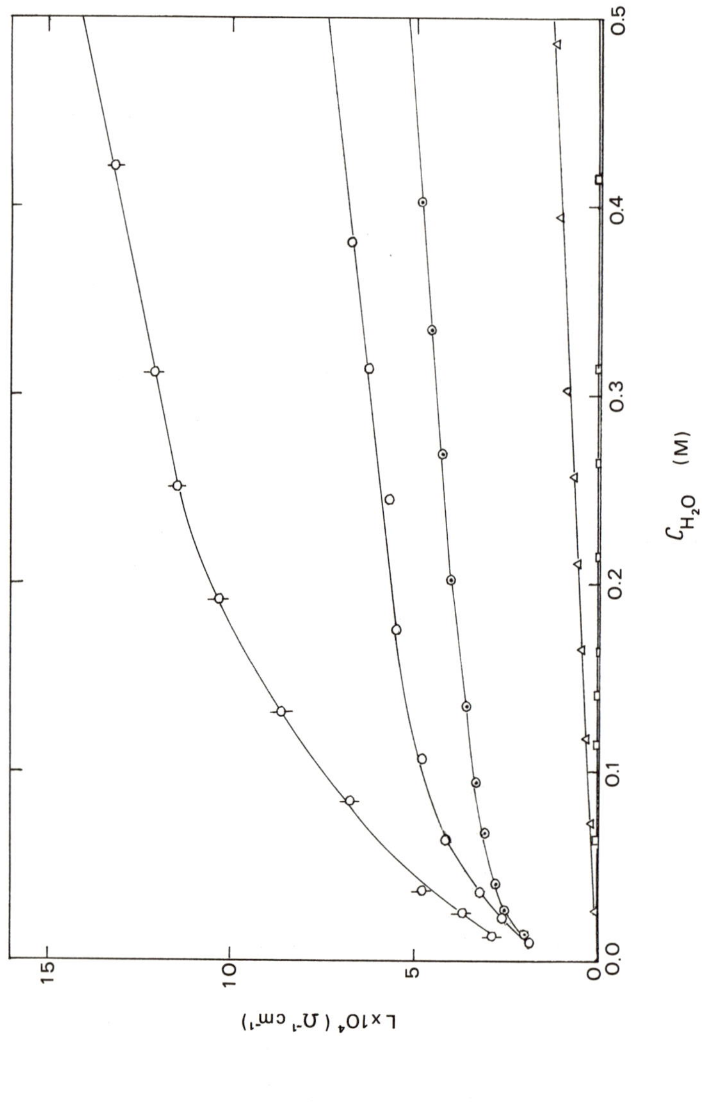

Figure 3. Specific conductance of sulfolane solutions of acids as a function of water concentration at 30°C: (□), *0.54M* HCF_3CO_2; (△), *0.53M* HCH_3SO_3; (ϙ), *0.50M* HCF_3SO_3; (○), *0.20M* HCF_3SO_3; (⊙), *0.10M* HCF_3SO_3.

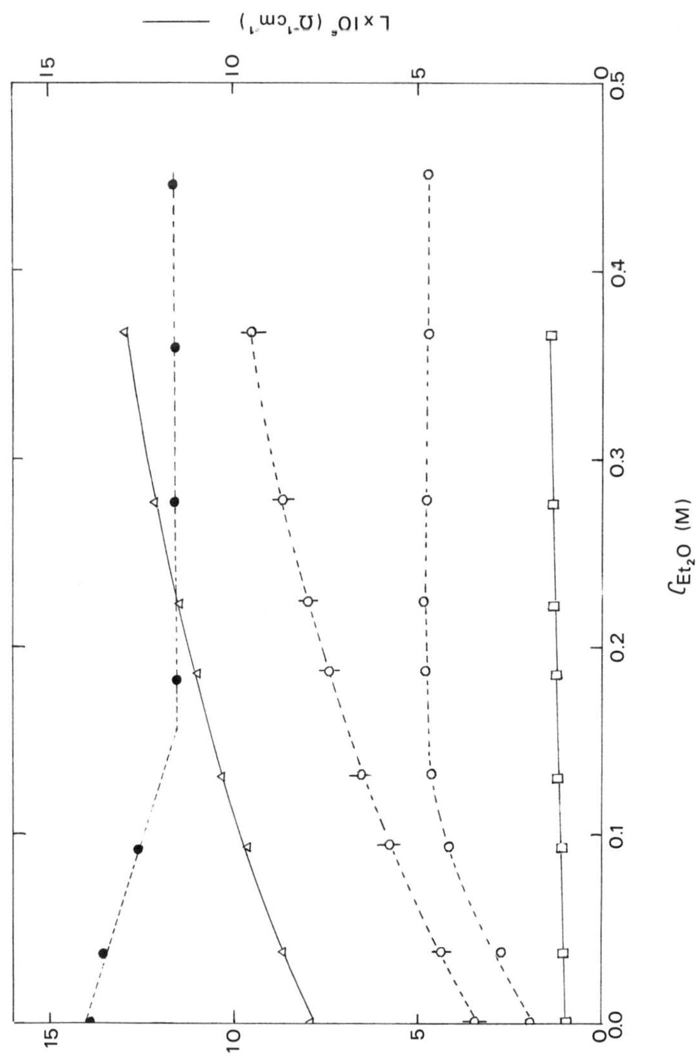

Figure 4. Specific conductance of sulfolane solutions of acids as a function of diethyl ether concentration at 30°C: (□), *0.48M* HCF_3CO_2; (△), *0.58M* HCH_3SO_3; (⦶), *0.58M* HCF_3SO_3; (○), *0.25M* HCF_3SO_3; (●), *0.15M* $HSbCl_6$ ($C_{SbCl_5}/C_{HCl} = 0.33$).

Table I. Heats of Mixing of Water with Sulfolane Solutions of Electrolytes at 30°C

HA	C_{HA} (M) *(initial)*	C_{H_2O} (M) *(initial)*	C_{H_2O} (M) *(final)*	−Q *(kcal) per mole of water added*
HCH_3CO_2	0.499	0.009	0.0401	−1.24
		0.009	0.0805	−1.21
		0.009	0.1214	−1.24
	0.498	0.009	0.0802	−1.26
HCF_3CO_2	0.500	0.0105	0.0413	0.420
		0.0105	0.0783	0.382
		0.0105	0.1198	0.333
HCH_3SO_3	0.501	0.016	0.0404	1.30
		0.016	0.0791	1.27
		0.016	0.1200	1.21
HCF_3SO_3	0.219	0.0074	0.102	8.71
		0.0074	0.183	7.82
		0.183	0.288	3.42
	0.498	0.010	0.088	8.84
		0.201	0.281	8.61
		0.201	0.384	7.79
		0.428	0.505	5.84
	0.522	0.540	0.772	3.32
	0.519	0.772	1.00	2.29
	0.516	1.00	1.23	1.54
	0.514	1.23	1.46	1.01
$HSbCl_6$	0.244	0.0072	0.081	9.90
		0.081	0.153	8.81
$\left(\frac{C_{SbCl_5}}{C_{HCl}} = 1.00\right)$	0.238	0.0027	0.096	9.36
		0.0027	0.098	9.28
		0.098	0.193	8.23
	0.478	0.519	0.752	2.76
	0.476	0.752	0.988	1.78
	0.478	0.519	1.009	2.26
	0.474	1.009	1.489	0.73
	0.470	1.489	1.967	0.15
$HSbCl_6$	0.160	0.008	0.90	9.94
$\left(\frac{C_{SbCl_5}}{C_{HCl}} = 0.16\right)$				
$HSbCl_6$	0.300	0.087	0.130	9.90
		0.130	0.260	9.50
$\left(\frac{C_{SbCl_5}}{C_{HCl}} = 0.37\right)$				
$HSbCl_6$	0.060	0.010	0.060	9.61
		0.060	0.120	2.36
$\left(\frac{C_{SbCl_5}}{C_{HCl}} = 0.05\right)$				

Table I. Continued

HA	C_{HA} (M) (initial)	C_{H_2O} (M) (initial)	C_{H_2O} (M) (final)	−Q (kcal) per mole of water added
$HSbCl_6$	0.250	0.250	0.310	2.95
$\left(\frac{C_{SbCl_5}}{C_{HCl}} = 0.21\right)$		0.310	0.370	2.62
		0.370	0.430	2.34
NEt_4Cl	0.50	0.010	0.050	0.45
		0.010	0.091	0.42
		0.010	0.131	0.40
$LiClO_4$	0.50	0.010	0.050	0.38
		0.010	0.090	0.36
		0.010	0.130	0.34

Table II. Heats of Mixing of Ether with Sulfolane Solutions of Acids HA at 30°C[a]

HA	C_{HA} (M) (initial)	C_{Et_2O} (M) (final)	−Q (kcal) per mole of ether added
HCF_3CO_2	0.500	0.038	−0.23
		0.076	−0.23
		0.110	−0.15
HCH_3SO_3	0.507	0.035	−0.24
		0.109	−0.13
HCF_3SO_3	0.225	0.017	6.42
		0.036	5.81
		0.056	5.41
		0.072	5.49
		0.090	5.37
		0.110	5.29
	0.251	0.065	5.95
		0.129	5.17
		0.179	4.72
	0.519	0.061	6.02
		0.116	6.07
		0.162	6.10
$HSbCl_6$	0.248	0.035	6.43
		0.072	6.62
$\left(\frac{C_{SbCl_5}}{C_{HCl}} = 1.00\right)$		0.119	6.58
	0.219	0.073	6.54
		0.131	6.51

[a] Initial concentration of Et_2O is nil.

Table III. Equilibrium Constants (L Mol^{-1}) for Reactions

$HA + R_2O \rightleftharpoons R_2O \cdot\cdot HA$	HA
	HCH_3CO_2
	HCF_3CO_2
	HCH_3SO_3
	HCF_3SO_3
$I(R_2O)^{\pm}_{n-1} + R_2O \rightleftharpoons I(R_2O)^{\pm}_n$	$I^{\pm}$
	H^+
	Li^+
	Cl^-

reduced further by a factor of 10. Therefore, we are justified in not considering Reactions 3 and 5 in our calculation of the association constant K for $H_2O \cdot\cdot HA$. Values of K and ΔH are in Tables III and IV, respectively.

The vapor pressure data (Figure 2) for the addition of ether to 0.5M solutions of HCF_3CO_2 and HCH_3SO_3 were treated as above to obtain the values of the constants, K, for $Et_2O \cdot\cdot HA$ in Table III. We note that the specific conductance, L, increases initially less rapidly when ether is added to HA solutions than when water is added. We can conclude that K_1 is larger for water than for ether. The conductivity data were not used to calculate the formation constants, K_1, of Et_2OH^+ because of the complexities involved. In addition to the previously emphasized difficulty in estimating Λ, further complications would arise from the need to consider the following supplementary equilibrium for HCH_3SO_3 and HCF_3CO_2:

$$HA + A^- \rightleftharpoons HA_2^- \qquad K_{HA_2^-} \tag{6}$$

Table IV. Enthalpy Changes (kcal mol^{-1}) for Reactions between Water, Diethyl Ether, and Acids (HA) and Ions ($I^{\pm}$)

		H_2O		Et_2O
$HA + R_2O \rightleftharpoons R_2O \cdot\cdot HA$	HA	$-\Delta H°$		$-\Delta H°$
	HCF_3CO_2	3.7		
	HCH_3SO_3	3.7		
	HCF_3SO_3	11		6.9
$I(R_2O)^{\pm}_{n-1} + R_2O \rightleftharpoons I(R_2O)^{\pm}_n$	$I^{\pm}$	$-\Delta H_1°$	$-\Delta H_2°$	$-\Delta H_1°$
	H^+	11.1	8.0	7.6
	Li^+	3.0		
	Cl^-	2.6		

between Water, Diethyl Ether, and Acids (HA) and Ions ($I^{\pm}$)

H_2O K	Et_2O K
~ 0.4	
2.5	~ 0.6
5.7	~ 0.3
~ 300	40

K_1	K_2	K_3	K_4	K_5	K_1
~ 10,000	50	7.5	5	4	120
4.5					
8.2					

Anions, A^-, such as $CH_3SO_3^-$ and $CF_3CO_2^-$, which correspond to weak acids and are therefore strong hydrogen-bond acceptors because of their poor solvation in dipolar aprotic solvents, tend to bind to undissociated HA to give HA_2^-. For example, HCl_2^- has been shown to be stable in sulfolane ($K_{HA_2^-} = 10^{2.5}$ (*5*)), $H(CH_3SO_3)_2^-$ is stable in acetonitrile ($K_{HA_2^-} = 10^{3.8}$ (*13*)), and $H(CF_3CO_2)_2^-$ is stable in the same solvent ($K_{HA_2^-} = 10^{3.5}$ (*14*)).

We now examine the vapor pressure results obtained when water and ether are added to solutions of the stronger acids, HCF_3SO_3 and $HSbCl_6$. As the first step we calculated, in the manner described before, the concentration (R_2O) of unbound H_2O or Et_2O corresponding to each concentration, C_{R_2O}, of added water or ether. We then computed values of

$$\bar{n} = \frac{C_{R_2O} - (R_2O)}{C_{HA}}$$

where $\bar{n}$ is the average number of R_2O bound per acid group, irrespective of the extent of HA dissociation. The values of $\bar{n}$ are plotted in Figure 5 against the concentration of unbound R_2O. This graph brings into focus the very different behavior of water from ether. With Et_2O, $\bar{n}$ increases with the concentration of unbound Et_2O up to about 1, with both HCF_3SO_3 and $HSbCl_6$. The species formed are therefore mainly $Et_2O \cdots HCF_3SO_3$ and Et_2OH^+. In contrast, with H_2O and HCF_3SO_3, $\bar{n}$ increases continuously with the unbound H_2O concentration, indicating that in addition to $H_2O \cdot \cdot HCF_3SO_3$, species $H^+(H_2O)_n$, with values of $\bar{n}$ up to at least 4, are formed. To interpret the vapor pressure curves in terms of Reactions 2, 3, 4, and 5, a program was written that enables us to calculate the concentration of all species present for a given set of constants $K, K_1 \ldots K_n, K_d$ and given total concentrations of acid (HA) and water or ether. We encountered some difficulties in determining the

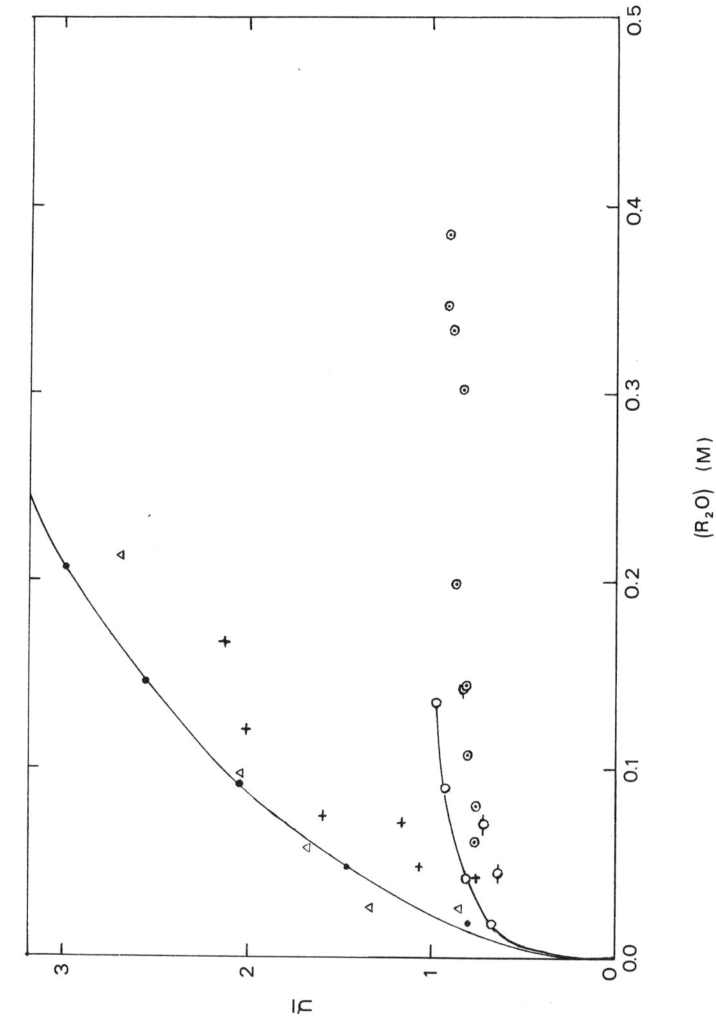

Figure 5. Average number (n) of R_2O groups attached per acid (HA and H^+) as a function of the unbound R_2O concentration: $R_2O = H_2O$; (●), 0.10M HCF_3SO_3; (△), 0.20M HCF_3SO_3; (+), 0.50M HCF_3SO_3; $R_2O = Et_2O$; (-○-), 0.25M HCF_3SO_3; (⊙), 0.51M HCF_3SO_3; (○), 0.20M $HSbCl_6$.

values of K and K_1 for the H_2O–HCF_3SO_3 system. The very low water vapor pressures for $0 < C_{H_2O}/C_{HA} < 1$ lead to imprecise values of K and K_1. However, the high value quoted for K_1 is in line with the conductivity and calorimetric data. The constants are listed in Table III. The vapor pressure data from the $HSbCl_6$–H_2O system resulted in lower values of K_n, which led us to suspect that the values of the vapor pressure were too high. This may be caused by difficulties in preparing $HSbCl_6$ solutions with a molar ratio (C_{SbCl_5}/C_{HCl}) equal to 1.00. Excess HCl over the ratio 1.00 gives a residual HCl pressure, while excess $SbCl_5$ reacts with water; further, a slight hydrolysis of $SbCl_6^-$ giving HCl may also take place when the water concentration increases. The enthalpy changes corresponding to Reactions 2, 3, and 4 were obtained from the experimental values, Q, of the heat involved and from program-calculated changes in the concentrations of all species before and after each calorimetric run. The values of the heat of mixing of water and ether in sulfolane used in the calculations were +1.63 and +0.60 kcal mol^{-1}, respectively. The enthalpy changes for Reactions 2, 3, and 4 are given in Table IV. The values of ΔH_n become inaccurate when $n > 2$ because of the cumulative effect of errors on K and K_1.

The vapor pressure curves obtained when water is added to 0.50*M* $LiClO_4$ and NEt_4Cl solutions in sulfolane are plotted in Figure 1. Both $LiClO_4$ and NEt_4Cl are strong electrolytes in sulfolane (*15, 16*). The vapor pressure data were used to calculate the hydration constants, K_1 (4.5 (Li^+H_2O) and 8.2 (Cl^-H_2O), respectively), under the assumption that hydration of the large ions ClO_4^- and NEt_4^+ is negligible (*3*). The ionic hydration enthalpies were computed from the corresponding calorimetric data (Table IV) and are −3.0 kcal mol^{-1} (Li^+H_2O) and −2.6 kcal mol^{-1} (Cl^-H_2O).

Discussion

In our treatment of the vapor pressure data we tacitly assume that the electrolytes that are usually present at low concentrations, except in $HSbCl_6$ solutions, have only a negligible salt effect on water and ether. This point has been discussed (*3*). A second assumption in our interpretation of the vapor pressure data for the H_2O–HCF_3SO_3 system is that the hydration of $CF_3SO_3^-$ is negligible when compared with the hydration of the proton and HCF_3SO_3. This assumption is justified on the basis of Kolthoff's (*10*) hydration constant of 3.6 for $CH_3SO_3^-$ in acetonitrile. From this value, the lower basicity of $CF_3SO_3^-$, and our data on the hydration of anions in dipolar aprotic solvents we estimate a value of less than 2 for the hydration constant of $CF_3SO_3^-$ in sulfolane. Although our data do not permit us to distinguish between the complexes $R_2O \cdot\cdot HCF_3SO_3$, on the one hand, and the ion pairs $R_2OH^+CF_3SO_3^-$, on the other hand, we feel that given the low hydrogen-bonding properties of $CF_3SO_3^-$ and existing data on ion-pair formation in sulfolane (*15, 16*), $H_2O \cdot \cdot HCF_3SO_3$ and $Et_2O \cdot \cdot HCF_3SO_3$ are the preferred species. However, it may be that with other anions (A^-) that have strong hydrogen-bonding properties, ion pairs such as $H_3O^+A^-$ could be formed, although this event would be competing with HA_2^- formation. In this respect, it is worth noting that in aqueous solutions of HF it has recently been advanced that "undissociated HF" is present as the ion pair $H_3O^+F^-$ and not as "hydrated HF" (*17*).

Our vapor and conductivity data show conclusively that addition of water to sulfolane solutions of HCF_3SO_3, HCH_3SO_3, and HCF_3CO_2, which are strong or moderately strong acids in the solvent water, causes only partial ionization of these acids. This is because, as shown by Reaction 5, hydration of the proton (which tends to increase ionization) is counter-balanced by formation of $H_2O \cdot \cdot HA$. The existence of such hydrogen-bonded molecular complexes $H_2O \cdot \cdot HA$ has already been inferred from physicochemical measurements in concentrated aqueous solutions of the strong acids, H_2SO_4 and HNO_3 (*18*). Further, direct spectroscopic evidence has been given for the existence of the complex $H_2O \cdot \cdot HCl$ in N_2 and Ar matrices (*19*). This latter species $H_2O \cdot \cdot HCl$ was also formed in sulfolane when water was added to HCl solutions, and K was estimated to be approximately 0.5 (*20*); however, we did not carry out any detailed study of the HCl–H_2O system because of the high volatility of HCl in sulfolane solutions (*5*). We can also surmise from our results that the projected study of the 1:1 mixture H_2O–HCF_3SO_3, which has recently been proposed as a preferred electrolyte (*21*) for improving the performance of fuel cells, will also reveal the presence of $H_2O \cdot \cdot HCF_3SO_3$ next to H_3O^+, $CF_3SO_3^-$, and other ionic species.

The stability of the $H_2O \cdot \cdot HA$ species increases with HA acid in the order $HCH_3CO_2 < HCF_3CO_2 < HCH_3SO_3 < HCF_3SO_3$, but the change is not very large because interactions between HA and the solvent sulfolane also increase with HA strength—i.e., H_2O and the solvent compete for HA. It is likely that in the gas phase, the stability of molecular complexes, $H_2O \cdot \cdot HA$, would increase much more rapidly with HA strength. Attempts to evaluate enthalpy changes for the formation of $H_2O \cdot \cdot HA$ in the gas phase (Reaction 2) from our data in sulfolane proved unsuccessful because of difficulties in finding a satisfactory model molecule with which to estimate the enthalpy of solvation of gaseous $H_2O \cdot \cdot HA$. Our results in Table III also show that for a given HA, the molecular complex $H_2O \cdot \cdot HA$ is somewhat more stable than $Et_2O \cdot \cdot HA$; however, the values of K less than 1 for $Et_2O \cdot \cdot HCF_3CO_2$ and $Et_2O \cdot \cdot HCH_3SO_3$ may not be meaningful and may just reflect the slight nonideality of the systems and not the formation of chemical species.

Comparing water and ether in their reactions with the solvated proton, our results in Tables III and IV show wide differences in the behavior of these related oxygen bases. Proton hydrates ($H^+(H_2O)_n$) containing at least four H_2O are formed, and the corresponding formation constants (K_n) and enthalpy changes (ΔH_n) decrease with increasing n. A similar situation prevails in the gas phase where H_3O^+ is not a preferred species (*22*). Our values of K_n for $H^+(H_2O)_n$ in sulfolane can be compared with those reported by Kolthoff (*23*) for $H^+(H_2O)_n$ in acetonitrile—namely, $K_1 = 160$, $K_2 = 50$, $K_3 = 7.5$, and $K_4 = 3.3$. These values were calculated from spectrophotometric measurements using Hammett indicators and dilute $HClO_4$ solutions but might require some correction because of the incomplete ionization of $HClO_4$ (*24*) and the possible formation of $H_2O \cdot \cdot HClO_4$. On the basis of the higher basicity of acetonitrile compared with that of sulfolane, the difference between the values of K_1 for $H^+(H_2O)$ in both solvents is of the expected order of magnitude. Considering next the reaction of the solvated proton with ether, $H^+(Et_2O)$ is less stable than $H^+(H_2O)$, as found by Kolthoff in acetonitrile (*23*), while $H^+(Et_2O)_2$ is not detected. It is significant that Kebarle (*25*) has recently shown that in the gas phase, the heat of binding successive dimethyl ether molecules, Me_2O to H^+, drops dramatically after the first two molecules are bound. This was attributed to the blocking of hydrogen bonding past the structure $Me_2OHOMe_2^+$; a similar pattern could, of course, be expected with Et_2O. The fact that we have not found $H^+(Et_2O)_2$ stable in sulfolane is attributed to competition between the weakly basic sulfolane molecules and Et_2O for $H^+(Et_2O)$.

A further comparison of the protonation of water and ether is best carried out on the basis of enthalpy data, both in sulfolane and in the gas phase. Consider the following thermodynamic cycle:

$$\begin{array}{ccccccc} H^+\,(g) & + & R_2O\,(g) & = & R_2OH^+\,(g) & \Delta H_1°\,(g) & \\ | & & | & & | & & \\ \Delta H_s°\,(H^+) & & \Delta H_s°\,(R_2O) & & \Delta H_s°\,(R_2OH^+) & & (6) \\ \downarrow & & \downarrow & & \downarrow & & \\ H^+\,(s) & + & R_2O\,(s) & = & R_2OH^+\,(s) & \Delta H_1°\,(s) & \end{array}$$

where (g) and (s) are used for the reaction that is taking place in the gas phase and in the solvent sulfolane. The values of $\Delta H_1°$ (g) are -170.3 kcal mol^{-1} (H_2O) and -197.4 kcal mol^{-1} (Et_2O), respectively (*26*); they indicate that Et_2O in the gas phase is much more basic than H_2O, which is in accordance with what is qualitatively expected from the inductive effect of the ethyl substituent. However, in sulfolane, the values of $\Delta H_1°$ (s), which are -11.1 kcal mol^{-1} (H_2O) and -7.6 kcal mol^{-1} (Et_2O), respectively, indicate that H_2O has now become more basic than Et_2O.

The use of the above thermodynamic cycle makes it possible to analyze the cause of this reversal of basicity order in sulfolane. Combination of the heat of solvation of gaseous R_2O, $\Delta H_s°$ (R_2O), −8.9 kcal mol^{-1} (H_2O) and −6.2 kcal mol^{-1} (Et_2O), respectively, with the heat of solvation of H^+ (g), $\Delta H_s°$ (H^+) = −252.6 kcal mol^{-1} (*27*), leads to the following values of $\Delta H_s°$ (R_2OH^+); −102.3 kcal mol^{-1} (H_3O^+) and −69.0 kcal mol^{-1} (Et_2OH^+) (Table V). It is therefore mainly the large heat of solvation of H_3O^+, some 33 kcal mol^{-1} more exothermic than that of Et_2OH^+, which more than compensates for the 27 kcal mol^{-1} difference in the proton affinity of H_2O and Et_2O and makes H_2O a better base in sulfolane than Et_2O by some 3 kcal mol^{-1}. Although electrostatic factors play some part in determining the 33 kcal mol^{-1} difference in the heat of solvation of H_3O^+ and Et_2OH^+, it is clearly the exceptional ability of H_3O^+ to hydrogen bond to the solvent which must be the determining factor. The recent enthalpy data of Scorrano (*28*) on the protonation of Et_2O in water, $\Delta H_1°$ (w) = −0.73 kcal mol^{-1}, show that this previous conclusion also holds when water is the solvent since we calculated $\Delta H_w°$ (H_3O^+) = −110.2 kcal mol^{-1} and $\Delta H_w°$ (Et_2OH^+) = −81.2 kcal mol^{-1}.

It is significant that this explanation of the reason for the much larger enthalpy of solvation of H_3O^+ compared with that of Et_2OH^+, whether in sulfolane, a DPA solvent, or water, is in line with our recent interpretation of our data on the basicity of a series of substituted amines, B, in DMSO and propylene carbonate (*29, 30*). We found the basicity order in DMSO to be $NH_3 \sim MeNH_2 > Me_2NH > Me_3N$, (i.e., an order that is at variance with that deduced from the proton affinity in the gas phase and predictions made on the basis of the alkyl substituent effect). We showed that just as for H_2O and Et_2O, it was the large decrease in the heat of solvation of the gaseous onium ions, $\Delta H_s°$ (BH^+), with the number of substituents, in the order $NH_4^+ > MeNH_3^+ > Me_2NH_2^+ > Me_3NH^+$, that was the main factor responsible for the basicity sequence observed in DMS. The values of $\Delta H_s°$ (BH^+) given in Table V also

Table V. Solvation Enthalpy of Gaseous Ions in

Solvent	H^{+a}	H_3O^+	Et_2OH^+
Sulfolane	252.6	102.3	69.0
Water	270.0	110.2	81.2
Propylene carbonate	259.5		
DMSO	276.1	119.9[b]	

[a] Ref. *27*.
[b] Calculated from data in Ref. *3*.
[c] Ref. *30*.

indicate a nearly constant decrease of some 8 kcal mol^{-1} per available hydrogen bond on the substituted ammonium ions in DMS (vs. 7 kcal mol^{-1} in propylene carbonate, a less basic solvent than DMS (*27*)). This regular decrease in the heat of solvation of the same substituted ammonium ions has also been observed in water (*30*), which is somewhat unexpected because water is a much more complex solvent than the DPA solvents we used. Data in Table V also show that heats of solvation for H_3O^+ and NH_4^+ are both more negative than that for K^+, an ion of similar size but without hydrogen-bonding properties. All these results give strong support to the conclusion that it is the number of available hydrogen bonds formed between the oxonium or ammonium ions and the basic DPA solvent molecules, more than electrostatic interactions between the ions and solvent (*32*), that is the main factor responsible for the regular decrease in the heat of solvation of a series of substituted onium ions. Results of our current study of the protonation of the MeOH and EtOH alcohols in sulfolane should be a further test of these views.

Finally, mention should be made of our exploratory investigation of the protonation of H_2S, a stronger base than H_2O in the gas phase since its proton affinity is 3.6 kcal mol^{-1} higher than that of H_2O (*26*). Vapor pressure experiments in sulfolane and acetonitrile using HCF_3SO_3 (HA) as acid, ($HSbCl_6$ reacts with H_2S) did not show any evidence of interactions between H_2S and H^+ or HA, but some decomposition of H_2S was observed in acetonitrile. Some proton NMR experiments using additional solvents showed only a small solvent effect ($0.8 < \delta < 1.3$ ppm) on the H_2S signal, and addition of HCF_3SO_3 only caused the collapse of proton peaks, while further addition of water led to the reappearance of the H_2S signal and appearance of a new signal caused by H_3O^+. We conclude that H_2S as a solute is a weaker base than H_2O despite its higher proton affinity, probably because H_3S^+ is a much poorer hydrogen-bond donor than H_3O^+. Nevertheless, H_2S is protonated at low temperature in HF–SbF_5 and HSO_3F–SbF_5 mixtures (*33*), which are obviously much more acidic than the media we used. In connection with the above results,

DPA Solvents and Water ($-\Delta H_s°$ (kcal mol^{-1}))

NH_4^{+c}	$MeNH_3^{+c}$	$Me_2NH_2^{+c}$	Me_3NH^{+c}	K^+
				92[d]
88.7	83.1	77.9	70.5	86.0
93.7	86.2	79.1	72.1	92[d]
98.2	90.2	82.3	73.7	94[d]

[d] Calculated from the solvation enthalpy of K^+ in water and enthalpies of transfer (*see* Ref. *36*).

Munson (*34*) recently showed that the proton affinity of HCF_3SO_3 is between that of H_2O and H_2S, as evidenced by the occurrence of gas-phase Reaction 7:

$$H_3O^+ (g) + HCF_3SO_3 (g) \rightarrow H_2O (g) + H_2CF_3SO_3^+ (g) \qquad (7)$$

while the similar reaction with H_3S^+ does not occur. Contrasting gas-phase Reaction 7 with Reaction 8 in sulfolane

$$H_2O (s) + HCF_3SO_3 (s) \rightarrow H_3O^+ (s) + CF_3SO_3^- (s) \qquad (8)$$

reveals dramatically the solvent effect on proton transfer reactions involving H_2O and H_3O^+ with an acid (HA).

Conclusion

This study has shown that the effect of water and diethyl ether ($X_{R_2O} < 0.1$) on the behavior of a series of acids (HA) in sulfolane can be accounted for by equilibrium Reactions 2–6. The species involved in these reactions are formed at different concentration levels according to the nature of R_2O (H_2O or Et_2O) and HA. First, if R_2O is H_2O, when the acid (HA) is very weak (HCH_3CO_2, HCF_3CO_2, HCH_3SO_3), the main species formed is the weak complex $H_2O \cdot\cdot HA$. When the acid is stronger (HCF_3SO_3), in addition to the formation of $H_2O \cdot\cdot HA$, formation of a series of proton solvates $H^+(H_2O)_n$ becomes important. H_3O^+ is the most stable solvate, while the stability of higher solvates decreases with n. Second, if R_2O is Et_2O, the picture is much simpler in that $Et_2O \cdot\cdot HCF_3SO_3$ and HEt_2O^+ are the only stable species formed. The reactions of H_2O and Et_2O with the proton are dissimilar largely because of the quite different hydrogen-bonding properties of both protonated species, H_3O^+ and HEt_2O^+. On the one hand, H_3O^+ is strongly stabilized with respect to HEt_2O^+ because of its superior ability to hydrogen bond to solvent molecules; on the other hand, H_3O^+, but not HEt_2O^+, can bond to several additional R_2O molecules through cooperative hydrogen bonding and give rise to stable solvates, $H^+(H_2O)_n$.

Our work gives insight into the many problems that would be met in trying to account for the influence of the concentration of water and acid both on reactions and physicochemical processes that take place in a solvent such as sulfolane. Our results also indicate some possible methods for solving such problems. For example, our present vapor–liquid equilibrium data on solutions of water and acid in sulfolane were correlated with solution composition along lines previously used for the system NH_3–Cu(II) salts in aqueous solution (*35*); in this latter system stepwise formation of $Cu^{2+}(NH_3)_n$ is akin to that of $H^+(H_2O)_n$ in the present system. Conceivably, therefore, distillation of both types of solutions could be treated in a similar manner.

Acknowledgments

The authors thank A. L. Beauchamp for writing the program and L. Bergeron for his valuable assistance in obtaining calorimetric data. Thanks are also due to the Editor of this book for his patience and understanding. Financial support of this work by the National Research Council of Canada and the Ministère de l'Education du Québec is gratefully acknowledged.

Literature Cited

1. Furter, W. F., Ed., "Thermodynamic Behavior of Electrolytes in Mixed Solvents," ADV. CHEM. SER. (1976) **155.**
2. Parker, A. J., *Quart. Rev.* (1962) **16,** 163.
3. Benoit, R. L., Lam, S. Y., *J. Am. Chem. Soc.* (1974) **96,** 7385.
4. Klages, F., Gordon, J. E., Jung, H. A., *Chem. Ber.* (1965) **98,** 3748.
5. Benoit, R. L., Rinfret, M., Domain, R., *Inorg. Chem.* (1972) **11,** 2603.
6. Benoit, R. L., Buisson, C., *Electrochim. Acta* (1973) **18,** 105.
7. Benoit, R. L., Buisson, C., Choux, G., *Can. J. Chem.* (1970) **48,** 2353.
8. Coetzee, J. F., Bertozzi, R. J., *Anal. Chem.* (1971) **43,** 961.
9. Ibid., (1969) **41,** 860.
10. Kolthoff, I. M., Chantooni, M. K., Jr., *J. Am. Chem. Soc.* (1973) **95,** 8539.
11. Haszeldine, R. N., Kidd, J. M., *J. Chem. Soc.* (1954) 4228.
12. Rossotti, F. J. C., Rossotti, H., "The Determination of Stability Constants," McGraw-Hill, New York, 1961.
13. Kolthoff, I. M., Chantooni, M. K., Jr., *J. Am. Chem. Soc.* (1965) **87,** 4428.
14. Louis, C., Ph.D. thesis, Université de Montréal, 1974.
15. Fernandez-Prini, R., Prue, J. E., *Trans. Faraday Soc.* (1966) **62,** 1257.
16. Della Monica, M., Lamanna, U., Senatore, L., *J. Phys. Chem.* (1968) **72,** 4329.
17. Giguère, P. A., Turrell, S., *Can. J. Chem.* (1976) **54,** 3477.
18. Gillespie, R. J., "Physicochemical Processes in Mixed Aqueous Solvents," F. Franks, Ed., Elsevier, New York, 1967.
19. Schriver, A., Silvi, B., Maillard, D., Perchard, J. P., *J. Phys. Chem.* (1977) **81,** 2095.
20. Domain, R., Ph.D. thesis, Université de Montréal, 1974.
21. Adams, A. A., Barger, H. J., Jr., *J. Electrochem. Soc.* (1974) **121,** 987.
22. Kebarle, P., "Modern Aspects of Electrochemistry," Vol. 9, J. O'M. Bockris, B. E. Conway, Eds., Plenum, New York, 1974.
23. Chantooni, M. K., Jr., Kolthoff, I. M., *J. Am. Chem. Soc.* (1970) **92,** 2236.
24. Kinugasa, M., Kishi, K., Ikeda, S., *J. Phys. Chem.* (1973) **77,** 1914.
25. Grimsrud, E. P., Kebarle, P., *J. Am. Chem. Soc.* (1973) **95,** 7939.
26. Wolf, J. F., Staley, R. H., Koppel, I., Taagepera, M., McIver, R. T., Jr., Beauchamp, J. L., Taft, R. W., *J. Am. Chem. Soc.* (1977) **99,** 5417.
27. Benoit, R. L., Louis, C., "The Chemistry of Nonaqueous Solvents," Vol. VA, J. J. Lagowski, Ed., Academic, New York, 1978.
28. Perdoncin, G., Scorrano, G., *J. Am. Chem. Soc.* (1977) **99,** 6983.
29. Mucci, A., Domain, R., Benoit, R. L., "Abstracts of Papers," 2nd CIC-ACS Joint Conference, May 29-June 2, 1977, ORGN n°44.

30. Mucci, A., Master's thesis, Université de Montréal, 1977.
31. Arnett, E. M., Jones, F. M., III, Taagepera, M., Henderson, W. G., Beauchamp, J. L., Holtz, D., Taft, R. W., *J. Am. Chem. Soc.* (1972) **94**, 4724.
32. Aue, D. H., Webb, H. M., Bowers, M. T., *J. Am. Chem. Soc.* (1976) **98**, 318.
33. Christe, K. O., *Inorg. Chem.* (1975) **14**, 2230.
34. Smith, D. E., Munson, B., *J. Am. Chem. Soc.* (1978) **100**, 497.
35. Benoit, R. L., *J. Chem. Eng. Data* (1961) **6**, 161.
36. Choux, G., Benoit, R. L., *J. Am. Chem. Soc.* (1969) **91**, 6221.

RECEIVED May 31, 1978.

11

Viscosity of Dilute Solutions of Alkali Halides in Methanol–Water Mixtures

ROBERT A. STAIRS

Department of Chemistry, Trent University, Peterborough, Ontario, Canada K9J 7B8

Viscosities of 0–0.1M solutions of lithium, potassium, and cesium chloride and potassium and rubidium iodide in water–methanol mixtures have been measured, and the results were fitted to the Jones–Dole Equation. The B *coefficients were interpreted as indicating slight to moderate preferential solvation of the ions by water vs. methanol. Values of the equilibrium constant representing the exchange at a single solvation site:* H_2O *(bound) + MeOH (free)* $\rightleftharpoons$ H_2O *(free) + MeOH (bound) were estimated as lying between 0.3 and 0.7 for all the salts, corresponding to differential solvation free energies,* $\Delta G°$ *(*H_2O*–MeOH) of 0.8–1.6 kJ/mol of bound solvent. Solvation numbers ranged from 10 to 16. The values of the* A *coefficients, although scattered, show trends that are in approximate agreement with a prediction based on Walden's rule, in the absence of the conductance data required for their calculation.*

In a previous symposium volume on this topic (*1*), we suggested that the viscosity *B* coefficient for the solution of a salt in a mixed solvent could indicate preferential solvation of the ions by one solvent component over the other. This suggestion was supported by some measurements that were interpreted qualitatively as showing preference of potassium iodide for water over methanol, for example. As a start toward more systematic study of the effect, this chapter reports the measurement of the viscosities of dilute solutions of several salts in a single solvent pair, water–methanol. We also report a small step toward a quantitative interpretation.

0-8412-0428-4/79/33-177-167$05.00/1

Let us define an equilibrium constant for the substitution of a molecule of solvent species 2 for species 1 at a single solvation site (assuming all sites are alike):

$$K = \frac{f_2 x_1}{f_1 x_2} \tag{1}$$

where f_1 and f_2 are the fractions of sites containing bound 1 and 2, respectively, and x_1 and x_2 are the mole fractions of the free components in the solvent. (For the present, 1 is water and 2 is methanol.)

Avoiding the water-rich region, where the special structure of water may cause problems, consider the B coefficient at $x_2 = 1/2$, $2/3$, and 1. From previous work we have

$$B_E = 2.5\,(V_o + nV_1 + \bar{q}(V_2 - V_1)) \tag{2}$$

where V_o, V_1, and V_2 are the molar volumes of the solute and solvent species (1 and 2), n is the number of available solvent sites per "molecule" of solute, and $\bar{q} = nf_2$, the mean number of moles of 2 bound per mole of solute. The subscript E in B_E indicates that this is the contribution to B calculated by Einstein (2), caused by the simple bulk of the solvated ions. If we assume that in this region of solvent composition the number of sites, n, remains constant and that the whole change in B is caused by the effect on B_E of the change in $\bar{q}$, we can write:

$$\beta = \frac{B_{1/2} - B_{2/3}}{B_{2/3} - B_1} = \frac{\bar{q}_{1/2} - \bar{q}_{2/3}}{\bar{q}_{2/3} - \bar{q}_1} \tag{3}$$

where the subscripts now indicate the value of x_2. From previous work,

$$\bar{q} = ny/(1 + y) \qquad (\text{where } y = f_2/f_1 = Kx_2/x_1) \tag{4}$$

Thus,

$$\bar{q}_{1/2} = nK/(1 + K),\ \bar{q}_{2/3} = 2nK/(1 + 2K),\ \bar{q}_1 = n$$

and

$$\beta = K/(1 + K) \qquad (\text{note: } 0 < \beta < 1)$$

or

$$K = \beta/(1 - \beta) \tag{5}$$

Now, with the same assumptions, from Equation 2:

$$B_1 - B_{1/2} = 2.5\ (V_2 - V_1)\ (\bar{q}_1 - \bar{q}_{1/2})$$
$$= 2.5\ (V_2 - V_1) n (1 + K)^{-1}$$

or

$$\mathrm{n} = \frac{(1 + K)\ (B_1 - B_{1/2})}{2.5\ (V_2 - V_1)} \tag{6}$$

Thus, measurement of B as a function of x_2 permits calculation of an apparent number of solvation sites from Equation 6 as well as the equilibrium constant (Equation 5) and, therefore, the free energy change, ΔG°, per mole of solvent 1 exchanged for 2 at an average site.

The viscosity A coefficient can be related to other measurable quantities through the expression (*3, 4*):

$$A = \frac{\beta^*}{320} \frac{\Lambda^\circ}{\lambda_1{}^\circ \lambda_2{}^\circ} \left[1 - 0.6863 \left(\frac{\lambda_1{}^\circ - \lambda_2{}^\circ}{\Lambda^\circ}\right)^2\right] \tag{7}$$

where

$$\beta^* = \frac{82.48}{\eta(\epsilon T)^{1/2}} \text{ and } \lambda_1{}^\circ, \lambda_2{}^\circ, \text{and } \Lambda^\circ\ (= \lambda_1{}^\circ + \lambda_2{}^\circ)$$

are the limiting equivalent conductance of the ions and the salt, respectively, and ϵ is the dielectric constant of the solvent. The lambdas and the viscosity can be related through Walden's rule (*5, 6*)

$$\lambda_i{}^\circ \eta = \text{constant} \tag{8}$$

but the constancy of this product depends on the constancy of the effective radius of the ion, through Stokes' law (*7, 8*).

This leads to the expectation that the Walden product might be inversely proportional to the effective radii. With one further assumption —i.e., that the transport numbers, $\lambda_i{}^\circ/\Lambda^\circ$, are insensitive to solvent composition—one may readily derive an expression for the dependence of the A coefficient on the mole fraction of solvent component 2:

$$A(x_2) = A(0) \left(\frac{\epsilon(0)}{\epsilon(x_2)}\right)^{1/2} \left(\frac{r(x_2)}{r(0)}\right) \tag{9}$$

To obtain the radius ratio, consider the expression for the hydrodynamic molar volume of the solvated salt that occurs in Equation 2 above, viz,

$$V = V_o + nV_1 + \bar{q}\,(V_2 - V_1) \qquad (10)$$

As reported below, K is not very different from unity; assuming that $K = 1$, we obtain $\bar{q} = nx_2$. The molar volume of the salt is likely to be small, and in fact the effective value may even be negative. We assume it to be negligible, and obtain:

$$\frac{r(x_2)}{r(0)} = \left(\frac{V(x_2)}{V(0)}\right)^{1/3} = \left[1 + x_2\left(\frac{V_2}{V_1} - 1\right)\right]^{1/3} \qquad (11)$$

Substituting appropriate values ($V_i = M_i/\rho_i^{25}$, $i = 1$ or 2) for V_1 and V_2 for water and methanol (the values for the pure bulk solvents were used) and using tabulated values (*9*) for the dielectric constant, we obtain numbers that can be approximated ($\pm 5\%$) by the linear expression

$$A(x) = A(0)\ (1 + 1.82x_2) \qquad (12)$$

Literature values (*10, 11*) tend to be somewhat higher, which is not surprising in view of the approximations used.

Experimental

The kinematic viscosity and density were measured by methods previously described (*1*); the only change was an improvement in the temperature stability of the air thermostat used for the viscometer. Short term control within 0.02°C and reproduction of setting to 0.05°C were achieved by substituting a thermistor device for the previously used bimetallic spiral thermoregulator. The improvement was not enough to enable satisfactory A coefficients to be obtained in all cases although the scatter has been reduced.

As before, the Jones–Dole Equation (*12*) was fitted to the data by a least squares procedure, in the form

$$\eta = \eta_o + aC^{1/2} + bC \qquad (13)$$

where η_o, $a(=A\eta_o)$ and $b(=B\eta_o)$ are all adjusted.

Results

The viscosity data are shown in Table I. Table II is a summary of results, showing the solvent properties (density and viscosity) and the solution parameters: the density–concentration coefficients and the constants A and B of the Jones–Dole Equation. The values of B for the four salts, for which new data are reported, are shown as functions of x_2 in Figure 1. (Some data from the literature are included as noted.)

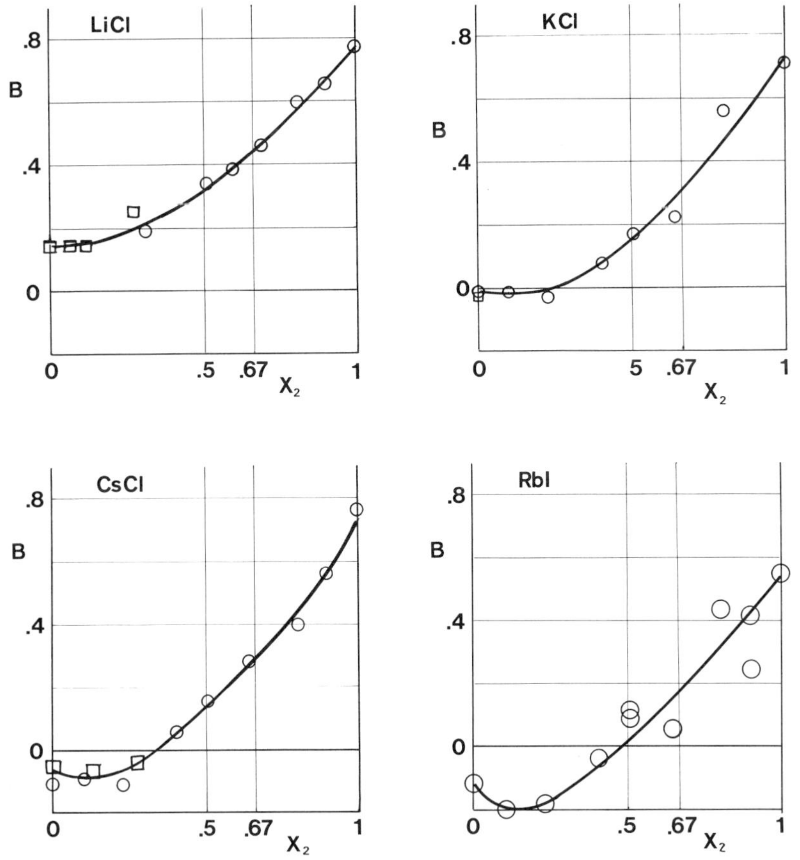

Figure 1. Viscosity B *coefficients for salt solutions in water–methanol mixtures, as functions of the mole fraction of methanol in the solvent,* x_2. (○, *this work;* □, *Refs.* 10, 11, 14)

The accuracy of the viscosity is estimated to be ±0.5%, but somewhat better precision was obtained within a single group of measurements on a particular solvent mixture and four to six solutions in that solvent. The precision of the measurements of relative viscosity (η/η_0) is accordingly claimed to ze ±0.03%. The uncertainty in the *A* values listed is around ±0.01 $L^{1/2}\ mol^{-1/2}$ in most cases, although a few are more uncertain. The negative values recorded are probably without physical significance. The precision of the *B* values is also variable, but when *A* looks reasonable, *B* is believed to be within ±0.015 L mol^{-1}. Assignment of "reasonable" *A* values would have led to less scatter in *B*, but this has not been done with the results reported here.

Table I. Viscosities of Solutions of Salts in

LiCl

x_2: *0.315* c	η	*0.502* c	η	*0.600* c	η	*0.686* c	η
0	1.5578	0	1.3046	0	1.1416	0	0.9822
4.9	1.5619	6.1	1.3076	5.0	1.1443	4.7	0.9853
14.5	1.5664	17.6	1.3136	9.0	1.1474	19.2	0.9937
29.9	1.5715	34.1	1.3220	19.2	1.1529	44.2	1.0067
50.8	1.5803	51.7	1.3298	26.3	1.1567	77.4	1.2049
		75.5	1.3408	34.2	1.1600	117.7	1.0430
				45.1	1.1655		
				69.9	1.1772		
				99.8	1.1917		

KCl

	η *at* x_2 =					
c	*0.0*	*0.100*	*0.229*	*0.401*	*0.510*	*0.641*
0	0.8711	1.2940	1.5586	1.4600	1.2953	1.0579
2.5	0.8703	1.2933	1.5604	1.4632	1.2970	1.0603
10.0	0.8706	1.2921	1.5612	1.4654	1.3016	1.0673
22.5	0.8700	1.2924	1.5642	1.4698	1.3055	1.0728
40.0	0.8701	1.2913	1.5651	1.4749	1.3137	1.0843
62.5	0.8699	1.2914	1.5650	1.4792	1.3173	1.0918
90.0	0.8692	1.2891	1.5660	1.4845	1.3281	1.1013

CsCl

	η *at* x_2 =				
c	*0.0*	*0.100*	*0.229*	*0.410*	*0.510*
0	0.8692	1.2831	1.5634	1.4668	1.3069
2.5	0.8702	1.2801	1.5626	1.4704	1.3091
10.0	0.8700	1.2798	1.5618	1.4695	1.3134
22.5	0.8697	1.2775	1.5611	1.4725	1.3149
40.0	0.8681	1.2734	1.5563	1.4761	1.3195
62.5	0.8676	1.2713	1.5546	1.4780	1.3283
90.0	0.8656	1.2677	1.5491	1.4818	1.3332

Water-Methanol at 25°C (η in cP, in mmol L^{-1})

x_2: *0.815*		*0.901*		*1.000*	
c	η	c	η	c	η
0	0.7813	0	0.6573	0	0.5198
5.9	0.7843	4.9	0.6613	11.1	0.5265
24.6	0.7947	22.4	0.6713	26.8	0.5341
50.8	0.8083	47.6	0.6844	50.4	0.5451
85.8	0.8247	81.5	0.7006	82.5	0.5590
144.8	0.8450	123.0	0.7205	128.2	0.5791

	η *at* x_2 =	
c	*0.801*	*1.000*
0	0.8173	0.5138
2.5	0.8189	0.5153
10.0	0.8242	0.5197
22.5	0.8315	0.5249
40.0	0.8411	0.5324
62.5	0.8521	

	η *at* x_2 =			
c	*0.641*	*0.801*	*0.897*	*1.000*
0	1.0832	0.8272	0.6710	0.5196
2.5	1.0856	0.8300	0.6719	0.5211
10.0	1.0906	0.8330	0.6752	0.5239
22.5	1.0934	0.8398	0.6812	0.5290
40.0	1.0982	0.8437	0.6907	0.5363
62.5	1.1092	0.8549	0.6978	
90.0			0.7094	

Table II. Summary of Viscosity and Density Parameters for Solutions of Salts in Water–Methanol Mixtures

Salt	x_2	ρ_o $(g\ mol^{-1})$	$\frac{d\rho}{dc}$	$\eta_o\ (cP)$	A $(L^{1/2}\ mol^{-1/2})$	B $(L\ mol^{-1})$
LiCl	0.32	0.9230	0.030	1.5579	0.021	0.185
	0.50	0.8808	0.033	1.3044	0.008	0.341
	0.60	0.8605	0.035	1.1414	0.018	0.384
	0.69	0.8424	0.037	0.9819	0.024	0.464
	0.81	0.8181	0.038	0.7810	0.018	0.595
	0.90	0.8035	0.040	0.6572	0.046	0.652
	1.00	0.7860	0.043	0.5198	0.042	0.772
KCl	0	0.9970	0.048	0.8709	0.0052	−0.0140*
	0.100	0.9693	0.072	1.2938	−0.006	−0.016
	0.299	0.9403	0.088	1.5584	0.028	−0.039
	0.401	0.9028	0.073	1.4600	0.034	−0.073
	0.510	0.8802	0.065	1.2951	0.029	0.181
	0.641	0.8537	0.054	1.0570	0.075	0.225
	0.801	0.8363	0.042	0.8170	0.033	0.561
	1.000	0.7846	0.035	0.5137	0.040	0.712
CsCl	0	0.9970	0.127	0.8963	0.017	−0.105
	0.100	0.9693	0.130	0.2815	−0.010	−0.088
	0.229	0.9403	0.133	1.5632	0.004	−0.110
	0.401	0.9028	0.137	1.4674	0.017	0.053
	0.510	0.8802	0.140	1.3072	0.021	0.154
	0.641	0.8537	0.143	1.0837	0.020	0.278
	0.801	0.8363	0.148	0.8275	0.030	0.394
	0.897	0.8033	0.151	0.6705	0.025	0.565
	1.000	0.7846	0.153	0.5197	0.006	0.764
RbI	0	0.9970	0.161	0.8696	0.008	−0.118
	0.10	0.9754	0.156	1.2764	−0.003	−0.200
	0.23	0.9403	0.150	1.5568	−0.003	−0.187
	0.40	0.9028	0.145	1.4699	0.002	−0.034
	0.51	0.8802	0.143	1.3075	0.021	0.066
	0.51	0.8802	0.143	1.2985	0.045	0.042
	0.65	0.8517	0.158	1.0644	0.016	0.053
	0.80	0.8363	0.164	0.8238	0.013	0.431
	0.90	0.8033	0.182	0.6626	0.060	0.242
	0.90	0.8033	0.182	0.6649	0.018	0.415
	1.00	0.7846	0.190	0.5196	0.029	0.550

Discussion

To consider the *A* coefficients briefly, discounting the negative and other aberrant values, the remainder show a trend that is in reasonable accord with the prediction outlined above. (So do those reported in a previous work (*1*).) A more detailed consideration requires better viscosity measurements and measurements in the mixed solvents of the limiting conductances and the transport numbers of the electrolytes.

The *B* coefficients are best considered in two regions of the solvent composition separately. The water-rich region is interesting because it has been suggested (*13*) that small amounts of hydroxylic solvents added to water enhance the special structure of water up to a point. The viscosity of water–methanol mixtures has a strong maximum near $x_2 = 0.27$. Structure-breaking salts—i.e., those with larger ions—have negative *B* coefficients in pure water. These same salts show pronounced minima in their *B* vs. x_2 dependence near $x_2 = 0.1$–0.15. The minimum is displaced to lower values of x_2 than the viscosity maximum. It is by no means clear that these two extremes of behavior have a common cause, but if they do, the displacement of B_{min} to lower x_2 may result from the strong upward trend in *B* discussed below.

LiCl, on the other hand, which has a positive *B* in water, shows no minimum; in fact, *B* changes very little with composition in the water-rich region. Thus, structure-breaking ions find more structure to break in water that contains a little methanol than in pure water. Structure-making ions, however, find little scope for their talents.

Once the water-rich region is passed, the *B* values all rise strongly. The interpretation of this rise in terms of the change in hydrodynamic volume of the ions seems reasonable. The extraction of a preferential solvation constant from it is perhaps a bit shaky in each individual case, but taken together they tend to confirm the original expectation. The values of *K* and corresponding values of $\Delta G°$, derived from the *B* coefficients, are recorded in Table III. They are sufficiently uncertain that they all must be considered equal within experimental error, i.e., $K = 0.5 \pm 0.2$, $\Delta G° = 1.6 \pm 0.8$ J mol^{-1}. That is, all these salts appear to show a slight preference for water over methanol.

The fact that $\Delta G°$ is so small is probably related to the rather large values of *n* found, large compared with those found by other methods, as noted earlier (*1*). Most of the energy change is probably associated

Table III. Salts in H_2O–MeOH at 25°C[a]

	K[b]	$\Delta G°$ *(kJ)*	n
LiCl	0.52	1.6	12.2
KCl	0.54	1.5	10.7
CsCl	0.54	1.5	17.4
KI	0.32	2.8	16.0
RbI	0.79	0.6	16.7

[a] Preferential solvation from viscosity *B* coefficient.
[b] Equilibrium constant for the reaction: H_2O (bound) + MeOH (free) ⇌ MeOH (bound) + H_2O (free).

with exchange of the innermost six or so solvent molecules, the remainder being hydrogen-bonded to the first few in a manner not very different from their condition in the nominally "free" part of the solvent. The actual significance of the n values recorded is not clear, although the approximate magnitudes are not unreasonable and are similar to those reported previously. Our expectation that larger values would be associated with the smaller ions was not fulfilled.

Acknowledgments

The author thanks Murray Fitzgerald for assistance with the measurements.

Literature Cited

1. Stairs, R. A., Adv. Chem. Ser. (1976) **155**, 332–342.
2. Einstein, A., *Ann. Phys. Leipzig* (1911) **34**, 591.
3. Falkenhagen, H., Vernon, E. L., *Phys. Z.* (1932) **33**, 140.
4. Harned, H. S., Owen, B. B., "The Physical Chemistry of Electrolyte Solutions," 3rd ed., p. 240, Reinhold, New York, 1958.
5. Walden, P., *Z. physik. Chem.* (1906) **55**, 249.
6. Harned, H. S., Owen, B. B., "The Physical Chemistry of Electrolyte Solutions," 3rd ed., pp. 283–285, Reinhold, New York, 1968.
7. Stokes, G. G., *Trans. Camb. Phil. Soc.* VIII (1845) 287.
8. Robinson, R. A., Stokes, R. H., "Electrolyte Solutions," 2nd ed., pp. 123–126, Butterworth, London, 1959.
9. Parsons, R., "Handbook of Electrochemical Constants," p. 11, Butterworth, London, 1959.
10. Jones, G., Fornwalt, H. J., *J. Am. Chem. Soc.* (1935) **57**, 2041–2045.
11. Ibid., (1936) **58**, 619.
12. Jones, G., Dole, M., *J. Am. Chem. Soc.* (1929) **51**, 2950–2964.
13. Werblan, L., Rotowska, A., Minc, S., *Electrochim. Acta* (1971) **16**, 41–49.
14. Kaminsky, M., *Discuss. Faraday Soc.* (1957) **24**, 171–179, 231.

Received May 5, 1978. Work supported by funds allotted from an institutional grant to Trent University by the National Research Council of Canada.

12

Ternary System Phase Diagram Determinations Concerning Potassium Electrolyte Influence on Aqueous Solutions of Dioxane or Tetrahydrofuran

JOHN A. McNANEY
Department of Chemistry, Fitchburg State College, Fitchburg, MA 01420

HOWARD K. ZIMMERMAN and PAUL H. GROSS
Department of Chemistry, University of the Pacific, Stockton, CA 95211

Phase diagrams, at 25°C, were determined for potassium acetate–water–dioxane, potassium acetate–water–tetrahydrofuran, and potassium chloride–water–tetrahydrofuran. Potassium acetate exceeded potassium chloride in its capacity to stratify aqueous solutions of either dioxane or tetrahydrofuran. Kobsev's (1) investigations revealed that the greater the solubility of an alkali metal salt, the greater its salting-out effect. The relative order of the water solubilities of the salts studied are potassium acetate >> potassium chloride. More potassium acetate is required to cause stratification in aqueous dioxane than is necessary to obtain the same results in aqueous tetrahydrofuran. It is proposed that, in comparison to dioxane, tetrahydrofuran forms a weaker association with water and, hence, the cations can more easily break these bonds causing liquid-phase separation.

The phenomena which gave rise to this study were first brought to the attention of H. K. Zimmerman, Director of Carbohydrate Research at the University of the Pacific, by research workers in peptide chemistry where the potassium hydroxide–water–dioxane system was used in the saponification of esters. He directed that this method be used by his

0-8412-0428-4/79/33-177-177$05.00/1

staff to saponify suger acetates and benzoates. They also found that aminosugar derivatives, even hydrochlorides of aminosugar glycosides, could be extracted from concentrated aqueous potassium chloride solutions.

In using this saponification procedure a salting-out of dioxane often was observed. Since such a salting-out effect can be quite troublesome, it became necessary to initiate a study of the three-compound system in order to learn how such a complication might be avoided.

It was also highly desirable to learn how to better control the salting-out effect in water–tetrahydrofuran systems so that extractions in this pair of mixed solvents could be performed. This present study of the phase relationships in some of the solvent systems used in aminosugar research includes the determination of the phase diagrams of the systems potassium acetate–water–dioxane, potassium acetate–water–tetrahydrofuran, and potassium chloride–water–tetrahydrofuran, and an attempt to provide a theoretical explanation for the experimental results.

Experimental

Chemicals. Potassium acetate from J. T. Baker Chemical Company (Baker Analyzed Reagent, 99.0% assay) was used in CO_2-free distilled water and was standardized with standard perchloric acid in glacial acetic acid, using methyl violet indicator (*2*). Potassium chloride of Baker Analyzed Reagent, 99.9% assay, was used as received. 1,4-Dioxane of technical grade from J. T. Baker Company was purified by distillation over metallic sodium as described by Vogel (*3*). Tetrahydrofuran (THF) of Baker Analyzed Reagent, density 0.881 g cm^{-3}, boiling range 65.8°–66.4°C, was used as received. Karl Fischer Reagent, obtained from Allied Chemical Company, was standardized to 1 mL of reagent to react with at least 5 mg of water. Fisher certified standard water in methanol (1 mL = 1 mg H_2O), used for standardizing Karl Fischer Reagent, was obtained from Fisher Scientific Company.

Apparatus. A circulating constant-temperature water bath equipped with a micro-set thermoregulator was used to supply the water for the jacket of a Pyrex water-jacketed reaction flask that was equipped with a magnetic stirrer. Burets used—automatic and conventional—were of Class-A grade and equipped with Teflon stopcocks.

General Procedure. The selected electrolyte solution was placed in the jacketed reaction flask which was held at constant temperature by the connected circulating water bath. The tip of a buret containing the cyclic ether was introduced through a stopper into the mouth of the reaction flask. With the electrolyte solution undergoing constant stirring the cyclic ether was delivered slowly from the buret into the reaction flask to the point of turbidity or precipitation, and the delivered volume of cyclic ether was noted. The densities of the cyclic ethers (dioxane and THF) and of the electrolyte solutions used had been determined previously; the normalities of the electrolyte solutions were known. With these data, the volume measurement of each of the components of the mixture was converted to weight data. Tie-lines were established by selecting mixtures whose component compositions would lie in a multiphase region of the phase diagram, and then analyzing each phase of the mixture for its components after equilibrium had been achieved

Specific Procedure. Table I provides a characteristic example of the data for the experimental observations in the system potassium acetate–water–dioxane, which data were used in plotting solubility curves at 25°C and 85°C on a three-component, equilateral triangular graph (Figure 1).

Table I. Solubility Curve and Tie-Line Data for the System Potassium Acetate–Water–Dioxane in Weight Percentage

Solubility Curve Data

25°C			85°C		
$KC_2H_3O_2$	H_2O	$C_4H_8O_2$	$KC_2H_3O_2$	H_2O	$C_4H_8O_2$
0.33	14.15	85.52	0.36	15.74	83.90
0.53	16.15	83.32	0.59	17.99	81.42
5.14	34.96	59.90	5.69	38.53	55.78
8.68	43.32	48.00	9.43	47.16	43.41
15.50	51.70	32.80	15.95	54.50	29.55
24.35	53.10	22.55	24.59	53.64	21.77
29.93	55.44	14.63			
72.00	27.40				
$71.00^{20°C}$ (*18*)					
$74.00^{30°C}$ (*18*)					
$72.93^{25°C}$ (*19*)					

Tie-Line Data 25°C

Denser Layer			Less Dense Layer		
$KC_2H_3O_2$	H_2O	$C_4H_8O_2$	$KC_2H_3O_2$	H_2O	$C_4H_8O_2$
22.70	55.15	22.15	1.33	9.35	89.32
27.00	56.00	17.00	0.06	7.47	92.47
36.15	53.20	10.65	0.03	4.48	95.49
45.00	49.00	6.00	0.04	6.07	93.89
50.95	43.25	5.80	0.00	1.96	98.04
63.90	33.10	3.00	0.00	1.03	98.97
67.80	29.40	2.80	0.01	0.85	99.14

Original Mixture

$KC_2H_3O_2$	H_2O	$C_4H_8O_2$
9.94	29.90	60.20
14.80	34.95	50.20
25.00	39.90	35.10
39.80	45.00	15.20
39.90	35.60	24.50
34.10	18.75	47.15
38.50	17.60	43.90

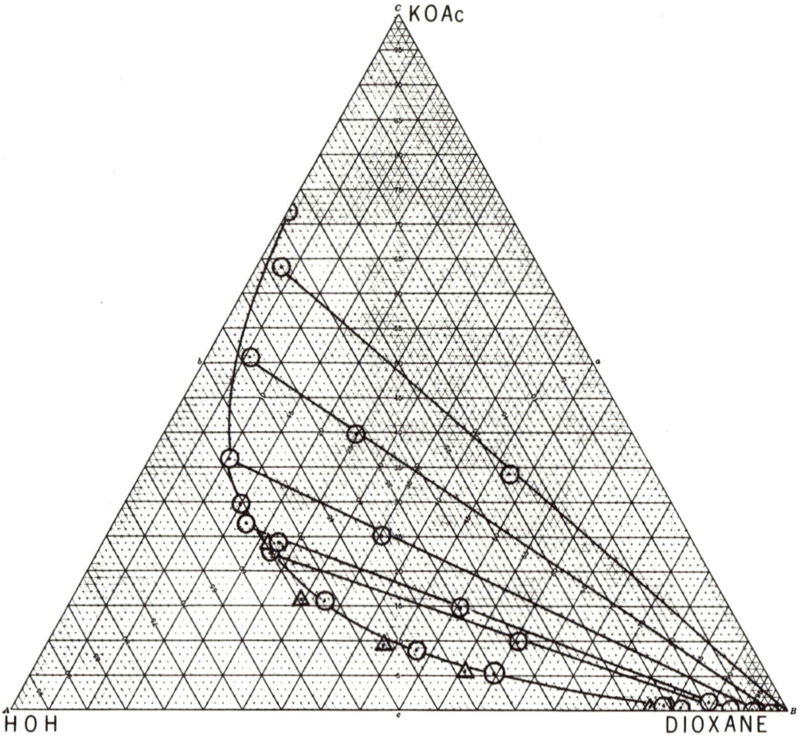

Figure 1. Isotherms of potassium acetate–water–dioxane in weight percent: ⊙, 25°C; △, 85°C.

Tie-lines were established by using a three-component mixture whose total composition was a point within the two-phase region of the graph. The resulting two-phase liquid mixture was separated with a separatory funnel, and two weighed samples were taken from each layer. One sample of each layer was analyzed for the potassium compound and the other sample was analyzed for water. In the systems potassium acetate–water–dioxane and potassium acetate–water–THF, a standard perchloric acid in glacial acetic acid was used in the potassium acetate determination (*2*). Water content was obtained by titration with standard Karl Fischer Reagent (*4*). Having the weight percentage of two of the three components, the cyclic ether percentage was calculated by difference in each case. The high volatility of THF precluded its use above 25°C.

For most of the potassium chloride–water–THF work, data were obtained in the same manner as the other two systems except that potassium chloride was analyzed by the Mohr method (*5*). At low concentrations of THF, however, a solid precipitate was observed. To ascertain its composition, a mixture of the three components with total mixture composition corresponding to a point within the solid–liquid, two-phase region of the preliminary graph was brought to equilibrium and the solid then was filtered off at experimental temperature. One sample from the liquid phase was analyzed for potassium chloride by the Mohr method (5), and another sample of the same liquid phase was analyzed for water with standard Karl Fischer Reagent. The amount of cyclic ether present then was calculated by difference. This procedure was repeated at various points within the solid–liquid, two-phase region. The phase behavior described here has as a consequence an invariant point in the solubility curve. In order to confirm the existence of this invariant point, synthetic mixtures of the three components were chosen arbitrarily to correspond to two points within the solid–liquid–liquid, three-phase area. After equilibrium was reached with these mixtures, the solid was allowed to settle out, and two weighed samples of the denser liquid were analyzed, the one for potassium by use of standard silver nitrate titration, and the other for water by Karl Fischer Reagent titration. This procedure then was repeated for the less dense liquid layer. Having determined the weight percentage of two of the three components, the THF concentration was calculated by difference for each layer.

The composition of the solid precipitate which was found in some of the mixtures of the potassium chloride–water–THF system was determined by using Schreinemakers wet residue method (*6*) as modified by Hill and Ricci (*7*).

Treatment of Data

From the recorded data it was possible to determine the solubility curves for each system, the liquid–liquid tie-lines, the solid–solid tie-lines, and to confirm the invariant point found in one of the solubility curves.

Solubility Curves. The solubility weight percentages are tabulated in Tables I, II, and III. Each experimental point on any of the solubility curves was obtained from the predetermined weights of the potassium compound and water plus the weight of the cyclic ether titrated into this solution.

All such points formed smooth curves when plotted on the three-component triangular graphs used for each system investigated except that of the potassium chloride–water–THF system. In the latter case, two solubility curves appeared which met in an invariant point. The solubility plots are shown in Figures 1, 2, and 3.

Liquid–Liquid Tie-Lines. The weight percentages of the mixtures, arbitrarily chosen as points within the two-phase, liquid–liquid regions and found in all of the systems studied, are tabulated in Tables I, II, and III. Tabulated in the same tables are the weight percentages of the three components found in each of the two liquid layers which resulted from each of the chosen mixtures.

Plots of the three points—original mixture, denser layer, and less dense layer—were made on the graph which already contained their solubility curve. Each set of these three points determined a straight line

Table II. Solubility Curve and Tie-Line Data for the System Potassium Acetate–Water–Tetrahydrofuran in Weight Percentage

Solubility Curve Data 25°C

$KC_2H_3O_2$	H_2O	C_4H_8O
0.38	26.87	72.75
0.87	37.00	62.13
1.85	52.80	45.35
2.90	62.60	34.50
4.80	68.70	26.40
10.10	72.10	17.80
15.30	73.10	11.60
19.95	71.25	8.80
34.40	61.50	4.10
44.80	63.30	1.90

Tie-Line Data 25°C

Denser Layer			*Less Dense Layer*		
$KC_2H_3O_2$	H_2O	C_4H_8O	$KC_2H_3O_2$	H_2O	C_4H_8O
24.80	68.00	7.20	0.13	6.00	93.87
16.10	72.10	11.80	0.05	8.87	91.08
10.00	72.97	17.03	0.10	12.70	87.20
42.70	54.10	3.20	0.02	2.66	97.32
61.20	36.00	2.80	0.10	1.00	98.90

Original Mixture

$KC_2H_3O_2$	H_2O	C_4H_8O
19.84	59.35	20.81
9.80	49.60	40.60
7.85	61.70	30.45
20.30	29.40	50.30
14.85	9.85	75.30

whose end points, corresponding to the denser and less dense layers, respectively, fell on the solubility curve. These tie-line plots are shown in Figures 1, 2, and 3.

Solid–Liquid Tie-Lines. Tabulated in Table III are the weight percentages of the originally prepared complexes which were chosen as points within the two-phase, solid–liquid region in the potassium chloride–water–THF system. The weight percentages of the saturated solutions for each of the original complexes are presented also in this table.

The data mentioned above were plotted on three-component triangular graphs. A slightly curved solubility line resulted from the plot of the saturated solution. Straight lines were drawn connecting each prepared complex point with its corresponding saturated solution point.

Table III. Solubility Curve and Tie-Line Data for the System Potassium Chloride–Water–Tetrahydrofuran (25°C) in Weight Percentage

Solubility Curve Data

Liquid–Liquid Equilibrium			Solid–Liquid Equilibrium		
KCl	H_2O	C_4H_8O	KCl	H_2O	C_4H_8O
0.63	29.85	69.52	21.45	71.50	7.05
1.39	42.00	56.61	23.00	70.70	6.30
3.65	62.35	33.90	24.15	72.10	3.75
7.28	72.80	19.95	25.60	73.00	1.40
11.05	73.80	15.15	26.80	72.90	0.00
14.85	74.35	10.80	26.34 (*20*)		
18.17	72.83	8.98			

Tie-Line Data *Liquid–Liquid Equilibrium*

Denser Layer			Less Dense Layer		
KCl	H_2O	C_4H_8O	KCl	H_2O	C_4H_8O
7.90	68.40	23.70	0.01	13.80	86.19
13.20	73.60	13.20	0.00	9.95	90.05
16.20	73.30	10.50	0.00	8.72	91.28

Original Mixture

KCl	H_2O	C_4H_8O
5.00	50.00	45.00
10.05	59.95	30.00
9.97	50.05	39.98

Solid–Liquid Equilibrium

Saturated Solution			Original Complex		
KCl	H_2O	C_4H_8O	KCl	H_2O	C_4H_8O
23.20	71.10	5.70	34.99	60.04	4.97
24.10	71.50	4.40	31.10	65.00	3.90
24.95	71.70	3.35	37.00	59.98	3.02

Solid–Liquid–Liquid Equilibrium

Denser Layer			Less Dense Layer		
KCl	H_2O	C_4H_8O	KCl	H_2O	C_4H_8O
22.60	71.00	6.40	0.00	6.00	94.00
22.70	70.93	6.37	0.00	6.10	93.90

Original Complex

KCl	H_2O	C_4H_8O
35.05	49.93	15.02
25.00	50.03	24.97

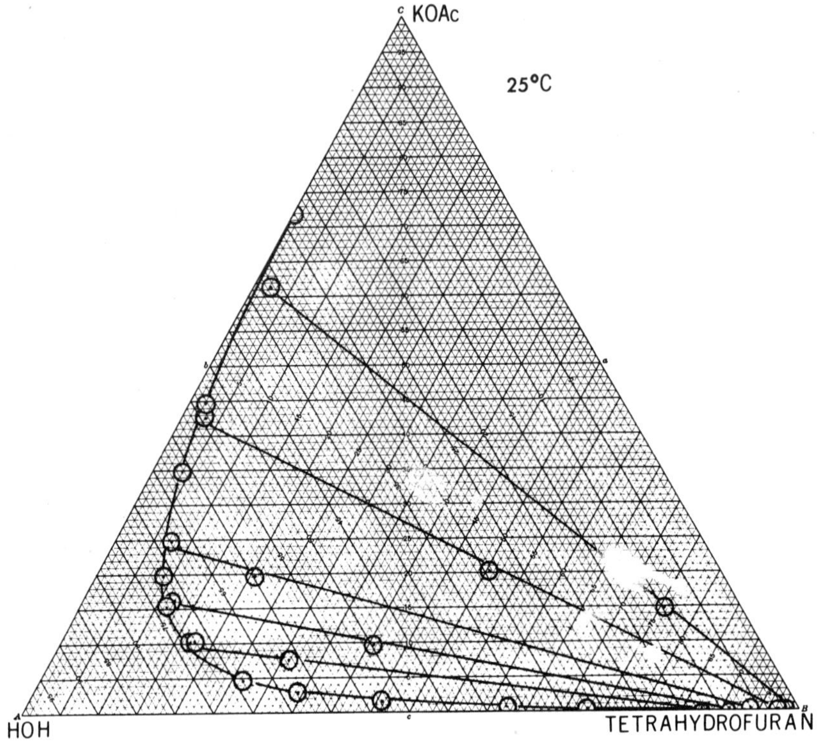

Figure 2. Isotherm of potassium acetate–water–THF in weight percent

When extrapolated, all of these straight lines met, approximately, in a point, the composition of the dissolved potassium chloride in the saturated solution (*6*). These plots are shown in Figure 3.

The modification of Schreinemakers wet residue method which was used in this part of the investigation is that instead of analyzing the wet residue, a complex is prepared of known composition and the solution only is analyzed. This again gives two points on the diagram: the solution point on the curve and the complex point which replaces the wet residue point. Hill and Ricci (*7*) claim that the complex method is as accurate or more accurate than the residue method if algebraic extrapolation of the tie-lines is used.

Assume the synthetic complex to be: Compound A—20%, Compound B (Water)—30%, dioxane—50%. The solution on analysis gives: Compound A—4%, Compound B—16%, dioxane—80%. The 80 parts dioxane contain 4A + 16B and 50 parts dioxane contain 2.5A + 10B. This amount of solution is subtracted from the complex giving:

$$\left.\begin{array}{l} 20 - 2.5 \;\; = 17.5\text{A} \\ 30 - 10.0 = 20.0\text{B} \end{array}\right\} \quad 37.5 \text{ total}$$

Thus, if all the dioxane could be removed from the complex as solution, the residue would be: 46.7% A and 53.3% B. This point, when marked on the edge of the diagram, gives the correct extrapolation of the complex point. It does not necessarily represent any existing solid phases, but the line joining this point to the solution point must pass through the true solid phase (*8*). In this investigation, however, all extrapolations ended with the solid phase as pure potassium chloride at one corner, i.e., A = 100%, B = 0%.

Invariant Point Confirmation. The weight percentages of the prepared complexes, arbitrarily chosen within the three-phase, solid–liquid–liquid region, are tabulated in Table III. Tabulated in the same table are the weight percentages of the denser liquid layers and those of the less dense layers.

Plots of each set of data—prepared complex, denser layer, and less dense layer—were made on the graph which already contained the solubility curve and tie-lines. A straight line was drawn from the prepared complex point to its denser layer point and, similarly, from the complex

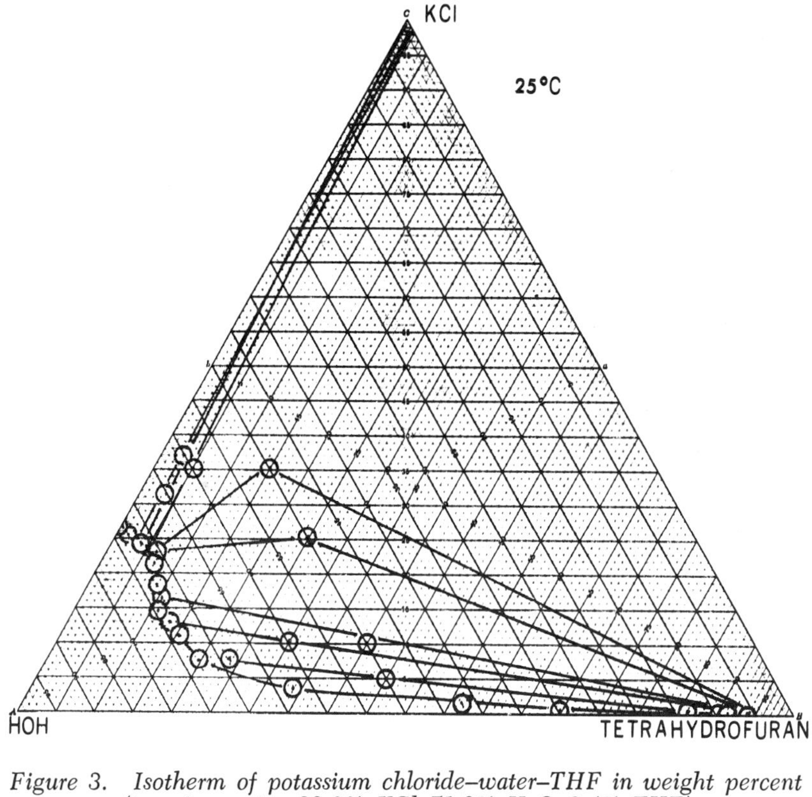

Figure 3. Isotherm of potassium chloride–water–THF in weight percent (invariant point: 22.6% KCl, 71.0% H_2O, 6.4% THF)

point to the less dense layer point. This formed an angle with the complex point as the vertex. The ends of the two sides, denser layer point and less dense layer point, fell on the solubility curve with the denser point approximately on the invariant position. A plot of the other set of data gave similar results. This denser layer point also coincided with the invariant point, confirming it, while the less dense layer point fell almost exactly on the previously plotted less dense layer point. These plots may be found in Figure 3.

Discussion of Results

Dioxane in an aqueous solution is associated with water (*9, 10, 11, 12, 13*). Evidence that THF behaves in a like manner has been reported also (*14*).

According to Godneva and Klocho (*13*), when an electrolyte is added to water–dioxane or water–THF, two phases result. In order to bring about this separation, the water–cyclic ether association must be broken down. These authors believe that the hydration of the electrolyte cations is accomplished by the dehydration of the cyclic ether molecules, causing stratification.

Salting-Out Effects Owing to Potassium Acetate. The system potassium acetate–water–dioxane (Figure 1, Table I) was investigated at temperatures of 25°C and 85°C. The salting-out isotherms are binodal curves and showed a very slight displacement toward the aqueous corner with an increase in temperature.

The liquid layer formation (at 25°C) of the system potassium acetate–water–THF is also a binodal curve (Figure 2). There is, however, a substantial difference between the layering effect on the water–dioxane by potassium acetate compared with its effect on the water–THF mixture. In the latter case, much less potassium acetate is required to bring about separation into two liquid phases than with the same percentage composition of water–dioxane. Even when plotted on a mole percentage composition basis (*see* Figure 4) this difference, while less pronounced, still exists. One could conclude that the reason for this behavior rests in the fact that THF has one-half as many oxygen sites for hydrogen bonding than does dioxane, and thus can hold less water in association. Previous experimenters, however, have produced results which seem to belie this. It has been postulated by different groups of investigators that the molecular ratio for water:THF is 17:1 (*15, 16*), as compared with 2:1, 3:1, and 4:1 for the molecular ratio of water:dioxane association, as put forth by others (*10, 11*). This study has determined that about three times as much potassium acetate, in mole percentage, is required to cause stratification in dioxane than is necessary to obtain

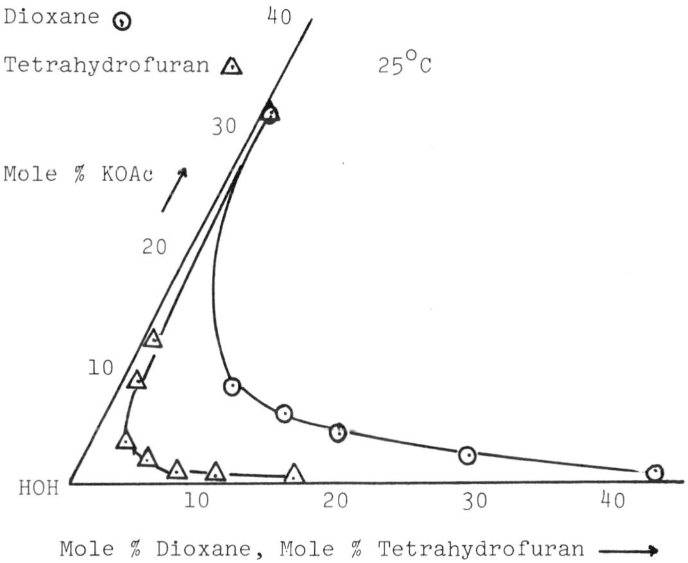

Figure 4. Comparison of potassium acetate–water–dioxane and potassium acetate–water–THF plots: ⊙, dioxane; △, THF

the same results in THF when both solvents are in aqueous solutions. If THF associates with more water than does dioxane, it would seem that the opposite should be true.

The investigators involved in this study feel that it is reasonable to conclude that compared with dioxane, THF forms a weaker association with water and, hence, the cations can more easily break these bonds causing liquid-phase separation. The basis for this judgment rests with the assumption that 17 water molecules would be held less firmly by one THF molecule than four water molecules by one dioxane molecule.

Salting-Out Effect Owing to Potassium Chloride. Kobzev (*1*) maintained, in a study of salts of potassium, sodium, lithium, rubidium, and cesium (all in aqueous solutions), that the solubility of a salt in water was related to its salting-out effect. He found that salts with the solubility from 6.4 to 2.8 g equiv wt per 100 mL of water at 25°C caused stratification in all systems except water–methanol and water–ethanol. Salts with solubilities below 2.82 g equiv wt did not cause salting-out.

The relative water solubilities of the two salts investigated are potassium acetate >> potassium chloride. As one might expect, then, potassium acetate has the greater capacity to salt-out aqueous solutions of dioxane and THF. Potassium chloride cannot bring about salting-out through the full percentage composition range of the ternary mixture.

Potassium chloride and potassium acetate were insoluble in both dioxane and THF. The solubility of these salts in water was ascertained and compared very closely with the literature values cited in Tables I and III.

E. Chulik has assisted in the foregoing work by developing analytical means for quantitative determination of acetate salts as modifications of the method of D. Ceausescu (*17*).

Literature Cited

1. Kobsev, V. V., *Sb. Nauchn. Rab. Aspir. L'vov. Politekh. Inst.* (1961) 21.
2. Fritz, J. S., "Acid Base Titrations in Nonaqueous Solvents," G. Frederick Smith Chemical Co., Columbus, Ohio, 1952.
3. Vogel, A. I., "A Text Book of Practical Organic Chemistry," 3rd ed., p. 177, Longmans, Green and Co., New York, 1957.
4. Mitchell, J., Smith, D. M., "Aquametry, Chemical Analysis," Vol. V, Interscience, New York, 1948.
5. Blaedel, W. J., Meloche, V. W., "Elementary Quantitative Analysis," p. 278, Row, Peterson and Co., White Plains, 1957.
6. Schreinemakers, F. A. H., *Z. Phys. Chem. Leipzig* (1893) **11**, 81.
7. Hill, A. E., Ricci, J. E., *J. Am. Chem. Soc.* (1931) **53**, 4305.
8. Purdon, F. F., Slater, V. W., "Aqueous Solutions and the Phase Diagram," p. 65, Edward Arnold and Co., London, 1946.
9. Zimmerman, H. K., *Bull. Soc. Chim.*, Beograd (1959) **24**, 1.
10. Schott, H., *J. Chem. Eng. Data* (1961) **6**, 19.
11. Tsypin, M. Z., Trifonov, N. A., *Tr. Kazan. Khim.-Tecknol. Inst. im S. M. Kirova* (1958) **22**, 120.
12. Grunwald, E., *Proc. Int. Symp., Electrolytes, Trieste, 1959* (1962) 62.
13. Godneva, M. M., Klocho, M. A., *Izv. Karel'. Kol'sk. Fil. Akad. Nauk SSSR* (1958) **5**, 122.
14. Montgomery, D., et al., *Proc. Okla. Acad. Sci.* (1949) **30**, 140.
15. Erva, J., *Suom. Kemistil.* (1956) **29B**, 183.
16. Pinder, K. L., *Can. J. Chem. Eng.* (1965) **43**(5), 271.
17. Ceausescu, D., *Rev. Chim., Bucharest* (1960) **11**, 174.
18. "International Critical Tables," Vol. 4, p. 260, McGraw Hill, New York, 1926.
19. Abe, R., *J. Tokyo Chem. Soc.* (1911) **32**, 980.
20. Shearman, R. W., *J. Am. Chem. Soc.* (1937) **59**, 185.

RECEIVED Feburary 6, 1978.

13

Solubility of Carbon Dioxide in Aqueous Mixed-Salt Solutions

AKIRA YASUNISHI[1], MASAKATSU TSUJI, and EIZO SADA

Department of Chemical Engineering, Kyoto University, Kyoto 606, Japan

The solubilities of carbon dioxide in aqueous solutions of mixed salts chosen from eight electrolytes (NaCl, KCl, Na_2SO_4, NH_4Cl, $MgSO_4$, $(NH_4)_2SO_4$, $CaCl_2$, KNO_3) were measured at 25°C and 1 atm by the saturation method. Experimental results for the mixed-salt system were not described easily by the modified Setschenow Equation. However, they were correlated very well by an empirical two-parameter equation. The parameters in the equation for the binary and ternary salt solutions could be estimated easily from these equations for the components of the mixed salts.

The solubility data of gases in aqueous mixed-salt solutions are fundamentally important in research on liquid–gas mass transfer and in designing gas–liquid contacting operations, especially gas absorption accompanied by chemical reaction.

The data have, however, been obtained for only a few systems by Onda et al. (*1*), Joosten and Danckwerts (*2*), and Hikita et al. (*3*). These authors also proposed methods for estimating the solubility of gases in aqueous mixed-salt solutions from the corresponding data for each salt in the systems. These studies are extensions of the empirical method proposed by van Krevelen and Hoftijzer (*4*) on the basis of the following modified Setschenow Equation.

$$\log(L_0/L) = K \cdot C_s \qquad (1)$$

[1] To whom correspondence should be addressed. Current address: Department of Environmental Chemistry and Technology, Tottori University, Tottori 680, Japan.

0-8412-0428-4/79/33-177-189$05.00/1

The range of applicability of the Setschenow Equation on the salt concentration in aqueous single-salt solutions varies with the system (gas plus an electrolyte) and is never confirmed clearly. Van Krevelen and Hoftijzer (*4*) showed the range to be up to 2 mol/L of ionic strength in all the systems, while Onda et al. (*5*) showed that the equation could be applied to the more concentrated solutions for some systems, such as up to 15 mol/L of ionic strength for carbon dioxide systems at the maximum.

However, from the measurements of solubilities of oxygen (*6, 7*) and carbon dioxide (*8*) in aqueous single-salt solutions over a wide range of salt concentration, it was found that data for many systems could not be correlated by the empirical Setschenow Equation. Therefore, the modified Setschenow Equation can not always be used to estimate satisfactorily the gas solubility data over a wide range of salt concentrations.

In the present investigation, the solubilities of carbon dioxide in aqueous mixed-salt solutions were measured at 25°C and 1 atm partial pressure of carbon dioxide, and the possibility of a method for estimating the solubility was discussed.

The following systems were studied: NaCl–KCl, NaCl–Na_2SO_4, NaCl–NH_4Cl, NaCl–$MgSO_4$, NH_4Cl–$(NH_4)_2SO_4$, KCl–$CaCl_2$, and NaCl–KNO_3 for aqueous solutions of binary mixed salts, and NaCl–Na_2SO_4–NH_4Cl, NaCl–$MgSO_4$–KNO_3, and NaCl–KCl–$CaCl_2$ for aqueous solutions of ternary mixed salts. These systems were selected to study the salt effect of the common ions on solubility.

Experimental

The apparatus and procedure were similar to those described by Tokunaga et al. (*9, 10*). The principle of this method is that a known volume of gas is brought into contact with a measured amount of gas-free liquid. The system was agitated enough to reach equilibrium rapidly, and the volume of remaining gas was measured. Correction for the vapor pressure was done, and the difference between initial and final measured volume of gas gave the solubility.

The solubility was expressed in terms of the Ostwald absorption coefficient, L, which is defined as the ratio of the volume of absorbed gas to the volume of absorbing liquid at the temperature and pressure of the measurement. This is because the Ostwald coefficient is really an equilibrium constant, and the method of correction to standard conditions does not need to be specified in comparison with the Bunsen coefficient.

Salt solutions were prepared as follows. A concentrated aqueous solution of mixed salts was prepared by weighing the proper amount of each component salt for the desired mole fraction in mixed salts and then dissolving the salt in deionized water. It was then diluted to desired concentrations and degassed by boiling with reflux and under vacuum to eliminate dissolved gases.

The total concentration of salts in a solution was determined by analyzing the concentration of chloride ion by the usual silver nitrate titration. Chlorides were present in all systems studied, and the mole fraction of the chloride was determined at the time the solution was prepared (*see above*).

Carbon dioxide was extra pure grade (99.96%), and its purity was confirmed by gas chromatography (GC). All salts were reagent grade and were used without further purification.

The vapor pressure of the mixed salt solution was estimated by the method of Robinson and Bower (*11*) from the corresponding data for each component salt solution.

The reliability and accuracy of the experimental apparatus and procedure were determined by the measurement of the solubility of carbon dioxide in pure water. The average value of the Ostwald coefficient, L_0, obtained in this work was 0.8264 and agreed quite well with values in the literature (*12*).

Results and Discussion

The solubility data of carbon dioxide in aqueous solutions of binary mixed salts obtained in this study are summarized in Table I; those for ternary mixed salts are summarized in Tables II, III, and IV. Figures 1 and 2 show the solubility data for the potassium chloride–calcium chloride and sodium chloride–sodium sulfate–ammonium chloride mixed solutions, respectively, which are representative of all the data. The salting-out effect was shown in all the systems studied.

Table I. Experimental Results of Carbon Dioxide Solubility in Aqueous Binary Mixed-Salt Solutions at 25°C and 1 atm

$x_1 = 0.00$[a]		$x_1 = 0.25$		$x_1 = 0.50$		$x_1 = 0.75$		$x_1 = 1.00$[a]	
C_s	L	C_s	L	C_s	L	C_s	L	C_s	L
KCl (1) + $CaCl_2$ (2)									
0.558	0.6477	0.426	0.7037	0.512	0.7012	0.429	0.7455	0.498	0.7576
1.237	0.4701	0.865	0.6011	1.079	0.6003	0.904	0.6735	0.980	0.7041
2.008	0.3380	1.611	0.4605	1.675	0.5062	1.845	0.5558	1.890	0.6177
2.809	0.2544	2.361	0.3704	2.605	0.3947	2.714	0.4691	2.702	0.5554
3.801	0.1876	3.233	0.2845	3.401	0.3253	3.501	0.4057	3.505	0.5063
$NaCl$ (1) + Na_2SO_4 (2)									
0.210	0.7165	0.233	0.7198	0.272	0.7283	0.266	0.7507	0.455	0.7343
0.657	0.5273	0.518	0.6098	0.595	0.6281	0.534	0.6844	1.000	0.6508
1.190	0.3627	0.854	0.5103	1.035	0.5201	1.005	0.5892	1.442	0.5842
1.710	0.2601	1.274	0.4038	1.502	0.4215	1.545	0.4907	1.945	0.5187
2.205	0.1828	1.744	0.3197	2.060	0.3360	2.073	0.4017	2.486	0.4653

Table I. Continued

$x_1 = 0.00$[a]		$x_1 = 0.25$		$x_1 = 0.50$		$x_1 = 0.75$		$x_1 = 1.00$[a]	
C_s	L	C_s	L	C_s	L	C_s	L	C_s	L
NaCl (1) + KCl (2)									
0.498	0.7576	0.537	0.7455	0.479	0.7502	0.545	0.7271	0.455	0.7343
0.980	0.7041	1.039	0.6813	1.051	0.6720	1.077	0.6511	1.000	0.6508
1.890	0.6177	1.765	0.6073	1.851	0.5742	1.830	0.5545	1.945	0.5187
2.702	0.5554	2.114	0.5705	2.297	0.5344	2.277	0.5070	2.486	0.4653
3.505	0.5063	3.104	0.4948	3.343	0.4439	3.434	0.4028	3.400	0.3727
4.131	0.4703	4.125	0.4243	4.396	0.3765	4.311	0.3415	4.216	0.3144
NaCl (1) + $MgSO_4$ (2)									
0.418	0.6480	0.264	0.7194	0.387	0.7001	0.263	0.7540	0.455	0.7343
0.724	0.5355	0.515	0.6329	0.723	0.6110	0.537	0.6917	1.000	0.6508
1.020	0.4513	1.028	0.4837	1.111	0.5147	1.079	0.5799	1.442	0.5842
1.643	0.3047	1.562	0.3766	1.616	0.4215	1.626	0.4811	1.945	0.5187
2.274	0.2049	2.056	0.2872	2.185	0.3315	2.179	0.4104	2.486	0.4653
NH_4Cl (1) + $(NH_4)_2SO_4$ (2)									
0.253	0.7378	0.282	0.7505	0.333	0.7568	0.254	0.7905	0.375	0.7958
0.519	0.6572	0.573	0.6785	0.658	0.6997	0.601	0.7451	0.811	0.7704
0.845	0.5794	0.998	0.5961	1.021	0.6460	1.018	0.6978	1.147	0.7460
1.235	0.4978	1.396	0.5322	1.330	0.5959	1.414	0.6594	1.508	0.7328
1.836	0.3952	1.946	0.4580	2.014	0.5223	1.985	0.6133	1.973	0.7102
2.486	0.3161	2.433	0.4015	2.673	0.4578	2.400	0.5800	2.338	0.6915
2.873	0.2802	2.806	0.3687			2.861	0.5524	2.816	0.6721
NaCl (1) + NH_4Cl (2)									
0.501	0.7880	0.426	0.7871	0.446	0.7655	0.945	0.6831	0.455	0.7343
0.980	0.7576	0.892	0.7405	0.931	0.7147	1.783	0.5834	1.000	0.6507
1.508	0.7328	1.709	0.6802	1.750	0.6309	2.648	0.4954	1.442	0.5824
2.338	0.6915	2.550	0.6252	2.233	0.5896	3.572	0.4243	1.945	0.5187
2.816	0.6721	3.398	0.5781	3.453	0.5024	4.284	0.3737	2.486	0.4653
3.682	0.6452	4.077	0.5467	4.157	0.4614	4.620	0.3552	3.400	0.3727
4.421	0.6258	4.459	0.5275	4.731	0.4296	5.281	0.3165	4.216	0.3144
5.480	0.6015	5.051	0.5059	5.176	0.4097			5.096	0.2584
NaCl (1) + KNO_3 (2)									
0.288	0.7994	0.298	0.7956	0.289	0.7857	0.264	0.7784	0.455	0.7343
0.801	0.7592	0.614	0.7663	0.711	0.7352	0.620	0.7315	1.000	0.6508
1.101	0.7399	1.007	0.7244	1.127	0.6854	0.913	0.6901	1.442	0.5842
1.412	0.7164	1.290	0.6983	1.601	0.6330	1.324	0.6367	1.945	0.5187
1.630	0.7023	1.664	0.6649	2.173	0.5820	1.723	0.5883	2.486	0.4653
						1.948	0.5666	3.400	0.3727

[a] From Ref. *8*.

Table II. Experimental Results of Carbon Dioxide Solubility in Aqueous Solution of Sodium Chloride (1)–Sodium Sulfate (2)–Ammonium Chloride (3) Mixed Salt at 25°C and 1 atm

$x_1 = 0.50$, $x_2 = 0.50$, $x_3 = 0.00$		$x_1 = 0.50$, $x_2 = 0.375$, $x_3 = 0.125$		$x_1 = 0.50$, $x_2 = 0.25$, $x_3 = 0.25$		$x_1 = 0.50$, $x_2 = 0.125$, $x_3 = 0.375$		$x_1 = 0.50$, $x_2 = 0.00$, $x_3 = 0.50$	
C_s	L	C_s	L	C_s	L	C_s	L	C_s	L
0.272	0.7283	0.404	0.7067	0.332	0.7447	0.407	0.7558	0.446	0.7655
0.595	0.6281	0.812	0.6068	0.720	0.6591	0.800	0.6900	0.931	0.7147
1.035	0.5201	1.211	0.5271	1.113	0.5899	1.207	0.6348	1.750	0.6309
1.502	0.4215	1.627	0.4565	1.576	0.5165	1.572	0.5867	2.233	0.5896
2.060	0.3360	1.978	0.4011	1.981	0.4613	2.008	0.5384	3.453	0.5042
				2.484	0.4017	2.549	0.4870		

Table III. Experimental Results of Carbon Dioxide Solubility in Aqueous Solutions of Potassium Chloride (1)–Sodium Chloride (2)–Calcium Chloride (3) Mixed Salt at 25°C and 1 atm

$x_1 = 0.50$, $x_2 = 0.50$, $x_3 = 0.00$		$x_1 = 0.50$, $x_2 = 0.25$, $x_3 = 0.25$		$x_1 = 0.50$, $x_2 = 0.00$, $x_3 = 0.50$	
C_s	L	C_s	L	C_s	L
0.479	0.7502	0.386	0.7505	0.512	0.7012
1.051	0.6720	0.722	0.6873	1.079	0.6003
1.851	0.5742	1.428	0.5901	1.675	0.5062
2.297	0.5344	1.938	0.5291	2.605	0.3947
3.343	0.4439	2.455	0.4704	3.401	0.3253
4.396	0.3765	2.762	0.4405		

Table IV. Experimental Results of Carbon Dioxide Solubility in Aqueous Solutions of Sodium Chloride (1)–Magnesium Sulfate (2)–Potassium Nitrate (3) Mixed Salt at 25°C and 1 atm

$x_1 = 0.50$, $x_2 = 0.50$, $x_3 = 0.00$		$x_1 = 0.50$, $x_2 = 0.25$, $x_3 = 0.25$		$x_1 = 0.50$, $x_2 = 0.00$, $x_3 = 0.50$	
C_s	L	C_s	L	C_s	L
0.387	0.7001	0.193	0.7859	0.289	0.7857
0.723	0.6110	0.498	0.7206	0.711	0.7352
1.111	0.5147	0.814	0.6577	1.127	0.6854
1.616	0.4215	1.131	0.6067	1.601	0.6330
2.185	0.3315	1.493	0.5506	2.173	0.5820
		1.910	0.4949		
		2.264	0.4497		

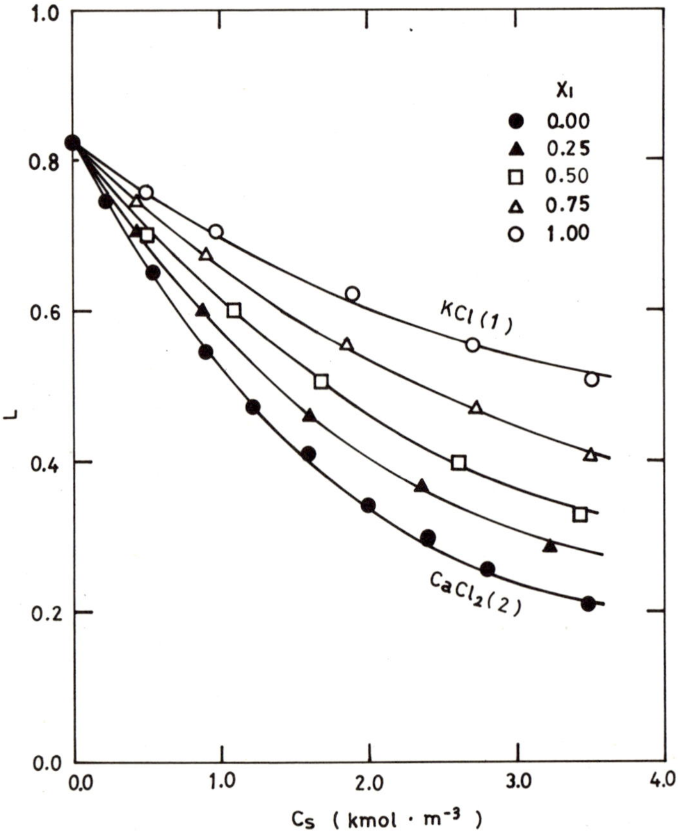

Figure 1. Solubility of carbon dioxide in aqueous potassium chloride (1)–calcium chloride (2) mixed solutions at 25°C and 1 atm

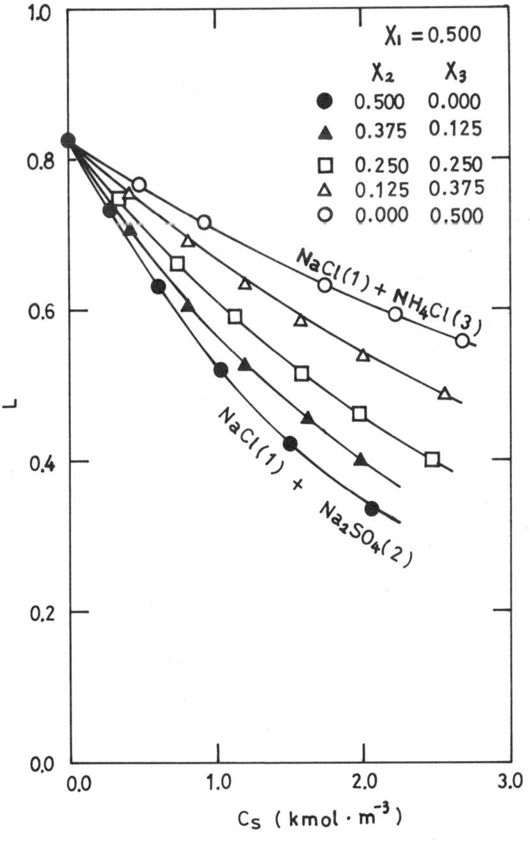

Figure 2. Solubility of carbon dioxide in aqueous sodium chloride (1)–sodium sulfate (2)–ammonium chloride (3) mixed solutions at 25°C and 1 atm

Examples of the variation of carbon dioxide solubilities with total salt concentrations in aqueous solutions are shown in Figures 3 and 4 in the form of the modified Setschenow plot for each salt composition.

Figure 3 shows the plot for potassium chloride–calcium chloride binary salt system. Figure 4 shows the plot for sodium chloride–sodium sulfate–ammonium chloride ternary salt system. As shown in these figures, the plots of $\log(L_0/L)$ vs. salt concentration all curve upward convexly, and the effects of these mixed salts on the solubility of carbon dioxide in the aqueous solutions do not show a direct correlation by the Setschenow Equation. These features are the same in all the mixed-salt systems considered here.

The estimating methods by van Krevelen and Hoftijzer (*4*) and Onda et al. (*1*) are based on the linear relationship between $\log(L_0/L)$ and salt concentration. When these methods are used to estimate the solubility of gases in aqueous salt solutions over a wide range of salt concentration, however, the estimates are sometimes in serious error, as shown in Figures 3 and 4.

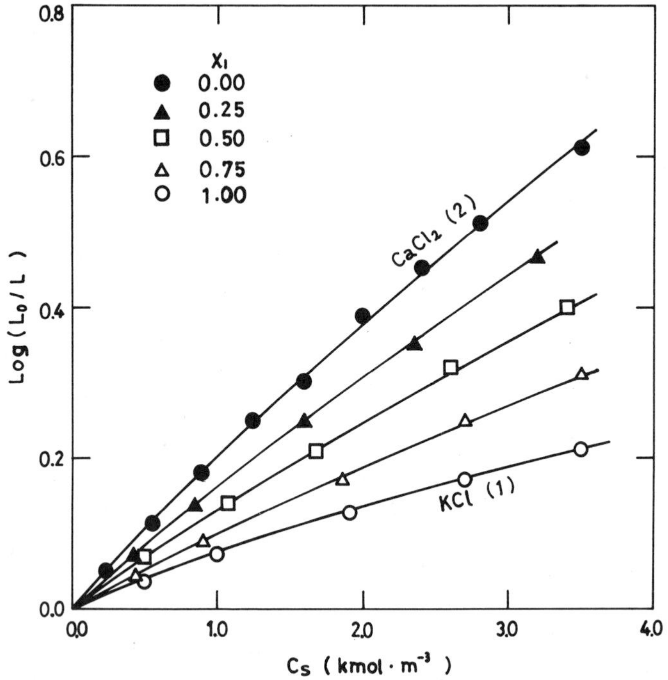

Figure 3. Effect of salt concentration on solubility of carbon dioxide in aqueous potassium chloride (1)–calcium chloride (2) mixed solutions at 25°C and 1 atm

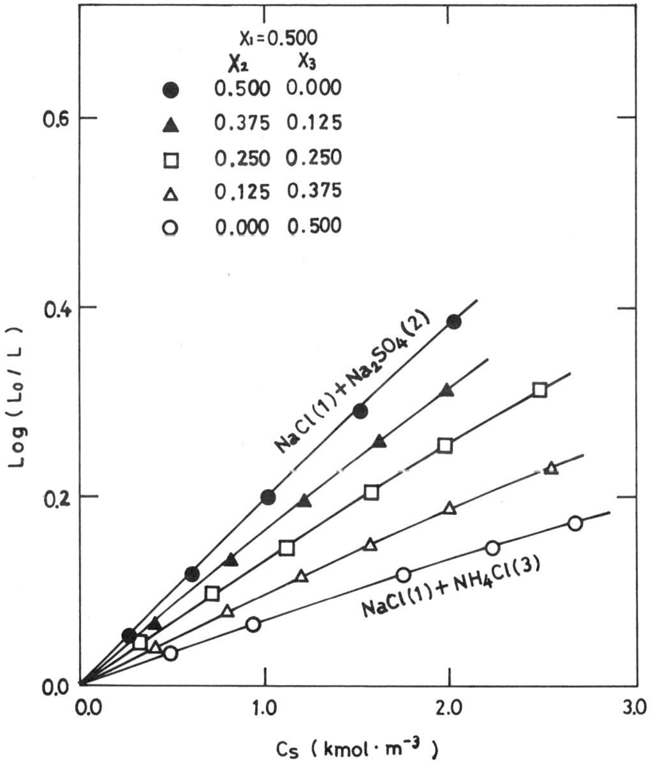

Figure 4. Effect of salt concentration on solubility of carbon dioxide in aqueous sodium chloride (1)–sodium sulfate (2)–ammonium chloride (3) mixed solutions at 25°C and 1 atm

For the solubility data of gases in aqueous single-salt solutions that could not be correlated by the empirical Setschenow Equation, Markham and Kobe (*13*) proposed an empirical two-parameter equation to correlate the data with the molality of the salt in solution.

One of the authors (*6*) proposed the following empirical two-parameter equation to correlate with those data,

$$\log(L_0/L) = A \cdot C_s/(1 + B \cdot C_s) \tag{2}$$

where A and B are arbitrary constants that can be determined from the observed data for each system. The molarity is used here as the salt concentration variable because it was used successfully as the variable in discussing the salt effect based on internal pressure concept (*12*) and perturbation theories (*14, 15*). The solubility data of oxygen (*6, 7*) and carbon dioxide (*8*) in aqueous single-salt solutions correlated quite well with the proposed Equation 2 over a wide range of salt concentration.

Table V. Values of *A* and *B* in Equation 2 for Binary Mixed-Salt Systems

	$x_1 = 0.00$	$x_1 = 0.25$	$x_1 = 0.50$	$x_1 = 0.75$	$x_1 = 1.00$
KCl (1) + $CaCl_2$ (2)					
A	0.2110	0.1691	0.1365	0.1014	0.0754
B	0.0572	0.0566	0.0428	0.0430	0.0671
$NaCl$ (1) + Na_2SO_4 (2)					
A	0.3000	0.2599	0.2031	0.1450	0.1050
B	0.0081	0.0560	0.0335	—0.0168	0.0124
$NaCl$ (1) + KCl (2)					
A	0.0754	0.0824	0.0899	0.0993	0.1050
B	0.0671	0.0435	0.0355	0.0268	0.0124
$NaCl$ (1) + $MgSO_4$ (2)					
A	0.2500	0.2236	0.1846	0.1487	0.1050
B	−0.0361	0.0030	0.0083	0.0278	0.0124
NH_4Cl (1) + $(NH_4)_2SO_4$ (2)					
A	0.1980	0.1528	0.1150	0.0793	0.0435
B	0.0752	0.0791	0.0751	0.1030	0.1330
$NaCl$ (1) + NH_4Cl (2)					
A	0.0435	0.0545	0.0715	0.0884	0.1050
B	0.1330	0.0576	0.0430	0.0236	0.0124
$NaCl$ (1) + KNO_3 (2)					
A	0.0478	0.0545	0.0746	0.0883	0.1050
B	0.0655	−0.0268	0.0274	0.0232	0.0124

In this study, Equation 2 was also used to correlate the solubility data of carbon dioxide in aqueous mixed-salt solutions. The values of *A* and *B*, obtained by the least squares method for all the systems, are summarized in Tables V, VI, VII, and VIII.

The solid lines in Figures 3 and 4 are the predictions of Equation 2. All the data could be correlated with Equation 2 within about 0.5% deviation.

In this study, the following equations were developed from Equation 2 to estimate the solubility of carbon dioxide in aqueous mixed-salt solutions from the solubility data in an aqueous solution of each salt component

$$\log(L_0/L) = A_M \cdot C_s/(1 + B_M \cdot C_s) \tag{3}$$

$$A_M = \sum_{i=1}^{n} A_i x_i \tag{4}$$

$$B_M = \sum_{i=1}^{n} B_i x_i \qquad (5)$$

where A_i and B_i represent the parameters A and B in Equation 2 for the ith component contained in a mixed salt, and x_i is the mole fraction of the ith component.

The variations of A and B with x_1 are shown in Figures 5, 6, and 7 for two binary mixed-salt systems and in Figure 7 for a ternary mixed-salt system. In these figures, the observed values of A and B deviate from the relationships represented by Equations 4 and 5.

To check the applicability of this estimation method, each observed solubility, L_{obsd}, was compared with the corresponding calculated value, L_{calcd}, by Equations 3, 4, and 5 using the parameters A_i and B_i for the salt components in the observed system. The deviation of L_{calcd} from L_{obsd} was calculated by Equation 6.

$$\% \text{ deviation} = \frac{|L_{calcd} - L_{obsd}|}{L_{obsd}} \times 100 \qquad (6)$$

The average deviation of each system is given in Table IX. It can be seen that the deviations are very small. This result shows that the deviation of the values of A and B from the linearly interpolated values, shown in Figures 5, 6, and 7, have little effect on the estimated values of

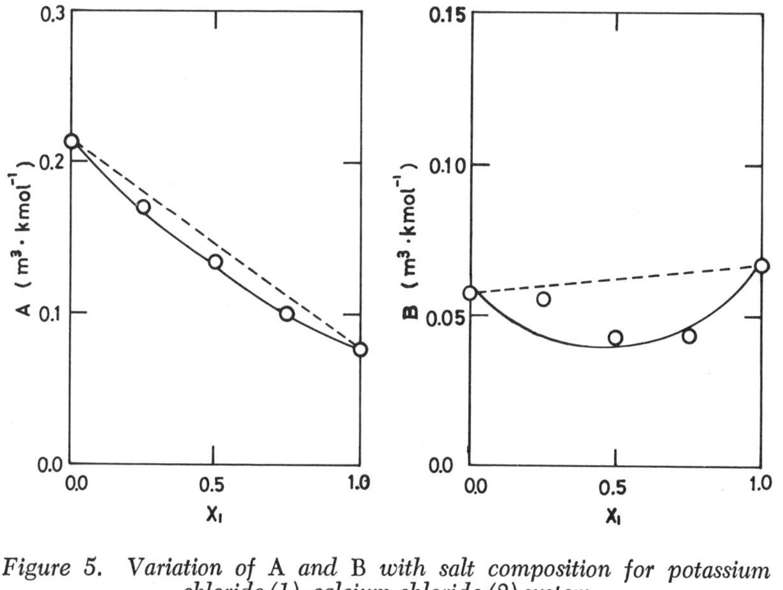

Figure 5. Variation of A *and* B *with salt composition for potassium chloride (1)–calcium chloride (2) system*

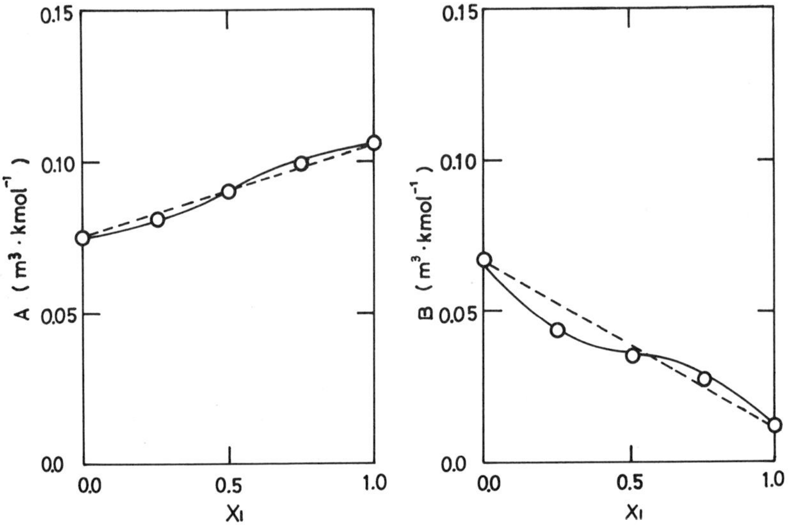

Figure 6. Variation of A and B with salt composition for sodium chloride (1)–potassium chloride (2) system

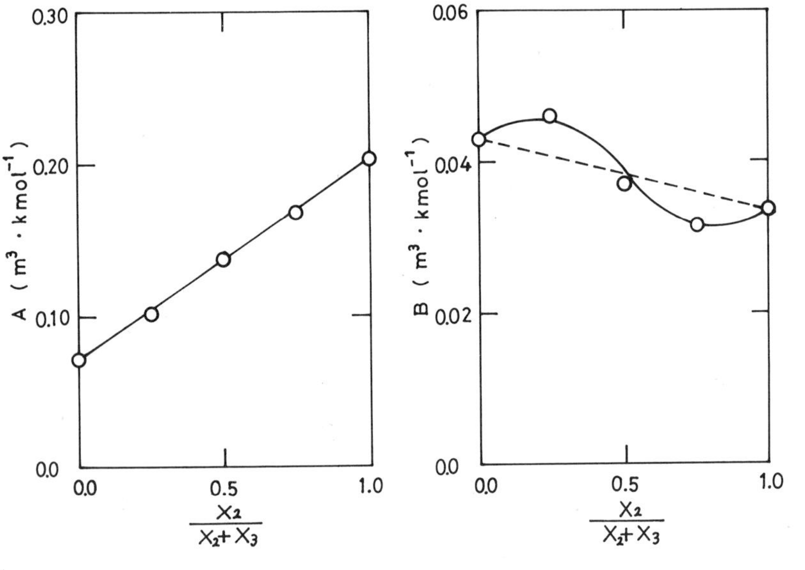

Figure 7. Variation of A and B with salt composition for sodium chloride (1)–sodium sulfate (2)–ammonium chloride (3) system

Table VI. Values of *A* and *B* in Equation 2 for Sodium Chloride (1)–Sodium Sulfate (2)–Ammonium Chloride (3) Mixed-Salt System

	$x_1 = 0.50$ $x_2 = 0.50$ $x_3 = 0.00$	$x_1 = 0.50$ $x_2 = 0.375$ $x_3 = 0.125$	$x_1 = 0.50$ $x_2 = 0.25$ $x_3 = 0.25$	$x_1 = 0.50$ $x_2 = 0.125$ $x_3 = 0.375$	$x_1 = 0.50$ $x_2 = 0.00$ $x_3 = 0.50$
A	0.2031	0.1679	0.1375	0.1009	0.0715
B	0.0335	0.0315	0.0372	0.0459	0.0430

Table VII. Values of *A* and *B* in Equation 2 for Potassium Chloride (1)–Sodium Chloride (2)–Calcium Chloride (3) Mixed-Salt System

	$x_1 = 0.50$ $x_2 = 0.50$ $x_3 = 0.00$	$x_1 = 0.50$ $x_2 = 0.25$ $x_3 = 0.25$	$x_1 = 0.50$ $x_2 = 0.00$ $x_3 = 0.50$
A	0.0899	0.1084	0.1365
B	0.0355	0.0361	0.0428

Table VIII. Values of *A* and *B* in Equation 2 for Sodium Chloride (1)–Magnesium Sulfate (2)–Potassium Nitrate (3) Mixed-Salt System

	$x_1 = 0.50$ $x_2 = 0.50$ $x_3 = 0.00$	$x_1 = 0.50$ $x_2 = 0.25$ $x_3 = 0.25$	$x_1 = 0.50$ $x_2 = 0.00$ $x_3 = 0.50$
A	0.1846	0.1215	0.0746
B	0.0083	0.0191	0.0274

Table IX. Average Deviations of Estimated Values L_{calc} from Observed Results

System	*Number of Determinations*	*Average Deviation (%)*
$KCl + CaCl_2$	15	1.35
$NaCl + Na_2SO_4$	15	1.21
$NaCl + KCl$	18	0.67
$NaCl + MgSO_4$	15	0.79
$NH_4Cl + (NH_4)_2SO_4$	20	0.62
$NaCl + NH_4Cl$	27	2.20
$NaCl + KNO_3$	16	0.45
$NaCl + Na_2SO_4 + NH_4Cl$	17	0.75
$KCl + NaCl + CaCl_2$	6	1.68
$NaCl + MgSO_4 + KNO_3$	7	1.77

solubility and that the presence of the common ion in the mixed salts has no particular effect on the solubility except for the effect of each salt component. These results suggest that the proposed method of estimation can be very useful in predicting the solubility of carbon dioxide in aqueous mixed-salt solutions from the practical point of view.

Conclusions

The solubilities of carbon dioxide in aqueous solutions of seven binary and three ternary mixed salts chosen from eight kinds of electrolytes were measured at 25°C and 1 atm partial pressure of carbon dioxide by the saturation method. The experimental results were not correlated easily by the modified Setschenow equation, but they were correlated very well by the empirical two-parameter equation. The parameters in the equation for the binary and ternary solutions could be estimated by assuming an additive rule for the parameters of the component salt systems. This method, therefore, is useful for predicting the solubility of carbon dioxide in aqueous mixed-salt solutions.

Nomenclature

A, B = parameters in Equations 2 and 3 ($m^3 \cdot kmol^{-1}$)
C_s = total concentration of mixed salt ($kmol \cdot m^{-3}$)
K = salting-out parameter ($m^3 \cdot kmol^{-1}$)
L = Ostwald absorption coefficient
n = number of components in mixed salt
x = mole fraction of a component in mixed salt

Subscripts

0 = free of salt
i = ith component
M = mixed salt
calcd = calculated value
obsd = observed value

Literature Cited

1. Onda, K., Sada, E., Kobayashi, T., Kito, S., Ito, K., *J. Chem. Eng. Japan* (1970) **3,** 137.
2. Joosten, G. E. H., Danckwerts, P. V., *J. Chem. Eng. Data* (1972) **17,** 452.
3. Hikita, H., Asai, S., Ishikawa, H., Esaka, N., *J. Chem. Eng. Data* (1974) **19,** 89.
4. van Krevelen, D. W., Hoftijzer, P. J., *Chim. Ind., 21st Congr. Int. Chim. Ind.* (1948) 168.
5. Onda, K., Sada, E., Kobayashi, T., Kito, S., Ito, K., *J. Chem. Eng. Japan* (1970) **3,** 18.
6. Yasunishi, A., *J. Chem. Eng. Japan* (1977) **10,** 89.
7. Yasunishi, A., Kagaku Kōgaku Ronbunshu (1978) **4,** 185.
8. Yasunishi, A., Yoshida, F., *J. Chem. Eng. Data,* in press.

9. Tokunaga, J., Nitta, T., Katayama, T., Kagaku Kōgaku (1969) **33**, 775.
10. Tokunaga, J., *J. Chem. Eng. Data* (1975) **20**, 41.
11. Robinson, R. A., Bower, V. E., *J. Res. Nat. Bur. Stand.* (1965) **69A**, 365.
12. Battino, R., Clever, H. L., *Chem. Rev.* (1966) **66**, 395.
13. Markham, A. E., Kobe, K. A., *J. Am. Chem. Soc.* (1941) **63**, 449.
14. Shoor, S. K., Gubbins, K. E., *J. Phys. Chem.* (1969) **73**, 498.
15. Gubbins, K. E., *AIChE J.* (1973) **19**, 684.

RECEIVED February 1, 1978.

14

Solubility of Sodium Chloride in an Inverse Micellar Solution of Pentanol

STIG FRIBERG

Chemistry Department, UMR, Rolla, MO 65401

IRENA BURASZEWSKA and EVA SJÖBLOM

The Swedish Institute for Surface Chemistry, Stockholm, Sweden

The solubilization of an aqueous sodium chloride solution by potassium oleate in the pentanol isotropic solution was determined. The presence of sodium chloride increased the minimum concentration for solubilization, reduced the maximum solubilization at high pentanol:potassium oleate ratios, and altered this ratio to lower values for maximal solubilization of the electrolyte solution. The increased minimum amount of electrolyte solution for solubilization arose from the fact that no micelles were present at the lowest fractions of water in the pentanol solution. The increased potassium oleate:pentanol ratio for maximal solubilization of the electrolyte was related to the destabilization of the lamellar liquid crystal with which the inverse micellar pentanol solution of high water content was in equilibrium.

One specific aspect of electrolyte solubility in nonaqueous media that so far has not been investigated widely is the solubilization of electrolytes in organic liquids using inverse micelles. These inverse micellar solutions are of interest in combination with microemulsions (*1, 2, 3, 4*); their structure has been investigated and the fundamental basis for their stability has been studied (*5*). The microemulsions contain inverse micelles (Figure 1) composed of an aqueous core (radius $\approx$ 50 Å) lined by polar groups of an ionic surfactant and a medium-chain-length alcohol (pentanol) with the nonpolar parts pointing towards the surrounding alcohol hydrocarbon solution.

0-8412-0428-4/79/33-177-205$05.00/1

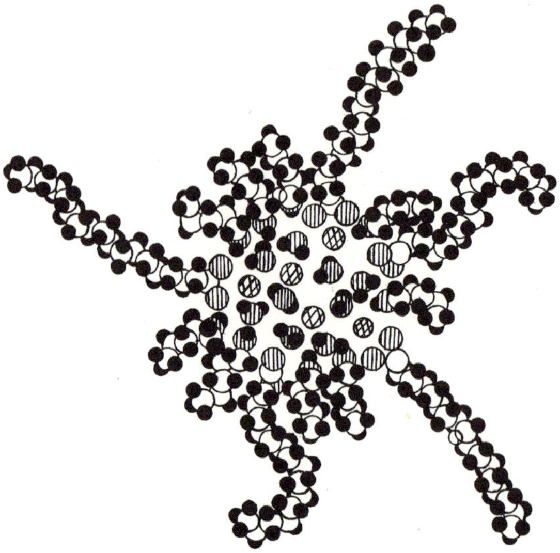

Figure 1. The inverse micelle as an aqueous core light with polar groups from the surfactant and the cosurfactant—a medium-chain-length alcohol. The hydrocarbon chains point radially outwards to the surrounding organic liquid.

These inverse micelles will solubilize electrolytes in their aqueous core but the presence of the electrolytes also will influence the stability of the inverse micelle. A change in the stability of the inverse micelle will be reflected in modifications of the solubility region of the inverse micellar solution. This chapter will relate the changes in solubility areas from addition of electrolytes to the water to the structure of inverse micelles and other association complexes in the pentanol solution.

Experimental

Materials. The following chemicals were used without further purification: *cis*-9-octadecanoic acid (Fluka A.G., 99%), 1-pentanol (Fluka A.G., 7 99%), potassium (Riedel and Hahn A.G.), sodium chloride (Merck p.a.), and twice-distilled water. The potassium oleate was prepared by titration with oleic acid of an ethanolic potassium ethoxide solution (*6*).

Solubilization Limits. The solubility regions were determined by titration with the sodium chloride solution until turbidity and the results checked by long-time storage of suitable compositions.

Association Structures. The onset of micellization was observed using light-scattering data and confirmed by electron microscopy of carbon replicas from freeze fracturing of samples frozen to − 150°C. The light-scattering unit was a Sophica Photo Gonia Diffusometer, Model

42,000, green light (5460 Å) was used, and the I_{90} scattering of benzene was given the intensity of 100. The Danaliker cells were cleaned with a solution of nitric and sulfuric acids, rinsed with distilled water, and finally cleaned by condensation of acetone vapor into the inverted cell. Removal of dust particles was achieved by centrifugation of the prepared solution at 25,000 *g* for 2 hr. Refractive indices were determined on an Abbe refractometer.

Platinum-shadowed replicas of freeze-fractured samples were prepared in a Denton Vacuum DFE-3 with a Denton DV-502 High-Vacuum Evaporator. An electron microscope Hitachi HU-11A with an optimum resolution of 7 Å was used to observe and photograph the replicas.

Results

The change of solubility areas with electrolyte content is shown in Figure 2. The solubility area for the pentanol solution of potassium oleate and water (—.—.—, Figure 2) shows a "minimum" water amount for solubilization. The minimum amount of water is proportional to the amount of soap present; the solubility limit to the right gives a molecular water:potassium oleate ratio of three, the minimum number of water molecules to bring the soap into solution. The solubility area is approximately triangular with maximum water solubilization at 12% (w/w) potassium oleate, 33% pentanol, and 55% water. The potassium oleate: pentanol ratio is .36 at that point.

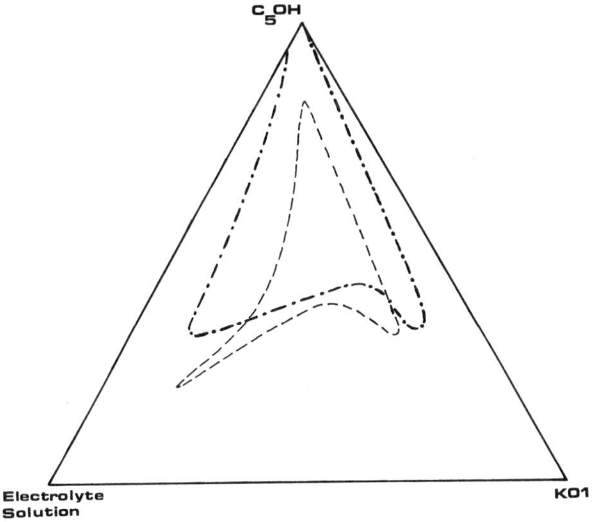

Figure 2. With added electrolyte (– – –) solution (0.5M NaCl) the solubility region of pure H_2O (– · –) will change.

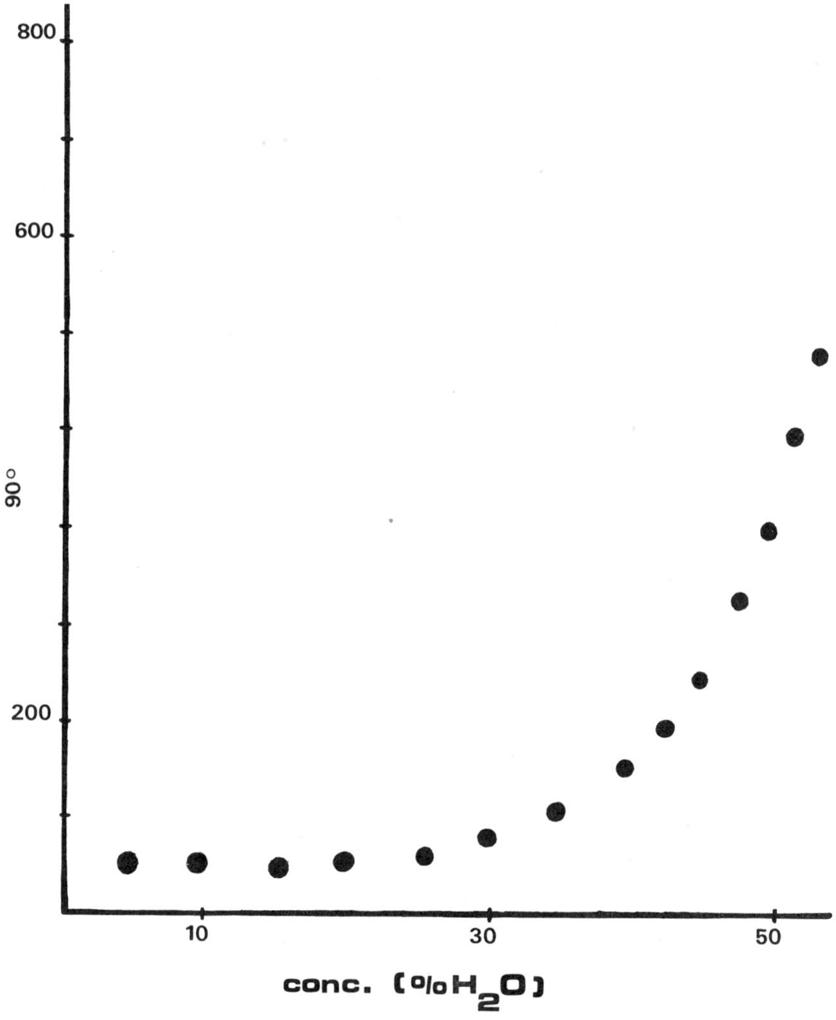

Figure 3. At low H_2O content (~ 25%) no association structures are present in the pentanol solution of potassium oleate and water. Pentanol:potassium oleate weight ratio = 2.9.

Addition of sodium chloride to the water caused pronounced changes in the solubility area. The minimum water amount to bring solubility of the soap was enhanced and the solubility region did not reach the pentanol corner. The solubility limit towards the water:pentanol axis (the left part of the solubility region) was moved to the right and the maximum water solubilization was shifted to a higher potassium oleate:pentanol ratio equal to 0.70. The maximum solubilization of the sodium chloride solution was greater than that for pure water; 64% of the solution was present.

The light-scattering results for the case involving pure water (Figure 3) showed no indication of association of surfactant at low water concentrations; the intensity of scattered light actually was reduced with added water. Above a certain water concentration (25%, Figure 3) the intensity gradually increased indicating stepwise association to larger aggregates. This rise of intensity began at higher concentrations of water for higher soap:alcohol ratios. Taking into account the solubility of water in pentanol the point at which the intensity began rising approximately corresponded to 10 water molecules per soap molecule.

The electron microscopy pictures showed no pattern at compositions for which the light-scattering data were at the low level (Figure 4, *right*); for compositions with higher scattered intensity the photos indicated the presence of colloidal structures (Figure 4, *left*).

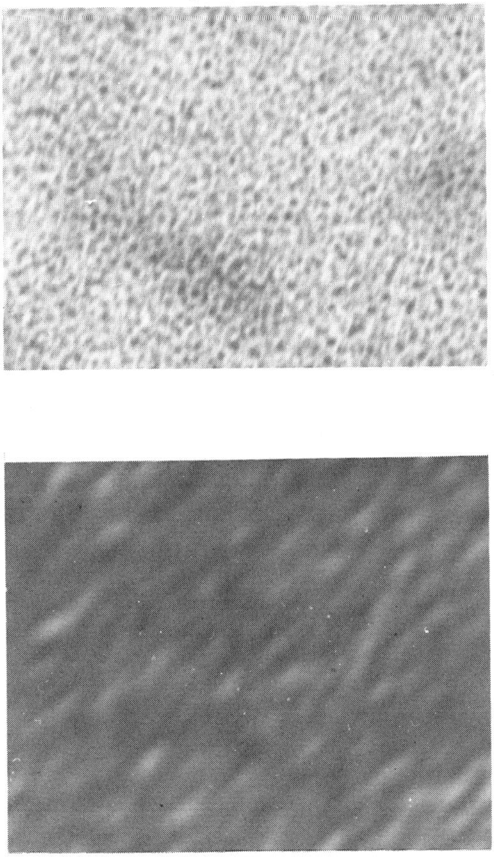

Figure 4. The presence of inverse micelles indicated by Figure 3 was confirmed by electron microscopy (top)*. Where light scattering indicated no inverse micelles, the electron microscopy photo was featureless* (bottom)*.*

Discussion

The results showed distinct and regular changes for the aqueous solubility region in pentanol:surfactant mixtures. With increased electrolyte content, the "minimum" amount of water for solubility was enhanced, the solubility limit towards the pentanol:water axis was shifted to higher soap concentrations, and the "maximum" solubility of the aqueous sodium chloride solution was obtained for higher surfactant:alcohol ratios (Figure 2).

The result showing an increase of the minimum water concentration is directly understood from the association structures in the aqueous solution without electrolytes. The light-scattering determinations indicated that no surfactant association took place at low water concentrations for the system without electrolytes. An association structure of one surfactant molecule, a few associated water molecules, and one or two alcohol molecules is a reasonable conclusion. The experimental results showed no electrolyte solubility in the part of the solubility region where these non-associated structures were found for the electrolyte-free solution. A small structure containing only 10 water molecules cannot be expected to accommodate electrolytes and the structural analysis offers a satisfactory explanation of the results.

The shift of the solubility limit against the pentanol:water axis (Figure 2, *left*) towards higher concentration of soap with an added electrolyte is mainly related to the solubility of the water in pentanol. The solubility of the aqueous solution in pentanol is reduced with increasing sodium chloride concentration; this fact accounts for the lowering of solubility at low surfactant:alcohol ratios.

The results showing augmentation of the surfactant: alcohol ratio for maximum aqueous solubility with added electrolytes are not amenable to a similarly simple explanation, and the influence of the presence of electrolytes must be discussed against the relative stability of the inverse micelles and of the lyotropic liquid crystalline phase with which the inverse micellar solution is in equilibrium (*7*).

The stability of inverse micelles has been treated by Eicke (*8, 9*) and by Müller (*10*) for nonaqueous systems, while Adamson (*1*) and later Levine (*11*) calculated the electric field gradient in an inverse micelle for a solution in equilibrium with an aqueous solution. Ruckenstein (*5*) later gave a more complete treatment of the stability of such systems taking both enthalpic (Van der Waals (VdW) interparticle potential, the first component of the interfacial free energy and the interparticle contribution of the repulsion energy from the compression of the diffuse part of the electric double layer) and entropic contributions into consideration. His calculations also were performed for the equilibrium between two liquid solutions—one aqueous, the other hydrocarbon.

However, experimental evidence has shown (*7*) that inverse micellar systems are rarely in equilibrium with aqueous micellar solutions but rather with a lamellar liquid crystalline phase. The presence of an electrolyte will influence the stability of both the inverse micelles and the lamellar liquid crystalline phase. This influence will be estimated now.

The discussion of the relative stability of solutions with inverse micelles and of liquid crystals containing electrolytes may be limited to the enthalpic contributions to the total free energy. The experimentally determined entropy differences between an inverse micellar phase and a lamellar liquid crystalline phase are small (*12*). The interparticle interaction from the Van der Waals forces is small (*5*); it is obvious that changes in them owing to added electrolyte may be neglected. The contribution from the compression of the diffuse electric double layer is also small in a nonaqueous medium (*11*) and their modification owing to added electrolyte may be considered less important. It appears justified to limit the discussion to modifications of the intramicellar forces.

Of these, the interfacial tension may be described by three terms, two of which have been described by Ruckenstein (*5*) who evaluated the noncoulombic and the coulombic part of the stretching force. The mechanical description (*13*) is

$$\gamma = \int_{\lambda_1}^{\lambda_2} p dz$$

in which the λ_v:*s* are coordinates of the *z*-axis perpendicular to the interface, and *p* is the excess pressure tensor parallel to the interface.

However, microemulsion droplets are sufficiently small that the second contribution to the surface free energy, the bending energy, is of the same magnitude as the pure stretching energy (*13*). The bending energy may be defined as

$$\gamma_2 = \int_{\lambda_1}^{\lambda_2} pz dz$$

For a flat interface or a droplet for which the interface transition zone is small compared with the radius the bending energy is insignificant compared with the purely stretching energy. A microemulsion droplet in this system has a radius of approximately 25 Å. The size of the transition zone is certainly not of negligible size in comparison with that value.

Addition of electrolyte will predominantly affect the coulombic term of the interfacial tension; some influence on the other terms may be present, but the significance must be comparatively less significant.

Analytical expressions for calculating the form of the inner double layer in an inverse micelle have been given (*11*). The following analysis will use the Debye–Hückel approach (5), an approach that will be justified by the results.

The energy of the inner electric double layer is

$$E_{\text{idl}} = \gamma_c \, 4\pi R^2 \tag{1}$$

in which R is the radius of the droplet and γ_c is the coulombic term of the surface free energy.

$$E_{\text{idl}} = -\int_0^{\psi_o} \delta d\psi \tag{2}$$

in which δ is the electrical surface charge and ψ_o is the surface potential.

For the spherical Debye–Hückel approximation, the Poission–Boltzmann Equation reduces to

$$\frac{d}{dr}\left(r^2 \frac{d\psi}{dr}\right) = \kappa^2 \psi r^2 \tag{3}$$

in which κ^2 is the square of the reciprocal Debye length and r is the varied radius for which ψ is sought.

$$\kappa^2 = 8\pi^2 e^2 n_o / \epsilon kT$$

in which e is the charge of an electron and ϵ is the dielectric constant.

For hydrocarbon or pentanol solutions, the potential gradient outside the inverse micelle is extremely small. Solving Equation 3 with the boundary conditions (5)

$$r = 0 \qquad \psi_o = \text{finite}$$
$$r = R \qquad \psi_R = \psi_R$$
$$r = \infty \qquad \psi = 0$$

and the energy for the electric double layer formation may be calculated from (2):

$$E_{\text{idl}} = \frac{-\psi_R{}^2}{8\pi R} \{\epsilon_{H_2O}(\kappa_{H_2O} R \text{ ctn h } \kappa_{H_2O} R - 1) + \epsilon_{\text{oil}}(1 + \kappa_{\text{oil}} R)\} \tag{4}$$

Equation 4 predicts a square dependence of the energy on the electric potential of the interface of the inverse micelle. Addition of electrolyte will not change the surface potential much; a slightly reduced stability may be expected. The higher surfactant:alcohol ratio that was observed (Figure 2) will increase the surface potential.

The energy of the electric double layer is directly dependent on the square of the surface potential (Equation 4) and the observed increase of the potassium oleate:alcohol ratio should enhance the stability of the inverse micelle. The stability of the inverse micelle is not the only determining factor. Its solution with a maximal amount of water is in equilibrium with a lamellar liquid crystalline phase (7) and the extent of the solubility region of the inverse micellar structure depends on the stability of the liquid crystalline phase.

An investigation of the conditions of the lamellar liquid crystalline phase is straight forward. The geometry of planar layers makes the estimation of the influence of electrolyte simple since the stability directly depends on the gradient of the electric field. For stability at high water content the gradient at large distances is decisive. If the effects of discreteness of charge are neglected, the repulsion force is

$$f = 64nkT\gamma^2 e^{-\kappa d}$$

in which $\gamma \sim \psi^o$ at low electric potentials and d is the total distance between the layers.

Increased electrolyte concentration will cause a reduction of the slope at distances for which $\kappa d > 1$. With an electrolyte concentration of 0.5*M* NaCl, $\kappa d = 1$ corresponds to a distance of 4.6 Å. A lamellar liquid crystal in [illegible] part of a system has an aqueous layer thickness of approximately 50 [illegible]nd the reduction of the repulsion force from the overlap of the electric double layer is obvious.

A reduction of the stability of the liquid crystalline phase means a reduced region where it is stable and a corresponding increase of the region for the inverse micellar solution. The present results agree with these predictions, and it is justifiable to relate the changes in stability areas mainly to modifications of the potential distribution within the electric double layers.

Summary

The changes in stability regions for inverse micellar solutions where added electrolytes appear were given a rational explanation using the associated structures determined in the inverse micelle solution with no electrolyte.

Acknowledgment

This research was supported by National Science Foundation Grant #NSF DMR76–23569.

Literature Cited

1. Adamson, A. W., *J. Colloid Interface Sci.* (1969) **29**, 261.
2. Gillberg, G., Lehtinen, H., Friberg, S., *J. Colloid Interface Sci.* (1970) **33**, 40.
3. Shinoda, K., Kunieda, H., *J. Colloid Interface Sci.* (1973) **42**, 381.
4. Ahmed, S. I., Shinoda, K., Friberg, S., *J. Colloid Interface Sci.* (1974) **47**, 32.
5. Ruckenstein, E., Chi, J. C., *J. Chem. Soc., Faraday Trans. II* (1975) **17**, 1690.
6. Mandell, L., Ekwall, P., *Acta Polytech. Scand., Chem. Incl. Metall. Ser.* (1968) **74**, 1.
7. Friberg, S., Buraczewska, J., *Progress in Colloid and Polymer Sci.*, in press.
8. Eicke, H. F., Christen, H., *J. Colloid Interface Sci.* (1974) **46**, 417.
9. Ibid. (1974) **48**, 281.
10. Muller, N., *J. Colloid Interface Sci.* (1977) **63**, 383.
11. Levine, S., Robinson, K., *J. Phys. Chem.* (1972) **76**, 876.
12. Steinius, P., private communication.
13. Murphy, S., Thesis, University of Minnesota (1966).

RECEIVED March 2, 1978.

15

Computational Techniques of Ionic Processes in Water-Organic Mixed Solvents

BUDDHADEV SEN—Department of Chemistry, Louisiana State University, Baton Rouge, LA 70803

RABINDRA NATH ROY and JAMES J. GIBBONS—Department of Chemistry, Drury College, Springfield, MO 65802

DAVID A. JOHNSON—Department of Chemistry, Spring Arbor College, Spring Arbor, MI 49283

LOUIS H. ADCOCK—Department of Chemistry, University of North Carolina, Wilmington, NC 28401

A curve-fitting computational technique for the computation of the standard potential of Ag–AgCl electrode in mixed solvents has been developed. The equations generated from experimental data parallel the mathematical form of Gronwall, LaMer, and Sandved's extended Debye–Hückel Equation for symmetrical electrolytes. Application of this technique to a large number of systems has yielded excellent results. The curve-fitting computational technique has been extended to ionic equilibrium processes. Generated equations seem to support Born's electrostatic model in a dielectric medium as a first approximation. The power of the curve-fitting technique rests on the fact that it does not assume any physical model: it only assumes that the experimental data are good and then proceeds to explore which mechanistic model best rationalizes the empirical equation.

During the past two decades the thermodynamics of chemical processes in mixed and nonaqueous solvents have been studied extensively by a large number of workers (*1*). Such studies have merit in their own right, aside from the fact that these studies have extremely important practical implications. In studying various types of chemical equilibria and in studying kinetics, it is sometimes necessary to use mixed-organic–

0-8412-0428-4/79/33-177-215$08.50/1

aqueous solvents in order to avoid complicating processes such as hydrolysis, undesired coördination, competing reactions, insolubility, etc. However, we are not certain if the use of mixed solvents does not, in its wake, create more problems than it solves (*2*). In fact, Amis and Hinton state "As perplexing as the search for a structural definition of water is, the eccentric physico-chemical properties of mixed-aqueous solutions is at times even more baffling" (*2*). We will not pursue the complex role of mixed solvents on reaction rates, and therefore on mechanisms, except in stating that the bulk dielectric constant of the mixed solvent is one of the important parameters of every proposed theory (*2*).

This chapter is concerned primarily with the computation of potentials of a cell using the hydrogen electrode as a probe for studying ionic equilibrium processes in mixed-organic–aqueous solvent systems. Computation of a number of other thermodynamic functions of the ionic process under investigation or of the solvent used is rather straightforward once the standard potential of the measuring cell has been calculated.

Therefore, from the experimental point of view the most important objective is to develop a method for computing the standard potential of the cell from experimental emf data. The first part of this chapter presents our endeavors in that direction.

Computation of the thermodynamic equilibrium constant (and $\Delta G°$) of an ionic process is quite simple once the standard potential of the relevant cell has been determined. Enthalpies and entropies of the process can be calculated when such equilibrium constant data are available over a range of temperatures. However the importance of such thermodynamic data, even when obtained by rigorous experimental procedures, is greatly reduced if the data cannot be referred to an arbitrarily accepted common reference state. The second part of this chapter presents computational methods for converting the thermodynamic data obtained in a mixed solvent to the corresponding thermodynamic data in a reference system which in this case is aqueous medium containing the equilibrium ionic process at infinite dilution. A large volume of the experimental data cited were obtained by the authors. In order to prevent the chapter from becoming unwieldy, we have been selective rather than comprehensive in reviewing the pertinent literature. Also, the background theoretical treatment has been kept brief with the assumption that the reader is more or less conversant with such material.

Computation of Standard Potential

The working cell is

$$\text{Pt, } H_2 \text{ (1 atm)} | \text{HCl } (m), \text{ S } (X), H_2O \ (Y) | \text{AgCl–Ag} \qquad \text{I}$$

in which S represents the organic component of the solvent, X and Y are weight percents or mole fractions of the two solvent components, and the other symbols have their usual significance. In their classic monograph (*3*) Harned and Owen have extensively summarized most of the earlier work on the cells of Type I including their own comprehensive work using the dioxane–water system. They found that as the dielectric constant of the solvent decreased, the contribution of Gronwall, LaMer, and Sandved's (GLS) (*4, 5*) extended terms for the Debye–Hückel Equation could no longer be ignored in extrapolating to obtain the limiting value of $E°$. Harned and Owen (for details *see* Ref. 5) have demonstrated graphically the effect of ignoring the higher-order terms by plotting $E°'$ and $E°' - E_{ext}$ against the molality of HCl in Cell I with varying percentages of dioxane (Figure 1). $E°' = E° + f(m)$ and E_{ext} is the extended terms contribution. Obviously, extrapolation of the $E°'$ vs m plot is unsatisfactory for a solvent system containing even 45 wt % dioxane (D = 38.48) owing to the curvature of the plots in the important

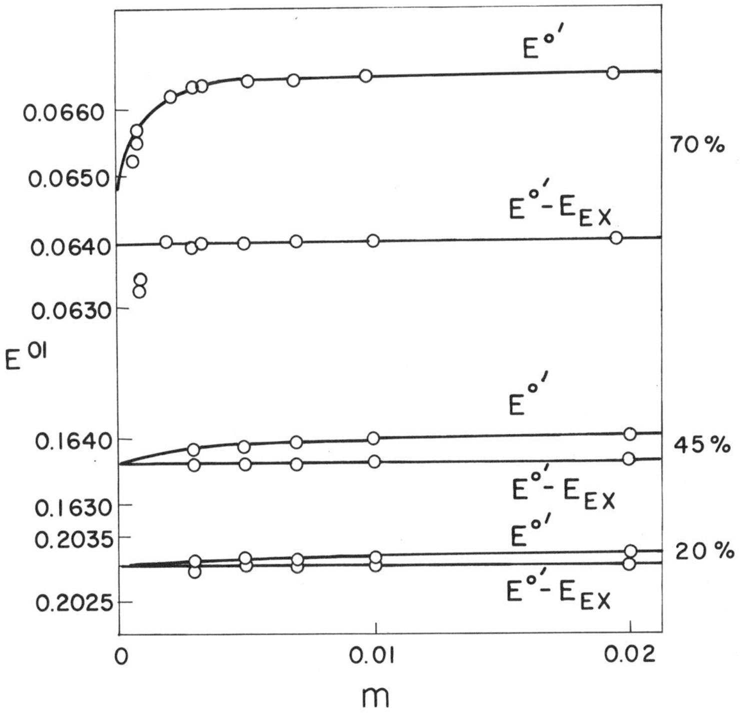

The Physical Chemistry of Electrolytic solutions

Figure 1. Extrapolation of emf (V) in dioxane–water mixtures computed by the Debye–Huckel extended equation. The percentages are those of weight percent of dioxane (5).

low molality region. $E^{\circ\prime} - E_{ext}$ plots seem to be quite satisfactory in solvents containing 45 wt % (D = 38.48) or more of dioxane; however, even these plots are unsatisfactory in a 70% dioxane medium. Harned and others attempted to improve the extrapolation method by introducing contributions from mass action and dissociation of the acid in low dielectric medium.

Let us now consider the explicit expression for E_{obs} of Cell I

$$E_{obs} = E^{\circ} - \frac{2RT}{F} \ln \gamma_{\pm} m \tag{1}$$

Rearrangement of Equation 1 leads to

$$E_{obs} + \frac{2RT}{F} \ln m = E^{\circ} - \frac{2RT}{F} \ln \gamma_{\pm} \tag{2}$$

Equation 2 was proposed by Lewis and Randall (*6*) even before the advent of the Debye–Hückel theory for the evaluation of E° by the extrapolation of LHS of Equation 2 against f(m) because by definition $\gamma_{\pm} \rightarrow 1$ as $m \rightarrow 0$. However, such extrapolations failed to yield precise values of E° because the plots did not produce the desired straight lines. The most common method of extrapolation is to substitute an expression for $\ln \gamma_{\pm}$ from the Debye–Hückel theory (*7*), and the two most frequently used substitutions are

$$\log \gamma_{\pm} = -S_{(f)} (\Gamma)^{\frac{1}{2}} + B'm \tag{3}$$

and

$$\log \gamma_{\pm} = -\frac{S_{(f)} (\Gamma)^{\frac{1}{2}}}{1 + A(\Gamma)^{\frac{1}{2}}} \tag{4}$$

where

$$S_{(f)} = \frac{1}{\nu} \sum_{1}^{p} \nu_j Z_j^{\,2} \frac{1.290 \times 10^6}{(DT)^{3/2}} \tag{5}$$

$$\Gamma = \sum_{1}^{s} C_i Z_i^2 \tag{6}$$

$$A = a^{\circ} \frac{35.56}{(DT)^{\frac{1}{2}}} \tag{7}$$

The term a° is frequently called the ion-size parameter or apparent ionic diameter, B′ is an empirically fitted coefficient. In reality, however, B′ has important theoretical significance in interionic attraction theory, and it is a higher-order function of the ion-size parameter.

Also a comparison of two alternative expressions for log $\gamma_\pm$ given by Equations 3 and 4 is of interest. A close examination of these equations will reveal certain inherent weaknesses of the linear extrapolation method for the evaluation of $E°$, even when using the extended terms. First, the linear extrapolation will require a precise value of D, the dielectric constant of the solvent, which in principle is experimentally measurable. Frequently, however, it is computed by using some empirical function or by graphical interpolation. Secondly, the uncertainty of the ion-size parameter is significant and no reliable method of computation exists, nor is it directly measurable experimentally.

This dilemma led us to investigate the feasibility of a nonlinear extrapolation for the evaluation of $E°$. For the sake of brevity, we will not reproduce here the detailed mathematical derivations of Gronwall, LaMer, and Sandved's extended terms of the Debye–Hückel theory. One can find these derivations in their original paper (*4*) or in Harned and Owen's classic monograph. For a description of the basic assumptions and the physical model of the interionic attraction theory one should consult Harned and Owen's monograph as well as the work of Gurney (*8, 9*).

For the electrolytes of symmetrical valence type, the extended equation for ln $\gamma_\pm$ is expressed as

$$\ln \gamma_\pm = -\frac{(\epsilon z)^2}{2DkT}\cdot\frac{\kappa}{1+\kappa a}+\left[\frac{(\epsilon z)^2}{DkTa}\right]^3\cdot\left[\frac{1}{2}X_3(\kappa a)-2Y_3(\kappa a)\right]+$$
$$\left[\frac{(\epsilon z)^2}{DkTa}\right]^5\cdot\left[\frac{1}{2}X_5(\kappa a)-4Y_5(\kappa a)\right]+\text{higher terms} \quad (8)$$

in which z is the valence (which for symmetrical electrolyte, $|z_+| = |z_-|$), ϵz is the ionic charge, D is the dielectric constant of the solvent, k is Boltzmann constant, T is absolute temperature, κ is a very important function in the interionic attraction theory and has the dimensions of a reciprocal distance, and $1/\kappa$ is related to the potential of the ionic atmosphere and it is explicitly given by

$$\kappa = \left(\frac{4\pi\epsilon^2 N}{1000DkT}\,\Gamma\right)^{\frac{1}{2}} \quad (9)$$

Upon substitution of the numerical values of the constants, Equation 9 for symmetrical electrolytes at 25°C becomes

$$\frac{1}{\kappa} = \frac{3.043 \times 10^{-8}}{\sqrt{c}}\ \text{cm} \quad (10)$$

in which c is the molar concentration of an ion. Obviously, in a normal solution $1/\kappa$ is of the order of a molecular diameter, and it varies inversely with the square root of c. In Equation 8, *a* is the mean distance of approach of the ions.

It is quite obvious that the first two terms of the right-hand side of Equation 8 represent Debye–Hückel's first approximation as expressed by Equation 4. That is,

$$\ln \gamma_{\pm} = -\frac{(\epsilon z)^2}{2DkT} \cdot \frac{\kappa}{1 + \kappa a} \qquad (11)$$

or

$$\log \gamma_{\pm} = -\frac{S_{(f)}(\Gamma)^{\frac{1}{2}}}{1 + A(\Gamma)^{\frac{1}{2}}} \qquad (4)$$

in which

$$A = \frac{a\kappa}{(\Gamma)^{\frac{1}{2}}} = \frac{35.56 \times 10^{-8}\, a}{(DT)^{\frac{1}{2}}},\ a \text{ in cm} \qquad (12)$$

$$= \frac{35.56 \times a^{\circ}}{(DT)^{\frac{1}{2}}},\ a^{\circ} \text{ in Ångstroms} \qquad (7)$$

The overall reaction in Cell I under the defined standard conditions is

$$HCl\ (1\ m) + Ag(s) = AgCl(s) + 1/2\ H_2(p = 1\ atm) \qquad (13)$$

The electrical work done by Cell I under standard and reversible conditions is a measure of the standard free energy change of Reaction 13. Thus

$$-\Delta G^{\circ} = NFE^{\circ} \qquad (14)$$

in which E°, the standard potential, is expressed by

$$E^{\circ} = E_{obs} + \frac{2RT}{F} \ln m + \frac{2RT}{F} \ln \gamma_{\pm} \qquad (15)$$

with E_{obs} being the measured potential of the working Cell I. It should be emphasized here that the standard cell is an imaginary device in which $m = a = \gamma_{\pm} = 1$. The interionic attraction theory essentially accounts for the nonideality of the solutions of strong electrolytes (i.e., completely dissociated) in a dielectric medium, and this nonideality is expressed by $\gamma_{\pm}$. In solutions of high dielectric constants ($D \geq 38$), Debye–Hückel's extended terms (in Equations 3 and 4) adequately account for this nonideality (cf. Figure 1). However, as the dielectric constant decreases, the contributions owing to the higher terms of Equation 8 can no longer be ignored.

At this stage we will omit all of the complicated and rather tedious mathematical derivations. Roy and Johnson (*10, 11, 12, 13*) have shown that GLS Equation 8 can be reduced to a polynomial in κ; thus

$$\ln \gamma_{\pm} = A''\kappa + B''\kappa^2 + C''\kappa^3 + D''\kappa^4 + \ldots \qquad (16)$$

in which the constants A″, B″, etc., emerge from the various series involved in the GLS equation. It has been shown earlier (Equations 9 and 10) that κ is proportional to the square root of ional strength or concentration (molar or molal). Equation 16 can be rewritten as

$$\ln \gamma_{\pm} = A'm^{\frac{1}{2}} + B'm + C'm^{3/2} + \ldots \qquad (17)$$

Substituting the value of $\ln \gamma_{\pm}$ from Equation 17 into Equation 2, one obtains

$$E_{\text{obs}} + \frac{2RT}{F} \ln m = E^{\circ} - \frac{2RT}{F} (A'm^{\frac{1}{2}} + B'm + C'm^{3/2} + \ldots) \qquad (18)$$

or

$$E_{\text{obs}} + \frac{2RT}{F} \ln m = E^{\circ} + Am^{\frac{1}{2}} + Bm + Cm^{3/2} + \ldots \qquad (19)$$

The coefficients of Equation 19 consist essentially of a number of constants among which are dielectric constant D of the medium and the ion-size parameter. It has been mentioned earlier that indeterminacy of the value of a makes prior computation of constants impossible. There are also doubts about the reliability of the values of D in many instances.

A number of important conclusions may be drawn from Equation 19. These are as follows.

1. ($E_{\text{obs}} + RT/F \ln m$) is a polynomial function in powers of $m^{\frac{1}{2}}$ which will cause a plot of the left-hand side vs. $m^{\frac{1}{2}}$ to be curved inwards, and will asymptotically approach E° as $m^{\frac{1}{2}} \rightarrow 0$.

2. It is only under special conditions that a plot of the left-hand side vs. $m^{\frac{1}{2}}$ produces a straight line; and all experimental evidence seems to indicate that such special conditions exist for dilute solutions of strong electrolytes (particularly 1–1 electrolytes) in solvents of high dielectric constant where the sum of contributions of terms containing powers of $m^{\frac{1}{2}}$ greater than 1 are insignificant. The Debye–Hückel extended equation includes the term containing the second power of $m^{\frac{1}{2}}$.

3. The contributions from higher order terms of $m^{\frac{1}{2}}$ rapidly become significant as $|z_+| = |z_-|$ becomes larger than 1, and as a becomes smaller with the decreasing value of the dielectric constant.

Because of the foregoing conclusions, we decided to abandon the linear extrapolation method and investigate the prospects of a curve-fitting technique for the evaluation of E° of Cell I. In order to use the curve-fitting technique, it is necessary to assume that a set of reliable and precise emf data for the cell reaction in the low-concentration region say from $10^{-1}m$ to $10^{-4}m$) is available. With the fine instrumentation available today, the foregoing expectation is experimentally attainable. It also is assumed that the set of experimental emf data can be best fitted to a polynomial of explicit form,

$$Y = A_o + A_1x + A_2x^2 + A_3x^3 + \ldots + A_nx^n \quad (20)$$

which is an analog of Equation 19. In using the polynomial for the computation of E°, y is replaced by $E_{obs} + RT/F \ln m$, and x is replaced by $m^{1/2}$. In the actual case, the value of the subscript n must be arbitrarily fixed, and its arbitrary value need not be fixed any higher than necessary in order to save computation time. However, computation must be continued and include higher- and higher-ordered terms until all the experimental data can be fitted to the appropriate polynomial producing a smooth plot. The first approximation of Equation 20 is $y = A_0 + A_1x$, which is an analog of the DH-extended equation as mentioned earlier. The coefficients A_i, which were determined by the method of least squares and the standard error of the dependent variable y, was compared with a predetermined maximum tolerance. Higher-order terms must be taken into consideration, if necessary, to have the error fall within the maximum tolerance. A modified Gaussian elimination technique was used to solve the resulting set of linear equations. In our computation, we used a standard IBM program (revised IBM 7.0.002).

Obviously, the curve-fitting polynomial for the computation of E° of Cell I may be written explicitly as

$$E_{obs} + \frac{2RT}{F} \ln m = A_o + A_1m^{1/2} + A_2m + A_3m^{3/2} + \ldots \quad (21)$$

in which $A_0 = E^\circ$, the standard potential of the cell. It was never necessary to use more than four terms for the systems we investigated. We would like to emphasize once again that it is important that Equation 21, with the computed coefficients, accommodates all experimental data; that is, the calculated $[E_{obs} + 2RT/F \ln m]$ is within the predetermined tolerance limit of experimental $[E_{obs} + 2RT/F \ln m]$ for all values of m in the given system at the given temperature. In fact, in our opinion, the systems for which such a fit cannot be obtained should be viewed with suspicion, and this criterion may be taken as an indication of presence of interactions other than coulombic interactions among ions and ions, and ions and solvent dipoles. During our investigations we experienced such deviations when acetonitrile was used as the organic component in Cell I. We will make a few brief comments about the use of acetonitrile as the solvent component later in this chapter.

Figures 2 and 3 represent typical plots of E' ($= E_{obs} + 2RT/F \ln m$) vs. $m^{\frac{1}{2}}$ in mixed solvents with a low (8.68%) and a high (89.00%) proportion of organic component (*10, 11, 12, 13*). The pronounced increase in the curvature of the plots with increased organic component (decreased bulk dielectric constant) is quite obvious.

Roy and his co-workers (*14–28*) have used the curve-fitting polynomial technique for the computation of $E°$ of Cell I containing a variety of mixed-solvent systems. Some of the values of $E_m°$ (molal scale) of Cell I (*29*) are compiled in Table I.

Extensive experimental data on bulk dielectric constants of mixed-aqueous–organic solvents are not available. Standard potentials of Cell I with limited dielectric data are compiled in Table II. Plots of standard potential vs. $(1/DT)^{\frac{1}{2}}$ are shown in Figure 4. The data were plotted on two different scales for a clear demonstration of the trend. The entire range of potential and dielectric data, $(1/DT)^{\frac{1}{2}}$, were plotted as open circles and the range potential data from 149–222 mV and the corresponding dielectric data were replotted as open squares on a more expanded scale (Figure 4). Obviously, for the case in which the bulk dielectric constant of the solvent is greater than 40, the standard potential of Cell I is a linear function of $(1/DT)^{\frac{1}{2}}$, as would be expected from the first-order approximation of DH-extended theory. Only three data points significantly deviate from the straight-line plot. These points correspond to 163 mV, 9338 $(1/\sqrt{DT})$; 149 mV, 8834 $(1/\sqrt{DT})$; and 203 mV, 7417 $1(/\sqrt{DT})$. The first of these points corresponds to 45 wt % dioxane ($D = 38$); it is the threshold region of failure of linearity; also the datum itself seems to be suspect. The other two points may not be considered beyond the realm of experimental error.

It would be interesting to compare the slopes of such experimental plots with those calculated from the theory. However, it is safe to conclude that in mixed solvents with a bulk dielectric constant greater than 40, first-order coulombic interactions are the only significant interactions among the ions and ions, and ions and solvent molecules assuming that there are no chemical interactions. In solvents with lower dielectric constants higher-order interactions, and perhaps even noncoulombic interactions, become increasingly significant. As experimentalists, we do not want to speculate any further, and in making the foregoing comments, it has been tacitly assumed that one of the components of the binary system is water.

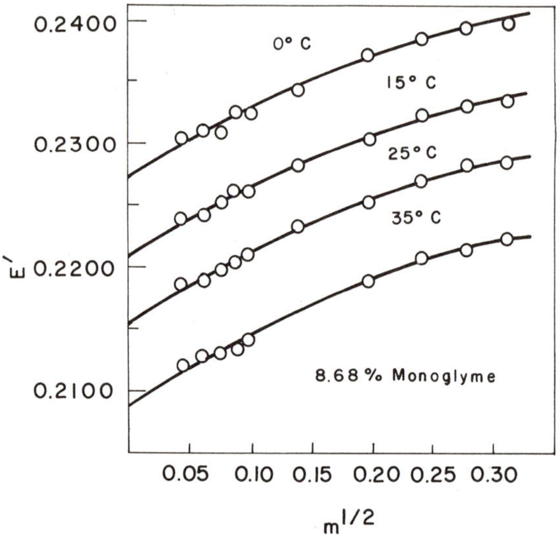

Figure 2. Variation of emf [E′ = E_{obs} + (2RT/F) ln m] with molality and temperature in 8.68% monoglyme

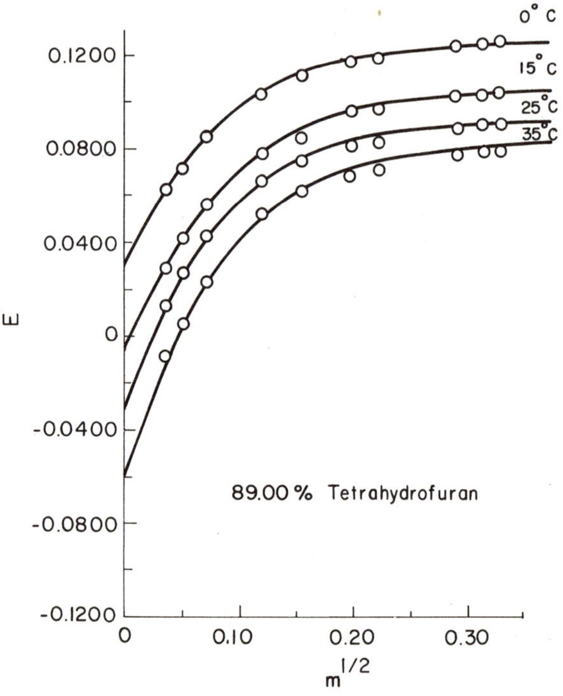

Figure 3. Variation of emf (E′) with molality and temperature in 89.00% tetrahydrofuran

Table I. Standard Potential ($E_m{}^\circ$ Volts) of Ag–AgCl Electrode (Cell I) Computed by the Curve-Fitting Method

Solvent Composition Wt % Organic Component	T (°C)					
	0	*5*	*15*	*25*	*35*	*45*
Organic Component: 1,2-Dimethoxyethane (11, 27) $D = 6.8$						
8.68	0.2272	—	0.2209	0.2155	0.2089	
17.81	0.2242	—	0.2158	0.2090	0.2009	
46.52	0.2053	—	0.1905	0.1794	0.1684	
67.03	0.1688	—	0.1449	0.1262	0.1099	
88.80	0.0250	—	−0.01125	−0.0395	−0.06039	
Organic Component: Tetrahydrofuran (12, 28) $D = 7.6$						
8.98	0.23467	—	0.22160	0.21368	0.20933	
18.21	0.21662	—	0.21062	0.20375	0.19828	
47.20	0.20365	—	0.18662	0.17060	0.16572	
73.03	0.12600	—	0.11163	0.09310	0.07330	
89.00	0.03397	—	−0.00279	−0.02582	−0.05689	
Organic Component: Isopropanol (15, 20) $D = 19.41$						
95	—	0.01878	−0.00435	−0.02269		
100	—	−0.1122	−0.1249	−0.1338		
Organic Component: 1-Butanol (22, 23) $D = 17.51$						
5	—	0.2168 (10°)	0.2101	0.2029	0.1958	0.1883 (40°)
100	—	−0.04027		−0.06863		−0.10802
Organic Component: 1-Butanol (14)						
5	—	0.21028	0.20864	0.20350	0.19488	0.18272
Organic Component: tert-Butanol (12) $D = 16.62$ (30°)						
10	—	0.2159	0.2102	0.2061	0.2002	0.1919
20	—	0.2083	0.2013	0.1957	0.1909	0.1862
40	—	0.1963	0.1897	0.1845	0.1764	0.1665
70	—	0.1444	0.1354	0.1251	0.1105	0.1025

Table I. Continued

Solvent Composition Wt % Organic Component	T (°C)					
	0	*5*	*15*	*25*	*35*	*45*
Organic Component: 1-Propanol (10, 21, 24, 25) D = 20.33						
10	—	0.2212	0.2157	0.2093	0.2034	0.2008
20	—	0.2099	0.2053	0.1998	0.1940	0.1862
40	—	0.1917	0.1849	0.1729	0.1689	0.1577
70	—	0.1568	0.1461	0.1362	0.1289	0.1116
90	—	0.0971	0.0777	0.0583	0.0461	0.0060
95	—	0.04542	0.03363	0.02105	−0.00319	−0.02047
100	—	−0.0854	−0.1043	−0.1200	−0.1333	−0.1500
Organic Component: Glycerol (17, 18) D = 40.1						
95	—	0.08569	0.07868	0.06674	0.05089	0.04149
100	—	0.04065	0.02605	0.02077	0.00956	−0.00776
Organic Component: Propylene Carbonate (19) D = 64.4						
5	—	0.2309	0.2260	0.2209	0.2150	0.2080
10	—	0.2289	0.2240	0.2188	0.2125	0.2049
20	—	0.2266	0.2190	0.2132	0.2082	0.1990
Organic Component: Dimethyl Sulfoxide (13) D = 46.7						
5	—	0.2314	0.2255	0.2179	0.2102	0.2036
10	—	0.2312	0.2210	0.2148	0.2093	0.2029
20	—	0.2271	0.2186	0.2117	0.2076	0.1990
Organic Component: Dimethylformamide (16) D = 36.71						
5	—	0.23081	0.22735	0.22139	0.21507	0.20476
10	—	0.22643	0.22307	0.21487	0.21063	0.20038
0 (pure water as solvent) (*29*)		0.2341	0.2286	0.2224	0.2156	0.2083

Incidentally, such plots (Figure 4) may be used to compute approximate values of the bulk dielectric constants of water–organic mixed solvents from the emf measurements provided that the bulk dielectric constant is higher than 40. In fact, for auto-ionizing solvents this may be a convenient method, and our calculations have shown that such computed values are within 5–7% of the experimental values.

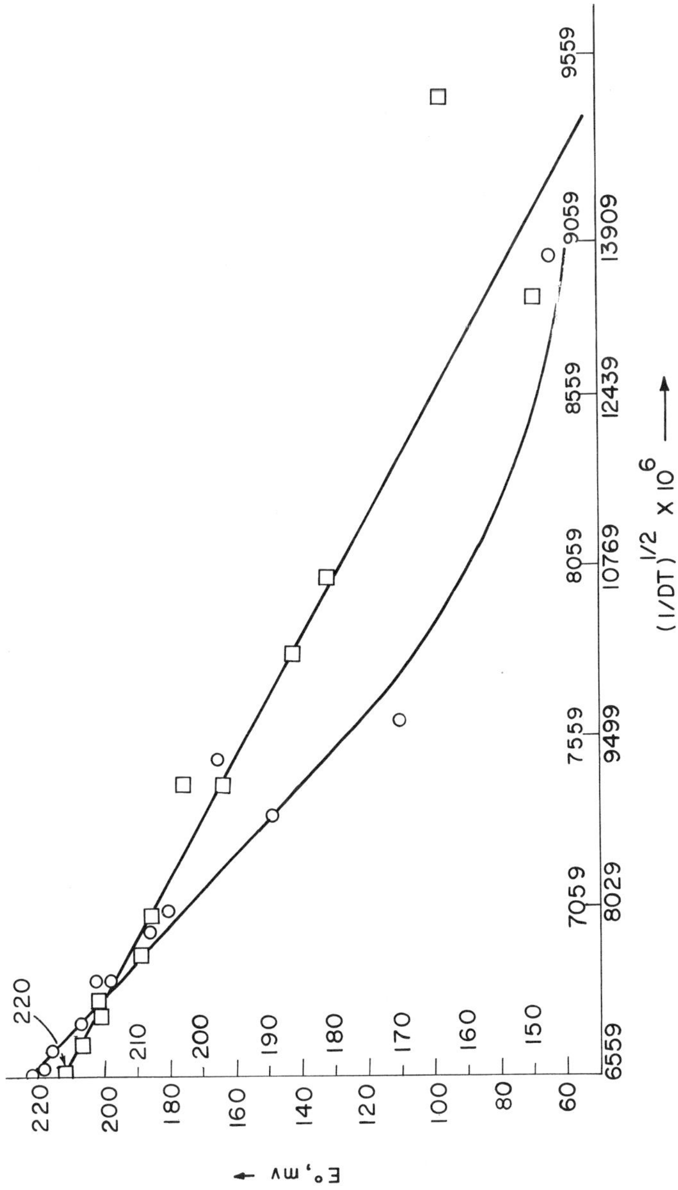

Figure 4. Variation of the standard potential of Cell I with $(1/DT)^{1/2}$ *at* 25°C. *Range of D*, 78–36. *Scale:* (○), *E*, 55–220 *mV*, $(1/DT)^{1/2}$, 6559–13909; (□), *E*, 100–220 *mV*, $(1/DT)^{1/2}$, 6559–9059.

Table II. Standard Potentials (Molal Scale) of Cell I and the Bulk Dielectric Constants of the Solvents[a]

Wt % of Organic Component	*D*	$E_1°$ *(V)*	$E_2°$ *(V)*	*No. of Observations*	*Std. Error* $\times 10^4$	*Deg. Equation*
5 Act	75.9	0.2190	0.21925	15	0.94	2
10 Me	74.1	0.21535	0.21532	18	0.47	3
10 Act	73.1	0.21565	0.2160	15	4.45	2
20 Me	69.99	0.2094	0.20914	7	0.82	3
20 Act	67.6	0.2095	0.20818	15	1.00	3
40 Me	60.94	0.1968	0.19682	8	1.59	3
20 Diox	60.79	0.20303	0.20301	10	0.70	3
40 Act	54.6	0.18595	0.18616	14	1.25	3
60 Me	51.67	0.1818	0.18120	9	0.5	4
80 Me	42.60	0.1492	0.14895	7	1.80	4
90 Me	35.76	0.1135	0.1148	9	3.20	2
45 Diox	38.48	0.16358	0.16331	10	0.57	5
70 Diox	17.69	0.06395	0.06408	10	0.39	5
82 Diox	9.53	−0.0415	−0.0339	7	1.24	3
100 Water	78.48	0.2224	0.22238	8	0.52	2
			0.2221		0.19	3

[a] $T = 25°C$. Act = acetone, Diox = dioxane, Me = methanol, D = dielectric constant, $E_1°$ and $E_2°$ respectively are the values of standard potential by the linear extrapolation and by the curve-fitting method. $E_1°$, dioxane (*61, 62, 63*); $E_1°$, acetone (*64*); $E_1°$, methanol (*65, 66*).

In our computations we found that second- or third-degree polynomials yielded values of standard potentials of sufficient accuracy for solvents of bulk dielectric constants higher than 40. For solvents with lower bulk dielectric constants, it was frequently necessary to use higher-degree (4th or 5th) polynomials for the computation of standard potentials. During the course of our work, we accumulated a large number of the coefficients of the polynomials for a wide variety of solvents; the coefficients obtained for water–tetrahydrofuran and water–monoglyme are compiled in Tables III and IV. The last column of these tables gives the values of the bulk dielectric constant of the mixed solvent at 25°C as obtained by interpolation of the linear plot $E_m°$ vs. $(1/DT)^{\frac{1}{2}}$ (Figure 4). It is not possible to make many comments regarding these coefficients at this time.

For the linear regions, the coefficients in both systems are of the same order of magnitude for nearly the same proportion of water in the mixture. This obviously implies that the H_2O molecules are the predominant species in the solvation sphere of the protons. But in the nonlinear region with a high percentage of organic component, the values of the coefficients for the two systems start to deviate significantly. This may imply different degrees and/or significantly different heats of solvation by the molecules of the organic component of the solvent. In that case, prediction regarding the value of the bulk dielectric constant may be in considerable error. However, the interpolated values for the linear region are perhaps very close to the actual values.

Once the standard potential of Cell I has been determined precisely, calculations of the mean activity coefficient, $\gamma_{\pm}$, of HCl and the primary and secondary medium effects using well-known relations are relatively simple tasks. Using empirical equations of the type $E = a + bT + cT^2$ and $E^\circ = a_0 + b_0T + c_0T^2$, it is possible to calculate the molal enthalpies and heat capacities. These types of calculations are demonstrated in many

Table III. The Coefficients of the Polynomial $E_{obs} + 2RRT/F \ln m = A_0 + A_1m^{1/2} + A_2m + A_3m^{3/2} + \ldots$ for Water–Tetrahydrofuran Mixtures[a]

T (°C)	A_0	A_1	A_2	A_3	Standard Error	Interpolated Value of D
			x = 8.98			
0	0.23467	0.055295	−0.060449	0	0.000246	
15	0.22160	0.060295	−0.056449	0	0.000191	
25	0.21368	0.062295	−0.053448	0	0.000267	72
35	0.20933	0.065395	−0.053450	0	0.000283	
			x = 18.21			
0	0.21662	0.056834	−0.054411	0	0.000227	
15	0.21062	0.059832	−0.051412	0	0.000376	
25	0.20375	0.062833	−0.047411	0	0.000306	67
35	0.19828	0.065864	−0.047667	0	0.000193	
			x = 47.20			
0	0.20365	0.082658	−0.075629	0	0.000520	
15	0.18662	0.090563	−0.092187	0	0.000494	
25	0.17060	0.095599	−0.113077	0	0.000403	48
35	0.16572	0.111072	−0.182976	0	0.000305	
			x = 73.03			
0	0.12600	0.353755	−0.592054	0	0.003224	
15	0.11163	0.354590	−0.516553	0	0.001500	
25	0.09310	0.407012	−0.624929	0	0.002459	
35	0.07330	0.434337	−0.615825	0	0.000526	
			x = 89.00			
0	0.03397	0.916603	−3.355034	0.328130	0.000687	
15	−0.00279	1.048794	−3.707654	0.591064	0.001346	
25	−0.02582	1.265785	−4.968096	0.719111	0.001142	
35	−0.05689	1.492791	−5.852357	0.880112	0.001220	

[a] The x = weight percent of tetrahydrofuran. $A_0 = E_m^\circ$.

Table IV. The Coefficients of the Polynomial $E_{obs} + 2RT/F \ln m = A_0 + A_1 m^{1/2} + A_2 m + A_3 m^{3/2} + \ldots$ for Water–Monoglyme Mixture[a]

T (°C)	A_0	A_1	A_2	A_3	Standard Error	Interpolated Value of D
			x = *8.68*			
0	0.22721	0.062099	−0.068714	0	0.00034	
15	0.22089	0.061487	−0.066973	0	0.00015	
25	0.21550	0.062737	−0.066091	0	0.00026	73
35	0.20889	0.061159	−0.058003	0	0.00031	
			x = *17.81*			
0	0.22415	0.047565	−0.031670	0	0.00025	
15	0.21581	0.057221	−0.050184	0	0.00025	
25	0.20903	0.065241	−0.068383	0	0.00020	69
35	0.20092	0.078325	−0.099941	0	0.00014	
			x = *46.52*			
0	0.20488	0.093269	−0.094175	0	0.00081	
15	0.19017	0.096545	−0.089622	0	0.00069	
25	0.17948	0.099162	−0.085895	0	0.00069	52
35	0.16855	0.098629	−0.078517	0	0.00097	
			x = *67.03*			
0	0.16368	0.171834	−0.174850	0	0.00075	
15	0.14161	0.193381	−0.205225	0	0.00082	
25	0.12604	0.219664	−0.261580	0	0.00031	
35	0.10907	0.244020	−0.309891	0	0.00025	
			x = *88.80*			
0	0.02502	0.814613	−2.612843	3.2842	0.00121	
15	−0.01125	0.826044	−2.445720	2.8735	0.00171	
25	−0.03095	0.849331	−2.466216	2.8993	0.00270	
35	−0.06039	0.979980	−3.081550	3.8657	0.00135	

[a] The x = weight percent of monoglyme. $A_0 = E_m{}^\circ$.

standard texts, and will not be pursued here. However, it should be pointed out that the constants a, b, c, a_0, b_0, and c_0 can be computed easily from a compilation of the polynomial coefficients.

We shall conclude this section after making a few brief comments regarding water–acetonitrile mixed-solvent systems. After spending a considerable amount of time and effort, we came to the conclusion that it is not possible to obtain reliable emf data at low hydrochloric acid concentration in this solvent system. Nitriles undergo two principal reactions:

$$CH_3 \cdot C \equiv N \xrightarrow{H_2O} CH_3 \cdot \overset{\overset{O}{||}}{C}\text{–}NH_2 \xrightarrow{H_2O} CH_3\text{–}\overset{\overset{O}{||}}{C}\text{–}OH + NH_3$$

However, in most circumstances, the first reaction is an intermediate step, and hydrolysis is both acid and base catalyzed. The relative rates of the two steps also depend on the structure of the substrate. Obviously, such hydrolysis of the solvent molecules will render emf data entirely unreliable. We also feel that the reliability of many rate and mechanistic experiments done in water–acetonitrile mixed solvents may be suspect.

Computation of the Ionization Constant of Weak Monoprotic Acids in Water–Organic Mixed Solvents

The work and results reported in Part I led us to believe that the first-order coulombic interactions among the ions and the ions and the solvent molecules are the significant interactions in solvents of bulk dielectric constant 40 or higher. This in turn led us to believe that a simple electrostatic model might be used for generating functions that would correlate the ionization processes in two solvents both of moderate bulk dielectric constants.

From a thermodynamic point of view, the ionization constants of a weak acid in two solvents are related to each other by primary medium effect. Thus,

$$Ka = \frac{{}_{S}{}_{W}\gamma_{oH^+} \cdot {}_{S}{}_{W}\gamma_{oA^-}}{{}_{S}{}_{W}\gamma_{oHA}} \cdot K_a^* = Q\gamma \cdot K_a^* \qquad (22)$$

in which the activity quotient term $Q\gamma$ represents the primary medium effect explicitly given by the second member of Equation 22. The primary and secondary medium effects and the relation between the two have been defined by Harned and Owen (*30*) in terms of transferring an ion from a solution of finite concentration in one solvent to a solution of finite concentration in another solvent. The transfer of the ion can be broken down into three imaginary steps:

(1) the transfer of an ion from a finite concentration in the first solvent to infinite dilution in the same solvent;

(2) the transfer of an ion at an infinite dilution in the first solvent to an infinite dilution in the second solvent; and

(3) the transfer of an ion at infinite dilution in the second solvent to a finite concentration in the second solvent.

Step 2 is defined as the primary medium effect and is obviously by definition the same as the free energy of transfer, ΔG_t°, of the ion from the infinitely dilute solution in one solvent to the infinitely dilute solution in another solvent.

Several theoretical models have been proposed (*31*) to rationalize the ionization process in a dielectric medium and to interpret the experimental data (*32*). Despite serious criticisms (*32*), the so-called Born electrostatic model, in the opinion of these authors, is perhaps still the best model. According to this model the free energy change for the ionization of a weak acid in a solvent of bulk dielectric constant ϵ is given by the equation

$$\Delta G^\circ = -RT \ln K_a = \Delta G^\circ{}_{nonel} + \frac{Ne^2}{2\epsilon}\left(\frac{1}{r_+} + \frac{1}{r_-}\right) \tag{23}$$

in which K_a is the ionization constant of the acid, r_+ and r_- are the radii of the proton and the anion respectively, and ϵ is the bulk dielectric constant of the solvent. Equation 23 can be refined further by expressing the quantity within the parentheses in terms of the process in the co-sphere region, and the process in the outer-sphere region. The quantity $Ne^2/2\epsilon\ (1/r_+ + 1/r_-)$ is commonly known as the free energy change owing to Born's solvation process. The free energy change of ionization of the acid in another solvent, mixed aqueous or pure, is given by

$$\Delta G^{\circ *} = -RT \ln K_a{}^* = \Delta G^{\circ *}{}_{nonel} + \frac{Ne^2}{2\epsilon^*}\left(\frac{1}{r_+} + \frac{1}{r_-}\right) \tag{24}$$

The primary medium effect may be expressed now by subtracting Equation 23 from Equation 24. Thus,

$$\Delta G_t{}^\circ = \Delta G^{\circ *} - \Delta G^\circ = -RT \ln K_a{}^* + RT \ln K_a = \Delta G_{t,}{}^\circ{}_{nonel} + \frac{Ne^2}{2}\left(\frac{1}{r_+} + \frac{1}{r_-}\right)\left(\frac{1}{\epsilon^*} - \frac{1}{\epsilon}\right) \tag{25}$$

Changing to common logarithm, the ionization constants of a weak acid in water and in a mixed-aqueous solvent are related by the expression

$$pK_a{}^* = pK_a + \frac{\Delta G_{t,}{}^\circ{}_{nonel}}{2.303RT} + \frac{Ne^2}{4.606RT}\left(\frac{1}{r_+} + \frac{1}{r_-}\right)\left(\frac{1}{\epsilon^*} - \frac{1}{\epsilon}\right) \tag{26}$$

In solvents of moderate-to-large dielectric constant, $\Delta G_{t,}{}^\circ{}_{nonel}$ can be neglected provided there are no quantum mechanical interactions and no secondary chemical processes; thus,

$$\Delta G_t{}^\circ = \frac{Ne^2}{2}\left(\frac{1}{r_+} + \frac{1}{r_-}\right)\left(\frac{1}{\epsilon^*} - \frac{1}{\epsilon}\right) \tag{27}$$

and

$$pK_a^* = pK_a + \frac{Ne^2}{4.606RT}\left(\frac{1}{r_+} + \frac{1}{r_-}\right)\left(\frac{1}{\epsilon^*} - \frac{1}{\epsilon}\right) \qquad (28)$$

Born's theoretical model has been criticized severely (*32*) because the plots and $\Delta G_t°$ vs. $(1/r_+)$ (for alkali metal ions) have negative slopes contrary to the requirement of Equation 27. We are not quite sure if such plots are legitimate, although in Equation 27 the electrical work involved in transferring the ions is given separately and explicitly: the experimental $\Delta G_t°$ refers to the transfer of the whole mole of the electrolyte. Secondly, the quantum mechanical interactions ($\Delta G°_{\text{nonel (ion)}}$) may no longer be neglected when individual ion–electrical work is being considered. In fact, Feakins (*32*) invoked quantum mechanical contributions in order to rationalize the negative slope of the $\Delta G_t°$ vs. $1/r_+$ plot. Equations 27 and 28 also do not include higher-order (outer-sphere) electrostatic terms. Finally, the strictly theoretical model calls for gaseous ionic radii which are different from crystallographic radii. On the other hand, solvated radii seem to give a good fit to Equations 27 and 28 in most cases. We ought to remember that Equations 27 and 28 are purely theoretical models, simple and primitive, but exceedingly useful as a starting point.

Our primary objective was to develop a computational technique which would correlate the ionization constant of a weak electrolyte (e.g., weak acid, ionic complexes) in water and the ionization constant of the same electrolyte in a mixed-aqueous solvent. Consideration of Equations 8, 22, and 28 suggested that plots of experimental pK_a^* vs. some linear combination of the reciprocals of bulk dielectric constants of the two solvents might yield the desirable functions. However, an acceptable plot should have the following properties: it should be continuous without any maximum or minimum; the plot should include the pK_a^* values of an acid for as many systems as possible; and the plot should be preferably linear. The empirical equation that fits this plot would be the function sought. Furthermore, the function should be analogous to some theoretical model so that a physical interpretation of the ionization process is still possible.

Through a series of parametrization, the most suitable linear combination was $(1/\epsilon^{**} - 1/\epsilon'')$ where $\epsilon^{**} = \epsilon^* + (\epsilon - \epsilon_{org})$, $\epsilon'' = 2\epsilon - \epsilon_{org}$, ϵ^* = bulk dielectric constant of the mixed solvent, ϵ_{org} = bulk dielectric constant of the pure organic component, and ϵ = bulk dielectric constant of water. (The dielectric constant data were obtained from literature; these references are not being cited in order to save space. Some of the ϵ^* values were obtained by these authors by interpolation. For the convenience of the reader we have compiled the values of bulk dielectric constants of a number of solvents frequently used in thermodynamic studies (cf. Table V). Some representative plots of pK_a^* vs. $(1/\epsilon^{**} - 1/\epsilon'')$ are shown in Figures 5 and 6. Many of the values of the ionization constants of the acids were obtained from literature (*33–48*), several of them were determined, and a large number of them rechecked by the authors (*49, 50*) using the cell

$$\text{Pt, H}_2\text{(1 atm)}\,|\,\text{HA}(m_1),\ \text{NaA}(m_2),\ \text{NaCl}(m_3),\ \text{organic}\ (x),\ \text{water}\ (y)\,|\,\text{AgCl–Ag} \qquad \text{II.}$$

For all systems investigated, except for the dimethyl sulfoxide–water system, the pK_a^* vs. $(1/\epsilon^{**} - 1/\epsilon'')$ plots for the entire range of dielec-

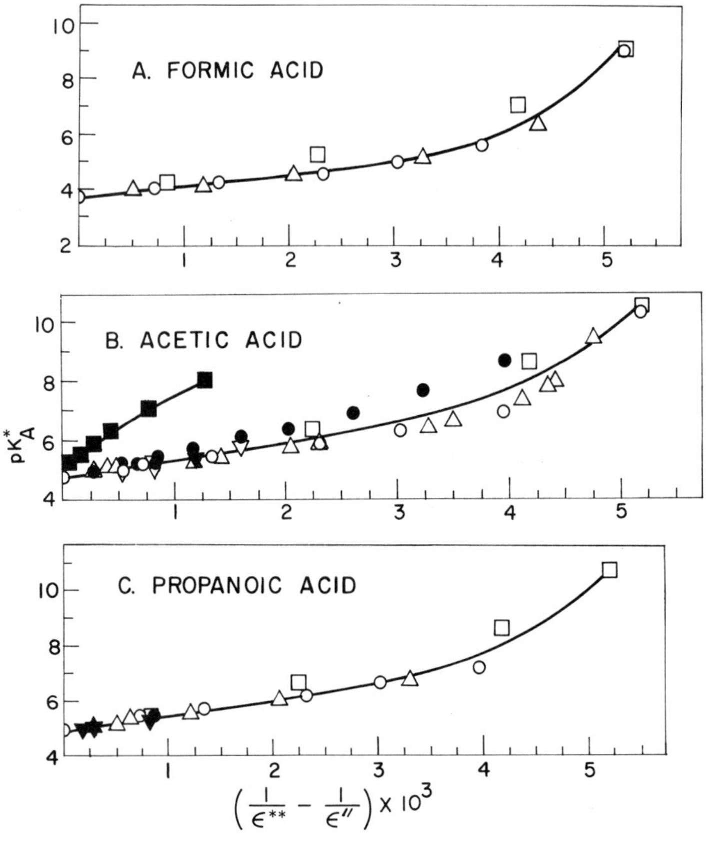

Figure 5. Ionization constants of acids in various solvents vs. the dielectric constant function at 25°C. A,B,C: ethanol–water (○), methanol–water (△), dioxane–water (□). B,C: acetone–water (●), glycerol–water (▼). B: 2-methoxyethanol–water (▽), dimethyl-sulfoxide–water (■).

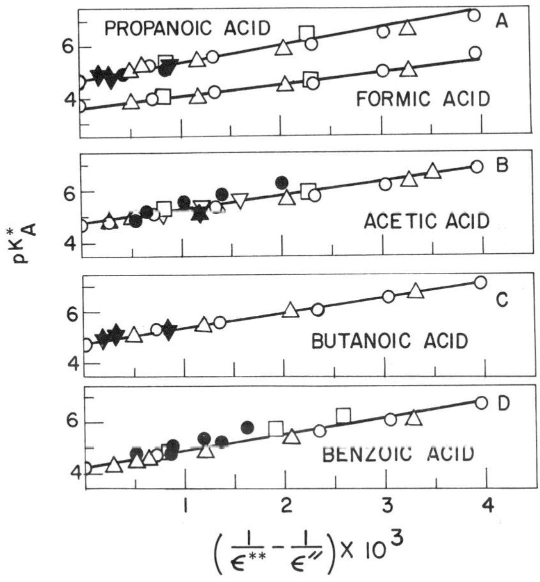

Figure 6. The linear range of the ionization constants of the acids vs. the dielectric constant function at 25°C. ethanol–water (○); methanol–water (△); 2-methoxyethanol–water; dioxane–water (□); acetone–water, glycerol–water, 2-propanol–water (●). Points owing to tetrahydrofuran–water, 2-ethoxyethanol–water, and 1,2-dimethoxyethane–water systems are not shown on the plots; they all lie on the linear plots.

tric function (0 to 0.005; corresponding range of bulk dielectric constant of the mixed solvent 78.5 − 9.5) yielded smooth curves (Figure 5) which were slightly concave upwards in the low-dielectric region (34 or less). These plots could be fitted to the equation

$$pK_a^* = pK_a + b\left(\frac{1}{\epsilon^{**}} - \frac{1}{\epsilon''}\right) + 10^{\left[c+d\left(\frac{1}{\epsilon^{**}} - \frac{1}{\epsilon''}\right)\right]} \qquad (29)$$

where pK_a^* = experimental ionization constant of the acid in the mixed solvent, pK_a = the intercept of the plot on pK_a^* axis, that is the computed value of the ionization constant of the acid in water. The empirical constants b, c, and d were obtained by a curve-fitting program requiring 90–99% goodness of fit. The experimentally determined values of pK_a^* in various solvents are compiled in Tables VIA-VIE; the experimentally determined and the computed values of pK_a, and the values of the constants b, c, and d are compiled in Table VII. The dimethyl sulfoxide–water system is the only system among those studied for which the

Table V. Bulk Dielectric Constants of Some Pure Solvents at 25°C

Compound	*Dielectric Constant*
Water	78.5
Propylene carbonate	64.4
Dimethyl sulfoxide	46.7
1,2-Ethanediol (ethylene glycol, glycol)	40.7
Glycerol	40.1
Dimethyl formamide	36.7
Acetonitrile	36.0
2-Ethoxymethanol	33.6
Methanol	31.5
Ethanol	24.3
1-Propanol	20.3
2-Propanol	19.4
Acetone	19.1
1-Butanol	17.5
2-Methoxyethanol	17.2
tert-Butanol (30°C)	11.6
Tetrahydrofuran	7.6
1,2-Dimethoxyethane	6.8
1,4-Dioxane	2.1

Table VIA. Values of Experimental pK_a*[a] (Molal Scale)

Wt % ethanol	0	8.04	16.26
Dielectric constant	78.54	73.5	69.2
pK_a* of			
Formic	3.75		
Acetic	4.76	4.81	4.90
Propanoic	4.87		
Benzoic	4.20		
Butanoic	4.82		
3-Methylbutanoic	4.78		
Chloroacetic	2.86		
Cyanoacetic	2.47		
Glycolic	3.83		
Lactic	3.86		
Malonic	2.75		
Succinic	4.13		
Glutaric	4.34		
Salicylic	3.00		

[a] Note on Tables VIA–VIE. pK_a* values and dielectric data are taken from various literature sources. Several pK_a* values are from our work, and many literature values were rechecked.

ionization constant data did not fall in line with the other plots, and these data could not be fitted into Equation 29. Dielectric constant data used by Morel (*43*) seemed to be particularly doubtful; both these and the ionization constant data should have been checked independently. It was deemed desirable to investigate the range of linearity of the function, $pK_a^* = f(/\epsilon^{**} - 1/\epsilon'')$, and accordingly the data was replotted to fit the equation

$$pK_a^* = pK_a + b'(1/\epsilon^{**} - 1/\epsilon'') \quad (30)$$

with goodness of fit between 90 and 99%. These plots are shown in Figures 6 and 7, and corresponding numerical values are compiled in Table VII. It became obvious from the excellent linearity of these plots and the goodness of fit to Equation 30 that the much simpler Born Equation should accommodate the data in the linear region. Accordingly pK_a^* data were replotted against $(1/\epsilon^* - 1/\epsilon)$, and excellent linear plots which fitted the equation

$$pK_a^* = pK_a + \beta\left(\frac{1}{\epsilon^*} - \frac{1}{\epsilon}\right) \quad (31)$$

were obtained (cf. Table VII).

of Various Acids in Ethanol–Water at 25°C[a]

20.30	34.90	50.10	65.10	79.90	100
66.8	58.3	48.9	40.4	32.8	24.3
4.02	4.24	4.60	5.01	5.64	9.15
5.13	5.43	5.84	6.29	6.87	10.32
5.33	5.68	6.13	6.63	7.17	
4.77	5.24	5.76	6.19	6.79	10.25
5.31	5.70	6.15	6.65	7.22	
5.29	5.75	6.22	6.76	7.35	
3.26	3.57	3.95	4.41	4.98	
2.78	3.07	3.39	3.81	4.39	
4.21	4.51	4.86	5.27	5.77	
4.14	4.44	4.80	5.26	5.77	
3.14	3.38	3.65	3.94	4.41	
4.54	4.86	5.21	5.64	6.16	
4.64	5.00	5.42	5.96	6.53	
3.23	3.62	3.99	4.46	5.03	

Table VIB. Values of Experimental pK_a^* (Molal Scale)

Wt % methanol	10.1	16.47	20.01	34.47	40.02	54.02
Dielectric constant	74.1	71.0	69.2	62.2	59.6	52.8
pK_a^* of						
Formic	3.91			4.14		4.56
Acetic	4.91	5.00	5.08	5.31	5.45	5.78
Propanoic	5.02	5.14	5.33	5.55		6.01
Benzoic	4.38	4.50	4.72	4.95		5.50
Butanoic		5.12		5.60		6.08

[a] *See* footnote, Table VIA.

Table VIC. Values of Experimental pK_a^* (Molal Scale)

Wt % acetone	10	20	25
Dielectric constant	73.02	66.98	63.70
pK_a^* of			
Acetic	4.90	5.09	5.22
Propanoic	4.98		5.40
Benzoic	4.44		4.96

[a] *See* footnote, Table VIA.

Table VID. Values of Experimental pK_a^* (Molal Scale)

Wt % dioxane	20	30	40
Dielectric constant	60.79	51.90	42.98
pK_a^* of			
Formic	4.18		
Acetic	5.92		
Propanoic	5.47		
Benzoic	4.87	5.28	5.79
Wt % 2-propanol	5	10	20
Dielectric constant	74.9	71.4	64.1
pK_a^* of			
Propanoic	4.98	5.09	5.33
Butanoic	4.95	5.05	5.34
Wt % Dimethyl Sulfoxide	10	20	30
Dielectric constant	78.2	77.9	77.2
pK_a^* of			
Acetic	4.89	5.04	5.26
Wt % glycerol	50		
Dielectric constant	64.0		
pK_a^* of			
Acetic	5.27		

[a] *See* footnote, Table VIA.

of Various Acids in Methanol–Water at 25°C[a]

60.05	75.94	80.03	90.02	93.47	95.02	99.99
50.1	41.9	40.1	35.7	34.1	33.6	31.5
	5.22			6.45		
5.90	6.45	6.64	7.31	7.77	8.00	9.52
	6.75					
	6.32			7.38		9.28
	6.82					

of Various Acids in Acetone–Water at 25°C[a]

30	40	50	60	70	80
61.04	54.60	48.22	41.80	35.70	29.62
5.61	5.83	6.41	6.96	7.64	8.78

of Various Acids in Several Organic–Water Solvents at 25°C[a]

45	50	70	82
38.48	34.36	17.69	9.53
5.29		7.02	9.14
6.31		8.61	10.51
6.55		8.61	10.75
	6.38		

40	50	60	70	80
76.4	75.2	73.3	69.5	64.7
5.48	5.84	6.31	7.03	8.06

Table VIE. Values of Experimental pK_a^* (Molal Scale) of Various Acids in Several Organic–Water Solvents at 25°C[a]

Vol % 2-Methoxyethanol	20	30	40	50	
Dielectric constant	68.90	63.93	58.49	52.81	
pK_a^* of					
Acetic	5.07	5.27	5.49	5.81	
Benzoic	4.60	4.86	5.16	5.52	
Vol % 1,2-Dimethoxyethane	20	30	40	50	
Dielectric constant	66.87	60.90	54.47	47.38	
pK_a^* of					
Acetic	5.13	5.37	5.65	6.05	
Benzoic	4.71	5.04	5.41	5.88	
Vol % 2-Ethoxyethanol	20	30	40	50	
Dielectric constant	68.37	62.69	56.63	49.83	
pK_a^* of					
Acetic	4.93	5.09	5.29	5.42	
Benzoic	4.49	4.78	5.07	5.34	
Vol % Tetrahydrofuran	10	20	30	40	50
Dielectric constant	73.40	65.70	58.83	50.82	42.58
pK_a^* of					
Acetic	4.89	5.14	5.39	5.70	6.08
Benzoic		4.87	5.33	5.78	6.23

[a] *See* footnote, Table VIA.

A legitimate question to ask is if there is any justification to curve-fitting computational technique aside from just fitting the experimental data. We wanted to investigate the significance of the slope β of Equation 31 (analogous to Born's Equation) by equating

$$\beta = \frac{Ne^2}{4.606RT}\left(\frac{1}{r_+} + \frac{1}{r_-}\right) \tag{32}$$

(where the constants are to be expressed in e.s. units; $e = 4.80 \times 10^{-10}$ esu, $R = 8.31 \times 10^7$ erg deg^{-1} mol^{-1}, $N = 6.02 \times 10^{23}$ mol^{-1}, $T =$ K). Theoretically, r_+ and r_- refer to gaseous ionic radii, and the right-hand member of Equation 32 represents the coulombic solvation work when multiplied by the reciprocal of the bulk dielectric constant of the solvent. However, the right-hand member totally neglects all noncoulombic work, and that is unrealistic. If we arbitrarily assume that r_+ and r_- in Equation

32 represent the solvated ionic radii, then knowing either r_+ or r_-, the other might be calculated.

There is considerable controversy over the value of the radius of the hydrated proton (*51–57*), and it has been reported from 0.945–4.5 Å. However, overwhelming experimental as well as theoretical work seems to suggest that the hydrated proton species is $H_9O_4^+$ with a radius of 3.0 Å. Hydrated anionic radii (r_-) were calculated using this value of the hydrated proton, and these are compiled in Table VIII. What do these radii mean? The term $(1/r_+ + 1/r_-)$ in Equation 32 really implies the sum of reciprocals of the minimum distances of approach of the two ions or some function of it. In the Dennison and Ramsey (*58*) treatment of the Born model this term is $1/r$, where r is the minimum distance of approach of the two ions. Implicit in the expression $(1/r_+ + 1/r_-)$ is the assumption that both ions are monatomic and spherical; therefore, the minimum distance of approach of the two ions may be apportioned as the hydronium ion radius and the anion radius. Although it might be possible to assign a value to the hydronium ion radius, the anionic radius

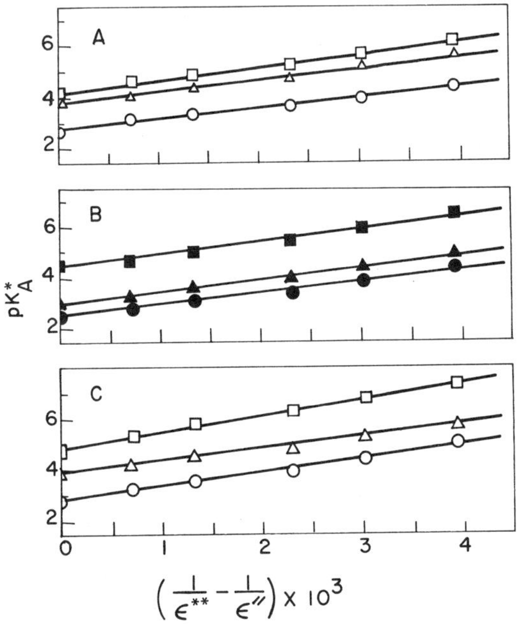

Figure 7. Ionization constants of acids in ethanol–water mixtures vs. the dielectric constant function at 25°C. A: Malonic acid (○), lactic acid (△), succinic acid (□). (B) Cyanoacetic acid (●), salicyclic acid (▲), glutaric acid (■). C: Chloroacetic acid (○), glycolic acid (△), isovaleric acid (□).

Table VII. Acids Studied, Calculated Constants of Empirical Solvents

Acid	pK_a[a] *Experimental*	*Eq. 29*	*Eq. 30*	*Eq. 31*	*No. of Composition of Each Mixed-Solvent System Studied Organic Component*
Formic	3.75	3.75	3.70	3.78	5 (MeOH)
					6 (EtOH)
					4 (Dioxane)
Acetic	4.76	4.76	4.76	4.87	13 (MeOH)
					8 (EtOH)
					4 (Dioxane)
					13 (Acetone)
					4 (Methyloxyethanol)
Propanoic	4.87	4.88	4.83	4.98	6 (MeOH)
					5 (EtOH)
					4 (Dioxane)
					2 (Acetone)
					3 (2-PrOH)
Butanoic	4.82	4.85	4.82	5.00	4 (MeOH)
					5 (EtOH)
					3 (2-PrOH)
3-Methylbutanoic (Isovaleric)	4.78		4.78	4.98	5 (EtOH)
Cyanoacetic	2.47		2.42	2.47	5 (EtOH)
Chloroacetic	2.86		2.86	2.98	5 (EtOH)
Glycolic	3.83		3.83	3.96	5 (EtOH)
Lactic	3.86		3.80	3.86	5 (EtOH)
Malonic	2.75		2.75	2.90	5 (EtOH)
Succinic	4.13		4.13	4.28	5 (EtOH)
Glutaric	4.34		4.26	4.40	5 (EtOH)
Benzoic	4.20	4.27	4.38	4.45	8 (MeOH)
					6 (EtOH)
					4 (Dioxane)
					6 (Acetone)
					4 (2-Methoxyethanol)
Salicyclic	3.00		2.91	3.04	5 (EtOH)

[a] First ionization constants for dicarboxylic acids.

could be quite misleading. In fact, Robinson and Stokes (52) reported that the carboxylate ion radius is of the order of 1.2 Å. Formate, benzoate, and glycolate cannot have the same radius, nor are they spherical. Such a radius would include a carbon chain "tail" (which is weakly hydrated) as well as the carboxylate "head" which is certainly strongly hydrated. This anionic radius, r_-, really is the thickness of the water sheath between the negative end of the anion and the outer periphery of the hydrated proton. Perhaps it is more realistic to report r rather than $(r_+ + r_-)$,

Equations, and Dielectric Constant Range of the Mixed Used (25°C)

Range of Dielectric Constant	*Eq. 29* b	c	d	*Eq. 30* b′	*Eq. 31* β
34–71	321	−1.80	463	477	110
24–67					
10–61					
31.5–74	530	−1.70	423	530	125
24–73.5					
10–61					
30–73					
53–69					
42–74	508	−2.00	484	676	128
33–67					
10–61					
64–73					
64–75					
42–71	482	−2.08	500	482	140
33–67					
64–75					
33–67				640	141
33–67				471	113
33–67				530	116
33–67				480	106
33–67				480	122
33–67				420	88
33–67				500	111
33–67				552	114
33–74	631	−2.80	620	627	152
24–67					
34–61					
53–73					
53–69					
33–67				514	115

and to report r_- as the anionic contribution to the minimum distance of approach. It is interesting to note that the values of minimum distance approach computed by us for the carboxylate ions are of the same order of magnitude as the minimum distance of approach obtained by Harned (59) for the halide ions (cf. Table VIII). We believe that the preceding discussion justifies the use of Equations 29, 30, and 31 for the computation of pK_a or pK_a^*, and that the experimental slope β rather than $(Ne^2/4.606RT)$ is the measure of the solvation energy. The values of β,

Table VIII. Values of Slope, β (Eq. 31), the Anionic Contribution to Minimum Distance of Approach (r_-) and the Minimum Distance Approach (r) for a Number of Acids (25°C)[a]

Acid	pK_a[b]	β	r_- (Å)	r (Å)
Formic	3.75	110	1.3	4.3
Acetic	4.76	125	1.1	4.1
Propanoic	4.87	128	1.1	4.1
Butanoic	4.82	140	1.0	4.0
3-Methylbutanoic	4.78	141	1.0	4.0
Cyanoacetic	2.47	113	1.2	4.2
Chloroacetic	2.86	116	1.2	4.2
Glycolic	3.83	106	1.3	4.3
Lactic	3.86	122	1.1	4.1
Malonic	2.75	88	1.7	4.7
Succinic	4.13	111	1.2	4.2
Glutaric	4.34	114	1.2	4.2
Benzoic	4.20	152	0.9	3.9
Salicylic	3.00	115	1.2	4.2
Hydrochloric				4.0
Hydrobromic				4.4
Hydroiodic				5.0

[a] The $r_+ = 3.0$ A.
[b] First ionization constants for dicarboxylic acids.

although of the same order of magnitude, will be different for different acids as they include small but significant contributions owing to noncoulombic interactions even in solvents of high-bulk dielectric constants (cf. Table VII). Too many factors such as the dipole moments, polarizability, structure, Lewis acidity and basicity of the solvent molecule, and long-range quantum mechanical interactions are involved in determining the values of β. Therefore, it will be futile to attempt a purely theoretical computation of β regardless of the sophistication of the model. As the value of the bulk dielectric constant becomes less than a certain critical value ($\epsilon^* < 40$), the contribution of noncoulombic interactions increases rapidly and Born's model becomes inadequate.

A few comments on the values of r_- and ($r_+ + r_-$) (cf. Table VIII) are necessary. Increasing hydrocarbon content decreases the hydrophilic property of the anion (cf. formic, acitic, propanoic, butanoic, 3-methylbutanoic, and benzoic acids) resulting in the decrease of hydration; the trend eventually levels off. Hydrophilic substitution increases hydration (cf. acetic, chloroacetic, cyanoacetic, glycolic acids; cf. glutaric, succinic, and malonic acids; cf. benzoic and salicyclic acids). Also note that r_- is smallest for benzoic acid and largest for malonic acid. These trends cannot be fortuitous.

Finally, it seems pertinent to make some comments on the trends of values of the constants b, b′, and β (cf. Table VIII) of the empirical

Equations 29, 30, and 31. For acids of the same class both b′ and β generally follow the trend of pK_a; they are influenced significantly by the substituent group as might be expected. No such trend is observed for the values of b. In computing b, the program did not attach any special weight to the points in the linear regions of the plots, hence b values are weighted by points in the curved region of the plots, whereas b′ and β values were computed by using data from the graphically selected linear region of the plots. It is interesting to note that by assuming r_- to be 1.2 Å, theoretical β value turns out to be 141 which is definitely of the same order values reported for β in Table VII. Small but significant deviations are obviously caused by noncoulombic interactions. One important difference between Equations 30 and 31 should be noted; Equation 30 is explicitly referenced to water, whereas in principle, Equation 31 is applicable for any two solvents.

Curve-fitting computation technique was applied for the calculation of formation constants of calcium lactate in methanol–water, ethanol–water, and glucose–water systems (*60*) with excellent results. In the calculation of the average formation constant $K_{av}(=(K_1 \cdot K_2)^{\frac{1}{2}})$, β turned out to be about 190 for the methanol–water system, 192 for the ethanol–water system, and 185 for the glucose–water system. The value of b′ was 204.

We believe that a compilation of b′ and β values for acids will enable one to calculate either pK_a or $pK_a{}^*$ knowing one and the solvent dielectric, which in turn may be determined graphically from $E_m{}^\circ$ vs. $(1/\sqrt{DT})$ plots for Cell I. It should also be possible to classify the acids on the basis of b′ or β values. These comments should apply also to weak ionic complexes.

Glossary of Symbols

A = a characteristic constant as defined by Equation 7

A, A′, A″ = constants of the polynomial

B, B′, B″ = constants of the polynomial

B′ = a constant of DH extended theory equation

C, C′, C″ = constants of the polynomial

c = molarity

D = bulk dielectric constant of the solvent in the first part of the chapter

DH = Debye–Hückel

E° = standard potential of the particular cell

E_{obs} = emf of the particular cell when partial pressure of hydrogen is one atmosphere

e = electronic charge in the second part of the chapter

F = Faraday

ΔG°, ΔG_t°, $\Delta G_t^{\circ *}$ = standard free energy change; asterisk indicates that the solvent contains an organic component or it may be nonaqueous

GLS = Gronwall, LaMer, and Sandved

k = Boltzmann constant

m = molality

N = Avogadro's number

pKa = ionization constant of an acid in pure water

pKa* = ionization constant of an acid in an organic–water mixed solvent or in a nonaqueous solvent

R = gas constant

$S_{(f)}$ = limiting theoretical slope of rational activity coefficient in interionic attraction theory; a function of γ, D, and T as expressed by Equation 5

T = temperature in Kelvin

X, Y = function of DH extended theory equation

z = valence of an ion

Γ = ional strength, $\sum_{1}^{s} C_i Z_i^2$

γ_t = mean molal activity coefficient

γ_o = mean molal activity coefficient at infinite dilution in the solvent indicated

ϵ = electronic charge in the first part of the chapter

ϵ, ϵ^* = bulk dielectric constant of the solvent in the second part of the chapter

ν = total number of ions from a single electrolyte

ν_i = number of ions of a kind

All other symbols are clearly defined where they appear in the text.

Literature Cited

1. Hogfeldt, E., Martell, A. E., "Stability Constants," Supplement No. 1, Special Publication No. **25**, The Chemical Society, London, 1971.
2. Amis, E. S., Hinton, J. F., "Solvent Effects on Chemical Phenomena," Vol. 1, Chap. 4, Academic, New York, 1973.
3. Harned, H. S., Owen, B. B., "The Physical Chemistry of Electrolytic Solutions," 3rd ed., Chap. 11, Reinhold, New York, 1958.
4. Gronwall, T. H., LaMer, V. K., Sandved, K., *Phys. Z.* (1928) **29**, 358.
5. Harned, H. S., Owen, B. B., "The Physical Chemistry of Electrolytic Solutions," 3rd ed., Chaps. 3, 4, 11, Reinhold, New York, 1958.
6. Lewis, G. N., Randall, M., "Thermodynamics," Revised by K. S. Pitzer, L. Brewer, McGraw-Hill, New York, 1961.
7. Harned, H. S., Owen, B. B., "The Physical Chemistry of Electrolytic Solutions," 3rd ed., Chaps. 5, 10, Reinhold, New York, 1958.

8. Gurney, R. W., "Ions in Solution," Dover, New York, 1962.
9. Gurney, R. W., "Ionic Processes in Solution," Dover, New York, 1962.
10. Roy, R. N., "The Activity and Other Thermodynamic Properties of Hydrochloric Acid in Tetrahydrofuran-Water Mixtures," Ph.D. Dissertation, Louisiana State University, 1966.
11. Johnson, D. A., "The Activity and Other Thermodynamic Properties of Hydrochloric Acid in Monoglyme-Water Mixtures," Ph.D. Dissertation, Louisiana State University, 1966.
12. Johnson, D. A., Sen, B., *J. Chem. Eng. Data* (1968) **13**, 376.
13. Roy, R. N., Sen, B., *J. Chem. Eng. Data* (1967) **12**, 584.
14. Roy, R. N., Vernon, W., Bothwell, A. L. M., *J. Chem. Soc. A* (1971) 1242.
15. Roy, R. N., Vernon, W., Bothwell, A. L. M., Gibbons, J., *J. Electrochem. Soc.* (1972) **119**, 694.
16. Roy, R. N., Vernon, W., Gibbons, J. J., Bothwell, A. L. M., *J. Electroanal. Chem.* (1972) **34**, 345.
17. Roy, R. N., Bothwell, A. L. M., *J. Chem. Thermodyn.* (1971) **3**, 769.
18. Roy, R. N., Vernon, W., Bothwell, A. L. M., *J. Chem. Soc. B* (1971) 2320.
19. Roy, R. N., Gibbons, J. J., Bothwell, A. L. M., *J. Electroanal. Chem.* (1972) **34**, 101.
20. Roy, R. N., Vernon, W., Bothwell, A. L. M., *J. Electroanaly. Chem.* (1971) **30**, 335.
21. Roy, R. N., Vernon, W., Gibbons, J. J., Bothwell, A. L. M., *J. Chem. Soc. A* (1971) 3589.
22. Roy, R. N., Vernon, W., Bothwell, A. L. M., *J. Chem. Soc., Faraday Trans. 1* (1972) **68**, 2047.
23. Roy, R. N., Vernon, W., Bothwell, A. L. M., *J. Chem. Thermodyn.* (1971) **3**, 883.
24. Roy, R. N., Bothwell, A. L. M., Gibbons, J. J., Vernon, W., *J. Chem. Soc., Dalton Trans.* (1972) 530.
25. Roy, R. N., Vernon, W., Bothwell, A. L. M., *Electrochim. Acta* (1972) **17**, 1057.
26. Ibid. (1972) **17**, 5.
27. Ibid. (1973) **18**, 81.
28. Roy, R. N., Vernon, W., Bothwell, A. L. M., Gibbons, J. J., *J. Chem. Eng. Data* (1972) **17**, 89.
29. Bates, R. G., Bowen, V. E., *J. Res. Natl. Bur. Stand.* (1954) **53**, 283.
30. Harned, H. S., Owen, B. B., "The Physical Chemistry of Electrolytic Solutions," 3rd ed., Chap. 15, Reinhold, New York, 1958.
31. Wiberg, K. B., "Physical Organic Chemistry," John Wiley, New York, 1964.
32. Feakins, D., "Physico-Chemical Processes in Mixed Aqueous Solvents," F. Franks, Ed., American Elsevier, New York, 1969.
33. Harned, H. S., et al., *J. Am. Chem. Soc.* (1936) **58**, 1912.
34. Ibid. (1939) **61**, 2374.
35. Ibid. (1941) **63**, 2579.
36. Harned, H. S., *J. Phys. Chem.* (1939) **43**, 275.
37. Dunsmore, H. S., Speakman, J. C., *Trans. Faraday Soc.* (1954) **50**, 236.
38. Grunwald, E., Berkowitz, B. J., *J. Am. Chem. Soc.* (1951) **73**, 4939.
39. Glover, D. J., *J. Am. Chem. Soc.* (1965) **87**, 5279.
40. Shedlovsky, T., Kay, R. L., *J. Phys. Chem.* (1956) **60**, 151.
41. Patterson, A., Felsing, W. A., *J. Am. Chem. Soc.* (1942) **64**, 1480.
42. Parton, H. N., Rogers, J., *Trans. Faraday Soc.* (1942) **38**, 238.
43. Morel, J. P., *Bull. Soc. Chim. Fr.* (1966) 2112.
44. Dippy, J. F. J., Hughes, S. R. C., Rozanski, A., *J. Chem. Soc.* (1959) 1442.
45. Reynaud, R., *C. R. Hebd. Seances Acad. Sci.*, (1966) **263C**, 105.
46. Feakins, D., French, C. M., *Chem. Ind.* (1954) 1107.
47. Moore, R. L., Felsing, W. A., *J. Am. Chem. Soc.* (1947) **69**, 2420.
48. Felsing, W. A., May, M., *J. Am. Chem. Soc.* (1948) **70**, 2904.

49. Adcock, H. L., "The Determination of the Ionization Constants of Weak Acids in Mixed Solvent Systems. A Study of Medium Effects," Ph.D. Dissertation, Louisiana State University, 1970.
50. Sen, B., Adcock, H. L., *Anal. Chim. Acta* (1970) **50,** 287.
51. Kielland, J., *J. Am. Chem. Soc.* (1937) **59,** 1675.
52. Robinson, R. A., Stokes, R. H., "Electrolytic Solutions," Academic, New York, 1959.
53. Tuck, D. G., Diamond, R. M., *Proc. Chem. Soc., London* (1958) 236.
54. Glueckauf, E., *Trans. Faraday Soc.* (1955) **51,** 1235.
55. Conway, B. E., Bockris, J. O'M., Linton, H., *J. Chem. Phys.* (1956) **24,** 834.
56. Sienko, J. M., Plane, R. A., Hestes, R. E., "Inorganic Chemistry," W. A. Benjamin, New York, 1965.
57. Wicke, E., Eigen, M., Ackermann, T., *Z. Phys. Chem. (New Series)* (1954) **1,** 340.
58. Dennison, J. T., Ramsey, J. B., *J. Am. Chem. Soc.* (1955) **77,** 2615.
59. Harned, H. S., Owen, B. B., "The Physical Chemistry of Electrolyte Solutions," 3rd ed., Chap. 12, Reinhold, New York, 1958.
60. Sen, B., Gibbons, J. J., *J. Chem. Eng. Data* (1977) **22,** 309.
61. Harned, H. S., et al., *J. Am. Chem. Soc.* (1938) **60,** 2130.
62. Ibid. (1936) **58,** 1908.
63. Ibid. (1939) **61,** 44.
64. Feakins, D., French, C. M., *J. Chem. Soc.* (1956) 3168.
65. Harned, H. S., Thomas, H. C., *J. Am. Chem. Soc.* (1936) **58,** 761.
66. Oiwa, I. T., *J. Phys. Chem.* (1956) **60,** 754.

RECEIVED February 13, 1978.

16

Thermodynamic Behavior of Hydrobromic Acid in Mixtures of Water and *N*-Methylacetamide from 5° to 45°C

RICHARD A. BUTLER and ROGER G. BATES

Department of Chemistry, University of Florida, Gainesville, FL 32611

The standard potential of the silver–silver bromide electrode has been determined from emf measurements of cells with hydrogen electrodes and silver-silver bromide electrodes in solutions of hydrogen bromide in mixtures of water and N-*methylacetamide (NMA). The mole fractions of NMA in the mixed solvents were 0.06, 0.15, 0.25, and 0.50, and the dielectric constants varied from 87 to 110 at 25°C. The molality of HBr covered the range 0.01–0.1 mol kg^{-1}. Data for the mixed solvents were obtained at nine temperatures from 5° to 45°C. The results were used to derive the standard emf of the cell as well as the mean ionic activity coefficients and standard thermodynamic constants for HBr. The information obtained sheds some light on the nature of ion–ion and ion–solvent interactions in this system of high dielectric constant.*

Much effort has been expended in attempting to elucidate the nature of the solute–solvent interactions that are responsible for the observed properties of solutions of ionic solutes. Because of its wide use as a solvent by both man and nature, water has been the solvent in the majority of such studies. The unique properties of water as a solvent, however, have made it difficult to extend knowledge of solute behavior, observed in aqueous solutions, to an understanding of the behavior of the same solutes in other media.

Most of the studies in solvents other than water have dealt with media of fairly low dielectric constant, mainly because most of the

0-8412-0428-4/79/33-177-249$05.00/1

common nonaqueous solvents have dielectric constants lower than that of water. Thermodynamic studies of solvent systems of higher dielectric constant are rare, presumably because of difficulties in obtaining and maintaining the solvents in a state of sufficient purity and in finding electrodes which are stable and reversible in these solvents.

As a solvent for electrolytes, however, considerable interest has been shown in *N*-methylacetamide (NMA) (*1, 2, 3, 4, 5*). This solvent is relatively easy to handle, and several electrodes have proven stable in its solutions. NMA has a dielectric constant of 165.5 at 25°C, or more than twice that of water. Although NMA is a solid below 30.5°C, it dissolves readily in water, giving a series of solvents with dielectric constants ranging from 78 to 110 at 25°C without exceeding an NMA mole fraction of 0.50.

In the preliminary stages of this study, some difficulty was experienced in obtaining stable potentials with the silver–silver chloride electrode in H_2O/NMA mixtures at a mole fraction of NMA of 0.5. However, the emf of cells with platinized platinum–hydrogen electrodes and silver–silver bromide electrodes reached values which were constant over a period of at least 4 to 5 hr, making this cell appear suitable for a study of the thermodynamics of HBr in H_2O/NMA mixtures with mole fraction NMA (x_2) up to at least 0.50.

Electromotive force measurements of HCl solutions in pure NMA and in NMA/dioxane solvent mixtures using the silver–silver chloride electrode have been reported by Dawson and his co-workers (*1, 2, 3*). The only other potentiometric studies in a solvent of dielectric constant higher than that of water appear to have been in formamide (*6, 7, 8, 9, 10*) and in *N*-methylpropionamide (NMP) (*11, 12, 13, 14, 15*).

We now have measured the emf of the cell without liquid junction

$$\text{Pt; } H_2 \text{ (g, 1 atm)} \mid \text{HBr } (m) \text{ in } H_2O/\text{NMA} \mid \text{AgBr; Ag} \qquad \text{(A)}$$

at NMA mole fractions (x_2) of 0.06, 0.15, 0.25, and 0.50. The standard emf of cell A (which is the standard potential of the silver–silver bromide electrode) has been determined for each of the mixed solvents at intervals of 5°C from 5°–45°C. From these data, the activity coefficient of HBr and the standard Gibbs energy,, enthalpy, and entropy changes on transfer of HBr from water to each solvent mixture were calculated.

Experimental

NMA was obtained from Eastman Kodak Co., fractionally crystallized once, and then continuously zone-refined 10–20 times. The zone refiner had three purification zones; the product from the first zone was used, while the one from the last two was discarded. The material so obtained had a specific conductivity of 1–5 $\times 10^{-7}$ S cm^{-1}.

The stock solution of HBr was prepared from HBr obtained by double distillation of commercial 48% HBr solution. In each distillation the first and third fractions were discarded. The product was diluted to a molality (m) of $\sim$ 1.0 mol kg^{-1} and the concentration determined to within $\pm$ 0.01% by weighing silver bromide.

Solutions for the cell measurements were made as follows: the approximate amount of NMA needed was weighed into a flask and the desired amount of HBr solution added from a weight buret. The additional amount of water needed to obtain the exact solvent composition then was calculated and added carefully. Distilled de-ionized water was used in preparing the solutions.

The cells were of the type used extensively in this laboratory for earlier studies. The essential features of the cell design have been set forth elsewhere (*16*). The emf was measured with a Hewlett–Packard digital voltmeter, Model 3460, checked occasionally against laboratory standards consisting of two saturated Weston cells maintained at a controlled temperature.

The silver–silver bromide electrodes were of the thermal type, made by heating a paste composed of 90 wt % silver oxide and 10% silver bromate, with a little water, to 600°C on a helix of platinum wire. In order to avoid solvent occlusion, the electrodes were not pretreated before use in the cells. After each run, the electrodes were cleaned and remade before immersion in a solution of a different composition.

The bases for the hydrogen electrodes were of platinum foil; they were platinized in a 2% solution of chloroplatinic acid in 1*M* HCl. The plating solution contained no lead. The finished electrodes were stored in water and rinsed with pure NMA just before being placed in the cells.

Dissolved air was removed from all of the cell solutions by bubbling purified hydrogen gas through them for 1–2 hr before the cells were filled. The emf reached equilibrium after 4–12 hr, depending on the composition of the cell solution. The criterion adopted for equilibrium was a drift in emf of less than 0.1 mV over a period of 6–12 hr. After a change of temperature, equilibrium was attained more rapidly. At the end of the series of measurements, the cells were returned to 25°C to check the constancy of the emf during the measurement interval. Irreversible changes amounted in general to several tenths of a millivolt. The temperatures of the cells were maintained to $\pm$ 0.01°C in a water bath.

It was necessary to measure the dielectric constant and density of each solvent mixture studied. Densities were determined in a constant-temperature bath maintained to within $\pm$ 0.02°C. Gay–Lussac pycnometers with a capacity of 25 mL were used for density measurements. Dielectric constants were determined with a Balsbaugh Model 2TN50 conductivity cell having a cell constant of $\sim$ 0.001. A Janz–McIntyre a–c bridge (*17*) was used. The dielectric constants and densities of the solvents are listed in Table I, along with the constants A and B of the Debye–Hückel theory.

Table I. Properties of H_2O/NMA

t(°C)	Density (g/mL)	Dielectric Constant	A[a]	B[a]
		$x_2 = 0.06$[b]		
5	1.00625	90.8	0.4560	0.3174
10	1.00465	90.0	0.4496	0.3158
15	1.00308	88.9	0.4457	0.3147
20	1.00128	87.8	0.4421	0.3137
25	0.99923	87.9	0.4300	0.3105
30	0.99714	86.3	0.4306	0.3105
35	0.99486	87.4	0.4117	0.3056
40	0.99250	88.8	0.3920	0.3004
45	0.98989	89.7	0.3765	0.2962
		$x_2 = 0.25$		
5	1.01705	101.9	0.3856	0.3012
10	1.01282	99.0	0.3912	0.3023
15	1.00991	96.8	0.3936	0.3025
20	1.00594	94.2	0.3988	0.3035
25	1.00218	92.6	0.3983	0.3030
30	0.99846	90.6	0.4005	0.3032
35	0.99466	89.1	0.4000	0.3027
40	0.99088	87.7	0.3991	0.3021
45	0.98694	86.1	0.3998	0.3019

[a] Debye–Hückel constants, molality scale.

Results

Isopiestic measurements showed that the water activity in H_2O/NMA solvent mixture of $x_2 = 0.5$ is the same, within experimental error, as that predicted by Raoult's Law. Hence, the emf was corrected to 1 atm partial pressure of hydrogen on the assumption that NMA is involatile and that all of the solutions obey Raoult's Law. The observed emf for each HBr solution at each solvent composition and temperature, corrected as described above, is listed in Table II.

It is expected that the mean activity coefficient of HBr in the molality range studied here can be represented by

$$-\log \gamma_{\mathrm{HBr}} = \frac{\mathrm{A}\, m^{1/2}}{1 + \mathrm{B}\mathrm{å}\, m^{1/2}} - \beta\, m \qquad (1)$$

where A and B are the Debye–Hückel constants given in Table I, while å (the ion-size parameter) and β are adjustable parameters. If Equation 1 is combined with the Nernst Equation for Cell A, one can define an apparent standard emf, $E^{\circ\prime}$, as follows:

Mixed Solvents from 5°–45°C

Density (g/mL)	*Dielectric Constant*	*A*	*B*
	$x_2 = 0.15$		
1.01553	95.9	0.4220	0.3103
1.01257	94.1	0.4221	0.3100
1.00958	92.5	0.4213	0.3095
1.00645	90.7	0.4222	0.3094
1.00338	90.0	0.4158	0.3075
1.00026	88.3	0.4167	0.3074
0.99698	86.8	0.4165	0.3070
0.99370	85.8	0.4130	0.3058
0.99018	83.8	0.4171	0.3065
	$x_2 = 0.50$		
1.00384	125.5	0.2803	0.2697
0.99984	121.1	0.2873	0.2716
0.99583	117.3	0.2930	0.2730
0.99189	113.8	0.2982	0.2742
0.98782	110.4	0.3037	0.2755
0.98387	107.2	0.3089	0.2767
0.97984	104.0	0.3148	0.2781
0.97583	101.3	0.3190	0.2789
0.97170	98.3	0.3252	0.2803

[b] Mole fraction of NMA.

$$E^{\circ\prime} \equiv E^{\circ} - 2k\,\beta\,m = E + 2k \log m - 2k \frac{A\,m^{1/2}}{1 + B\mathring{a}\,m^{1/2}} \qquad (2)$$

where k is written for 2.3026RT/F, values which have been tabulated elsewhere (*18*). If Equation 1 represents the activity coefficient of HBr satisfactorily and the proper value of å is chosen, $E^{\circ\prime}$ will be a linear function of m with intercept at $m = 0$ corresponding to the standard emf E°. The standard emf of Cell A is also the standard potential of the silver–silver bromide electrode.

A value of 5.2Å for the ion-size parameter å yielded straight-line plots of $E^{\circ\prime}$ vs. m at each temperature and solvent composition. The intercepts were obtained by standard linear regression techniques. A graphical representation of the data for each of the H_2O/NMA solvent mixtures at 25°C is shown in Figure 1. The calculations were performed with the aid of a PDP-11 computer with a teletype output. The intercepts (E°) and the standard deviations of the intercepts are summarized in Table III.

A tendency for the cell emf to drift slowly at temperatures above 35°C, indicative of some sort of irreversible side reaction, was noted.

Table II. Electromotive Force of the Cell: Pt; H_2 (g, 1 atm)|HBr (*m*) 5°–45°C,

Molality (m)	*5°C*	*10°C*	*15°C*	*20°C*
		$x_2 = 0.06$		
0.009096	0.33056	0.33337	0.33594	0.33842
0.02059	0.29272	0.29483	0.29686	0.29872
0.02983	0.27572	0.27750	0.27925	0.28081
0.04035	0.26194	0.26347	0.26494	0.26626
0.04606	0.25556	0.25721	0.25853	0.25976
0.05740	0.24578	0.24705	0.24829	0.24931
0.07196	0.23521	0.23635	0.23740	0.23829
0.09729	0.22172	0.22263	0.22338	0.22397
		$x_2 = 0.15$		
0.009546	0.35120	0.35341	—	—
0.019953	0.31620	0.31783	0.31934	0.32074
0.02868	0.30007	0.30150	0.30277	0.30378
0.04052	0.28363	0.28456	0.28547	0.28634
0.04072	0.28307	0.28421	0.28518	0.28606
0.04601	0.27715	0.27818	0.27911	0.27990
0.05920	0.26570	0.26658	0.26729	0.26786
0.06854	0.25882	0.25959	0.26019	0.26057
0.07745	0.25323	0.25391	0.25439	0.25475
0.09075	0.24616	0.24648	0.24691	0.24719
		$x_2 = 0.25$		
0.019773	0.33607	0.33689	0.33762	0.33814
0.03031	0.31521	0.31574	0.31609	0.31633
0.03986	0.30325	0.30352	0.30368	0.30406
0.04959	—	—	—	—
0.05239	0.28973	0.28974	0.28968	0.28939
0.06020	0.28370	0.28361	0.28350	0.28356
0.07023	0.27601	0.27586	0.27559	0.27517
0.08000	0.26983	0.26960	0.26922	0.26871
0.10076	0.25885	0.25843	0.25791	0.25714
		$x_2 = 0.50$		
0.010031	0.38835	0.38781	0.38729	0.38682
0.02008	0.35489	0.35395	0.35317	0.35219
0.03004	0.33577	0.33457	0.33334	0.33210
0.04030	0.32169	0.32026	0.31890	0.31744
0.05542	0.30646	0.30488	0.30320	0.30144
0.06055	0.30233	0.30059	0.29892	0.29717
0.06996	0.29527	0.29345	0.29161	0.28974
0.08074	0.28808	0.28653	0.28500	0.28311
0.09855	0.28002	0.27795	0.27580	0.27373

in H_2O/NMA|AgBr; Ag for Four Mixed-Solvent Compositions, from
in Volts

25°C	*30°C*	*35°C*	*40°C*	*45°C*
		$x_2 = 0.06$		
0.34111	0.34297	0.34509	0.34698	0.34898
0.30046	0.30249	0.30405	0.30578	0.30739
0.28223	0.28410	0.28532	0.28676	0.28813
0.26725	0.26899	0.26998	0.27119	0.27229
0.26081	0.26178	0.26261	0.26329	0.26401
0.25013	0.25155	0.25224	0.25315	0.25398
0.23900	0.24016	0.24068	0.24114	0.24208
0.22441	0.22480	0.22500	0.22509	0.22520
		$x_2 = 0.15$		
0.35959	—	—	—	—
0.32231	0.32343	0.32454	0.32593	0.32686
0.30513	0.30578	0.30659	0.30744	0.30823
0.28716	0.28782	0.28837	0.28917	0.28961
0.28687	0.28752	0.28806	0.28887	0.28929
0.28059	0.28111	0.28153	0.28182	0.28218
0.26837	0.26878	0.26904	0.26955	0.26970
0.26104	0.26124	0.26139	0.26156	0.26168
0.25496	0.25515	0.25520	0.25551	0.25545
0.24735	0.24742	0.24737	0.24717	0.24704
		$x_2 = 0.25$		
0.33891	0.33888	0.33951	0.33987	0.34017
0.31665	0.31652	0.31677	0.31686	0.31685
0.30372	0.30365	0.30328	0.30334	0.30325
0.29316	—	—	—	—
0.28942	0.28901	0.28860	0.28824	0.28788
0.28300	0.28254	0.28208	0.28163	0.28124
0.27475	0.27408	0.27367	0.27312	0.27244
0.26819	0.26743	0.26697	0.26627	0.26549
0.25661	0.25576	0.25485	0.25403	0.25317
		$x_2 = 0.50$		
0.38661	0.38627	0.38596	0.38575	0.38568
0.35122	0.35045	0.34950	0.34857	0.34773
0.33068	0.32961	0.32838	0.32720	0.32620
0.31594	0.31461	0.31310	0.31165	0.31026
0.29962	0.29818	0.29645	0.29480	0.29332
0.29536	0.29373	0.29189	0.29014	0.28841
0.28776	0.28614	0.28426	0.28245	0.28080
0.28073	0.27877	0.27677	0.27478	0.27270
0.27072	0.26853	0.26633	0.26415	0.26188

Table III. Standard Emf of

(t°C)	$x_2 = 0$	$x_2 = 0.06$
5	0.07961[a]	0.1011 (0.12)[b]
10	0.07773	0.0998 (0.09)
15	0.07572	0.0984 (0.08)
20	0.07349	0.0968 (0.06)
25	0.07106	0.0954 (0.09)
30	0.06856	0.0937 (0.24)
35	0.06585	0.0919 (0.27)
40	0.06310	0.0909 (0.45)
45	0.06012	0.0894 (0.55)

[a] Data from Ref. *19*.

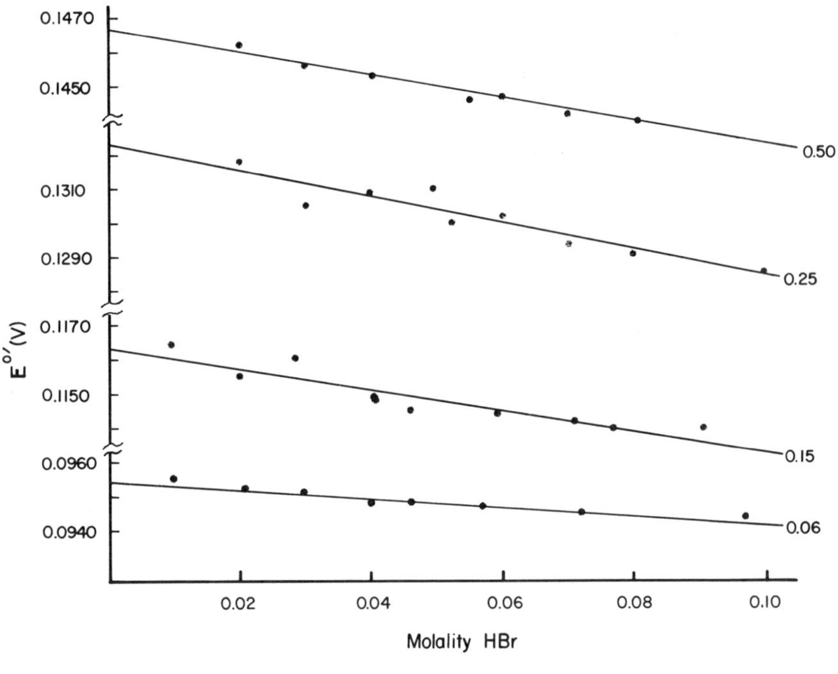

Figure 1. Plots of the quantity E°′ (Equation 2) as a function of the molality of HBr for four mixed solvents at 25°C. Numbers at the right are the mole fractions (x_2) of NMA.

The drift was most pronounced at higher NMA concentrations, yet it was at the lower NMA compositions that the highest standard deviations were found.

The values obtained for the standard emf of Cell A at the nine temperatures (t in °C) were fitted to power series equations in $t - 25$, with the following results:

Cell A from 5°–45°C, in Volts

$x_2 = 0.15$	$x_2 = 0.25$	$x_2 = 0.50$
0.1241 (0.26)	0.1434 (0.30)	0.1651 (0.22)
0.1223 (0.23)	0.1407 (0.30)	0.1605 (0.17)
0.1200 (0.27)	0.1379 (0.31)	0.1561 (0.10)
0.1179 (0.24)	0.1351 (0.42)	0.1515 (0.08)
0.1163 (0.25)	0.1323 (0.34)	0.1467 (0.17)
0.1137 (0.24)	0.1288 (0.33)	0.1426 (0.14)
0.1113 (0.24)	0.1258 (0.27)	0.1381 (0.15)
0.1092 (0.28)	0.1228 (0.28)	0.1334 (0.17)
0.1066 (0.28)	0.1197 (0.32)	0.1287 (0.19)

[b] Figures in parentheses are standard deviations in millivolts.

$$x_2 = 0.06: \quad E° = 0.0953–2.99 \times 10^{-4} (t - 25) \tag{3}$$

$$x_2 = 0.15: \quad E° = 0.1159–4.36 \times 10^{-4} (t - 25) - 1.45 \times 10^{-6} (t - 25)^2 \tag{4}$$

$$x_2 = 0.25: \quad E° = 0.1320–5.97 \times 10^{-4} (t - 25) - 1.32 \times 10^{-6} (t - 25)^2 \tag{5}$$

$$x_2 = 0.50: \quad E° = 0.1470–9.06 \times 10^{-4} (t - 25) - 1.6 \times 10^{-7} (t - 25)^2 \tag{6}$$

The standard deviations of the measured E° from the calculated curves were 0.1 mV in each case.

The standard Gibbs energy ($\Delta G_t°$) for the transfer of HBr from water to each mixed solvent was calculated from the relationship

$$\Delta G_t° = \mathrm{F}(E_{\mathrm{w}}° - E_{\mathrm{s}}°) \tag{7}$$

where w and s refer to water and the mixed solvent, respectively. The results for each solvent composition are listed in Table IV. Likewise, from the identity

$$\Delta S_t° = -\frac{d\Delta G_t°}{dt} \tag{8}$$

and the relationship among $\Delta G_t°$, $\Delta S_t°$, and $\Delta H_t°$, the standard enthalpy and entropy changes for the transfer process also were calculated at 25°C.

Table IV. Standard Gibbs Energy Changes, $\Delta G_t°$, for the Transfer Process $HBr(H_2O) \rightarrow HBr(H_2O/NMA)$, in Calories

t (°C)	$x_2 = 0.06$	$x_2 = 0.15$	$x_2 = 0.25$	$x_2 = 0.50$
5	−496	−1026	−1471	−1971
10	−509	−1028	−1452	−1909
15	−523	−1021	−1434	−1854
20	−538	−1024	−1421	−1799
25	−561	−1043	−1412	−1744
30	−580	−1041	−1389	−1707
35	−601	−1048	−1383	−1666
40	−641	−1063	−1377	−1621
45	−675	−1072	−1374	−1582

Table V. Standard Enthalpy and Entropy Changes for the Transfer Process $HBr(H_2O) \rightarrow HBr(H_2O/NMA)$ at 298.15 K

x_2	$\Delta H_t°$ (*kcal mol*$^{-1}$)	$\Delta S°$ (*cal K*$^{-1}$ *mol*$^{-1}$)
0.06	0.73	4.3
0.15	−0.69	1.2
0.25	−2.16	−2.5
0.50	−4.62	−9.7

Table VI. Mean Ionic Activity Coefficients (Molality Scale) of HBr in H_2O/NMA Solvent Mixtures at 25°C

m (*mol kg*$^{-1}$)	$x_2 = 0$[a]	$x_2 = 0.06$, $\beta = 0.101$	$x_2 = 0.15$, $\beta = 0.253$	$x_2 = 0.25$, $\beta = 0.313$	$x_2 = 0.50$, $\beta = 0.287$
0.005	0.929	0.940	0.944	0.948	0.959
0.01	0.906	0.920	0.926	0.932	0.947
0.02	0.879	0.896	0.906	0.914	0.933
0.03	—	0.881	0.894	0.904	0.926
0.04	—	0.869	0.885	0.897	0.921
0.05	0.838	0.860	0.879	0.893	0.918
0.06	—	0.852	0.875	0.890	0.916
0.07	0.823	0.846	0.872	0.888	0.916
0.08	—	0.841	0.869	0.888	0.916
0.09	—	0.836	0.868	0.888	0.916
0.10	0.808	0.832	0.867	0.888	0.917

[a] Values in water from Ref. *19*.

These are listed in Table V. The uncertainties in the primary data did not justify the calculation of these derived quantities at temperatures other than 25°C.

Discussion

It is evident that smoothed values of the mean ionic activity coefficient of HBr (γ_{HBr}), scale of molality, can be derived by Equation 1; A and B are given in Table I, and å is 5.2 Å. Values of β (the slope of the $E^{\circ\prime}$ vs. molality plot divided by $-$ 2k) are listed in Table VI, along with γ_{HBr} at molalities from 0.005–0.1 for each solvent mixture at 25°C. For comparison, the corresponding values of the activity coefficient in pure water are listed (*19*).

The activity coefficients in the mixed solvents are all higher than those in pure water. This suggests that HBr is dissociated completely and does not form ion pairs in mixtures of water and NMA of these compositions. From electrostatic considerations, this is in accord with expectation since mixed solvents have higher dielectric constants than water. This observation is consistent with earlier studies, which showed that the activity coefficient of HCl is higher in the pure solvents formamide (*6, 7*), NMA (*1, 2*), and NMP (*15*) than in water. Similarly, our own preliminary results show the same to be true of HBr in pure NMA at 35°C.

Nonetheless, the activity coefficient is not determined by the dielectric constant alone. In this connection, it is interesting to note that acetic acid is much weaker in NMP than in water (*13*). When NMP is added to the aqueous solvent, the dissociation of the protonated form of tris-(hydroxymethyl)aminomethane is enhanced initially (*12*). In pure NMP, however, this acid is weaker than in water (*14*), despite the greatly increased dielectric constant ($\epsilon = 176$ at 25°C). These results point to the controlling influence of solute–solvent interactions on the behavior of these weak electrolytes.

In Figure 2, the values of ΔG_t°, ΔH_t°, and $T\Delta S_t^\circ$ (where T is the thermodynamic temperature in Kelvin) are plotted as a function of the mole fraction of NMA (x_2). Since all of the ΔG_t° values are negative, it appears that there is a further stabilization of HBr in the solvent mixtures, as compared with pure water. Both ΔH_t° and $T\Delta S_t^\circ$ decrease in the same fashion, with $T\Delta S_t^\circ$ decreasing less rapidly than ΔH_t°; this leads to continuously negative values for ΔG_t°. As these quantities are zero at $x_2 = 0$, the results suggest a pronounced solvent effect of small amounts of NMA. Nevertheless, Table III reveals a large uncertainty in the temperature coefficient at $x_2 = 0.06$.

Negative values are to be expected, on electrostatic grounds, for a transfer from a medium of lower dielectric constant to one of higher.

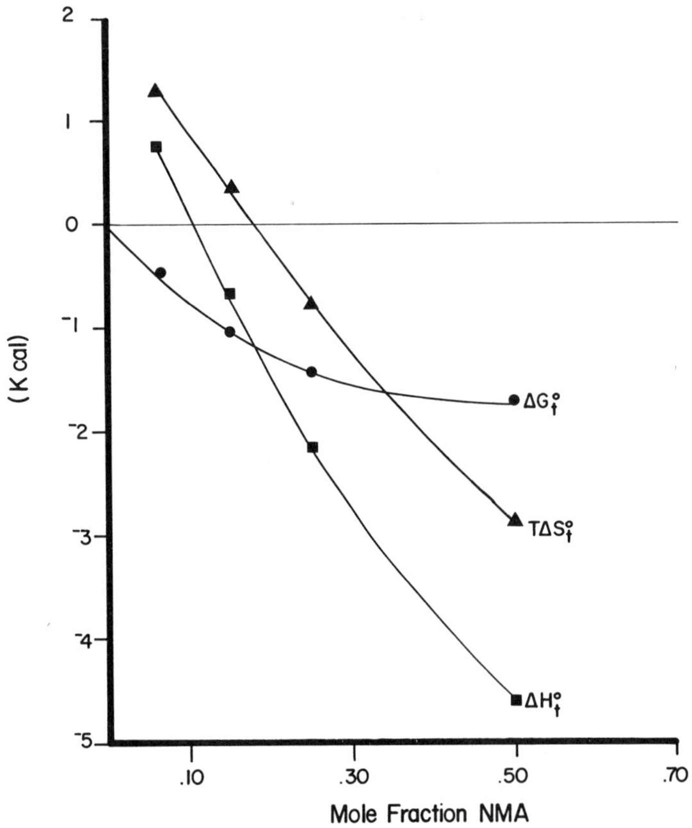

Figure 2. Plots of the transfer functions $\Delta G_t°$, $\Delta H_t°$, and $T\Delta S_t°$ at 25°C as a function of the mole fraction (x_2) of NMA in the mixed solvent

Thus, the Born Equation predicts a decrease in Gibbs energy of 0.53 kcal mol^{-1} for the transfer of HBr from water to H_2O/NMA, $x_2 = 0.5$, if the radii of the hydrated hydrogen and bromide ions are taken to be 2.8 and 1.96 Å, respectively. That the observed decrease is more than three times this value may have a variety of explanations; among them are deficiencies in the Born calculation, increased basicity of the medium through addition of the amide component, and solvent–solvent interactions of an undetermined nature.

Literature Cited

1. Dawson, L. R., Sheridan, R. C., Eckstrom, H. C., *J. Phys. Chem.* (1961) **65**, 1829.
2. Dawson, L. R., Zuber, W. H., Jr., Eckstrom, H. C., *J. Phys. Chem.* (1965) **69**, 1335.

3. Dawson, L. R., Kim, K. H., Eckstrom, H. C., *J. Phys. Chem.* (1966) **70,** 775.
4. Wood, R. H., Wicker, R. K., Kreis, R. W., *J. Phys. Chem.* (1971) **75,** 2313.
5. Bonner, O. D., Bunzl, K. W., Woolsey, G. B., *J. Phys. Chem.* (1966) **70,** 778.
6. Mandel, M., Decroly, P., *Nature* (1958) **182,** 794.
7. Mandel, M., Decroly, P., *Trans. Faraday Soc.* (1960) **56,** 29.
8. Agarwal, R. K., Nayak, B., *J. Phys. Chem.* (1967) **71,** 2062.
9. Agarwal, R. K., Nayak, B., *J. Phys. Chem.* (1966) **70,** 2568.
10. Dash, V. N., Nayak, B., *Aust. J. Chem.* (1975) **28,** 1657.
11. Bokra, Y., Bates, R. G., *Anal. Chem.* (1975) **47,** 1110.
12. Bates, R. G., Falcone, J. S., Jr., Ho, A. Y. W., *Anal. Chem.* (1974) **46,** 2004.
13. Etz, E. S., Robinson, R. A., Bates, R. G., *J. Solution Chem.* (1972) **1,** 507.
14. Etz, E. S., Robinson, R. A., Bates, R. G., *J. Solution Chem.* (1973) **2,** 405.
15. Etz, E. S., Ph.D. thesis, Clarkson College of Technology (1974).
16. Gary, R., Bates, R. G., Robinson, R. A., *J. Phys. Chem.* (1964) **68,** 1186.
17. Janz, G. J., McIntyre, J. D. E., *J. Electrochem. Soc.* (1961) **108,** 272.
18. Bates, R. G., "Determination of pH," 2nd ed., Wiley, New York, 1973, Appendix Table I.
19. Hetzer, H. B., Robinson, R. A., Bates, R. G., *J. Phys. Chem.* (1962) **66,** 1423.

RECEIVED February 13, 1978. This work was supported in part by the National Science Foundation under Grant CHE76 24556.

Thermodynamic Behavior of Tetraalkylammonium Bromide and Hydrobromic Acid in Aqueous Media at 298.15 K

R. N. ROY, J. J. GIBBONS, J. MOELLER, and B. SNELLING

Department of Chemistry, Drury College, Springfield, MO 65802

Electromotive force measurements of the cell (without liquid junction) of the type Pt; H_2(g, 1 atm)|HBr(m_1), $(C_2H_5)_4$-NBr(m_2)|AgBr; Ag have been made at several constant total molalities (m = 0.05, 0.1, 0.25, 0.5, 1.0, and 1.5 mol-kg^{-1}, where $m = m_1 + m_2$), at 298.15 K. The trace activity coefficients of HBr in HBr + $(C_2H_5)_4$NBr + H_2O mixtures, the Harned interaction coefficients (α_{12} and α_{21}), the Bronsted–Guggenheim parameter ($\beta_{(C_2H_5)_4N^+,Br^-}$), the Pitzer parameters Θ_{MN} (cation–cation doublet interaction), and Ψ_{MNX} (cation–anion–cation triplet interaction), along with the excess Gibbs free energies of mixing, ΔG^E, at 298.15 K, are reported. Interpretation of these results has been made in light of recent work by Pitzer. The results are discussed also in relation to the structural changes of water caused by the presence of the large-sized hydrophobic cation [$(C_2H_5)_4N^+$].

The determination of the activity coefficient of an electrolyte in binary mixtures of another electrolyte with a common ion has prompted considerable interest, both in the development of the underlying theory, as well as precise experimental measurements. This kind of study is useful in applications that involve metallurgical and biological systems. The ion interaction parameters (doublet and triplet, among others) responsible for an understanding of the effects of cation size on the thermodynamic

0-8412-0428-4/79/33-177-263$05.00/1

behavior of tetraalkylammonium salt–water mixtures are well recognized (*1–12*). There has been some progress towards unravelling the complex nature of these interactions, but the results are still far from quantitative. As a continuation of our previous investigations to study the thermodynamic behavior of mixtures of strong aqueous electrolytes, such as HCl + NH_4Cl + H_2O (*13*), HBr + NH_4Br + H_2O (*14*), HBr + $(C_3H_7)_4NBr$ + H_2O (*15*), HBr + $(C_4H_9)_4NBr$ + H_2O (*16*), and HCl + $LaCl_3$ + H_2O (*17*), we have made electromotive force measurements on the HBr + $(C_2H_5)_4NBr$ + H_2O system at 298.15 K in the total molality range $m = 0.05–1.5$ mol-kg^{-1}. The data for the activity and osmotic coefficients of pure $(C_2H_5)_4NBr$ are available in the literature (*9, 10*), based on either emf or gravimetric isopiestic vapor pressure techniques, but no thermodynamic data for various mixtures of the HBr + $(C_2H_5)_4NBr$ + H_2O system have been reported in the literature. Emf measurements were made therefore at 298.15 K using a cell of the type (without transference):

$$\text{Pt; } H_2(\text{g, 1 atm})|\text{HBr}(m_1),\ (C_2H_5)_4\text{NBr}(m_2)|\text{AgBr, Ag} \qquad (1)$$

Experimental Procedure

The hydrobromic acid used was twice-distilled, constant-boiling acid. Triplicate gravimetric analyses of the stock solution of aqueous HBr (about 3*M*) agreed to within ± 0.01%. The tetraethylammonium bromide (Eastman Kodak Co.) was recrystallized twice from benzene–ligroin mixtures, as recommended by Unni et al. (*18*). The molality of the stock solution (about 4*M*) was standardized by gravimetric determination of bromide and was accurate to within ± 0.02%.

Emf measurements were made with a Leeds and Northrup K-5 potentiometer equipped with a Leeds and Northrup DC null detector (Model 9829). The temperature of the bath was regulated to within 0.02 K. Details of the experimental procedure, including preparation of the electrodes (*19*), cell design, preparation of solutions, purification of the hydrogen gas, and other experimental aspects, have been reported elsewhere (*13, 14*).

Preliminary emf measurements were made on Cell I, and the standard potential of the Ag–AgBr electrode was determined as 0.07106 V from data taken in 0.01000 mol kg^{-1} hydrobromic acid. This value of E_m° was identical with that given in the literature (*20*). The emf values were reproducible up to $m = 1.0$ mol kg^{-1}. There was some evidence of irreversible behavior for $m = 1.5$ mol kg^{-1}. In order to avoid this kind of drift in the emf values at the highest constant total molality tested, the cell with the hydrogen electrode was allowed to equilibrate for ~ 45 min before the Ag–AgBr electrode (which was kept in a separate standard-joint test tube containing a solution of the same composition) was transferred to the electrode compartment. The equilibrium emf value was recorded every 5 min until no deviation was noticed.

Table I. Experimental Emf Data of the Cell Pt; H_2 (g, 1 atm)|$HBr(m_1)$, $(C_2H_5)_4NBr(m_2)$|AgBr; Ag at 25°C, in Volts, for Various Values of α_{12} Obtained from Equation 6

m = *0.05 mol kg⁻¹*		m = *0.1 mol kg⁻¹*		m = *0.25 mol kg⁻¹*		m = *0.5 mol kg⁻¹*	
y_2	E,V	y_2	E,V	y_2	E,V	y_2	E,V
α_{12} =	0.3504	α_{12} =	0.2894	α_{12} =	0.2793	α_{12} =	0.2578
$\sigma(\alpha_{12})$ =	0.00045	$\sigma(\alpha_{12})$ =	0.0014	$\sigma(\alpha_{12})$ =	0.00025	$\sigma(\alpha_{12})$ =	0.00023
$\sigma(E)$ =	0.00001	$\sigma(E)$ =	0.00008	$\sigma(E)$ =	0.00003	$\sigma(E)$ =	0.00007
0	0.23424	0	0.20063	0	0.15512	0	0.11837
0.24592	0.24197	0.24300	0.20873	0.09292	0.15837	0.23548	0.12901
0.47618	0.25183	0.49490	0.22066	0.28464	0.16613	0.49526	0.14365
0.63776	0.26162	0.65570	0.23033	0.49512	0.17670	0.63534	0.15397
0.75506	0.27193	0.75210	0.23914	0.68704	0.19063	0.73706	0.16400
0.86248	0.28690	0.87430	0.25689	0.87928	0.21673	0.86290	0.18264

m = *1.0 mol kg⁻¹*		m = *1.0 mol kg⁻¹*		m = *1.5 mol kg⁻¹*		m = *1.5 mol kg⁻¹*	
y_2	E,V	y_2	E,V	y_2	E,V	y_2	E,V
α_{12} =	0.2156	α_{12} =	0.3044	α_{12} =	0.1083	α_{12} =	0.2272
		$\sigma(\alpha_{12})$ =	0.0007			$\sigma(\alpha_{12})$ =	0.0011
$\sigma(\alpha_{12})$ =	0.0015	β_{12} =	−0.1021	$\sigma(\alpha_{12})$ =	0.0021	β_{12} =	−0.0954
		$\sigma(\beta_{12})$ =	0.0008			$\sigma(\beta_{12})$ =	0.0009
$\sigma(E)$ =	0.0007	$\sigma(E)$ =	0.00007	$\sigma(E)$ =	0.0023	$\sigma(E)$ =	0.0003
0	0.07765	0	0.07765	0	0.05000	0	0.05000
0.10650	0.08125	0.10650	0.08125	0.08720	0.05599	0.08720	0.05599
0.25920	0.08602	0.25920	0.08602	0.28447	0.06858	0.28447	0.06858
0.46970	0.09206	0.46970	0.09206	0.45000	0.07860	0.45000	0.07860
0.67370	0.09624	0.67370	0.09624	0.63473	0.08471	0.63473	0.08471
0.87920	0.09996	0.87920	0.09996	0.85020	0.11514	0.85020	0.11514

Results and Calculations

The experimental emf of Cell I, corrected to a hydrogen pressure of 1 atm (101,325 Pa), is summarized in Table I as a function of the molality fraction y_2 of tetraethylammonium bromide (which is given by $y_2 = m_2/m$).

Harned's Rule. The logarithm of the activity coefficient of each electrolyte in a mixture of constant total ionic strength can be expressed mathematically by the equations:

$$\log \gamma_{HBr} = \log \gamma^{\circ}{}_{HBr} - \alpha_{12}m_2 - \beta_{12}m_2^2 \tag{1}$$

and

$$\log \gamma_{(Et)_4NBr} = \log \gamma^{\circ}{}_{(Et)_4NBr} - \alpha_{21}m_1 - \beta_{21}m_1^2 \tag{2}$$

where γ°_{HBr} is the activity coefficient of hydrobromic acid in a pure HBr solution (i.e., $y_2 = 0$) at the same total molality, γ_{HBr} is the activity coefficient of HBr in the mixture, and α_{12}, β_{12}, etc., are the Harned coefficients, which may be functions of the constant total ionic strength, but are not dependent on the composition at a particular ionic strength. The linear forms of Equations 1 and 2 are known as Harned's Equations.

The values of the parameter α_{12}, and the standard deviations $\sigma(\alpha_{12})$ and $\sigma(E)$, are given in Table I, and were obtained in the following manner: first, from combination of the Nernst Equation,

$$E = E^\circ - \mathrm{k} \log m_1 (m_1 + m_2) \gamma_1^2 \quad (3)$$

with the linear form of Equation 1, one obtains

$$E + \mathrm{k} \log m_1 = E^\circ - \mathrm{k} \log m - 2\,\mathrm{k} \log \gamma_1^\circ + 2\,\mathrm{k}\, \alpha_{12}\, m_2 \quad (4)$$

which one can then express (after proper rearrangement) in both a linear and a nonlinear form:

$$E + \mathrm{k} \log m_1 = a + b\, m_2 \quad (5)$$

and

$$E + \mathrm{k} \log m_1 = a + b\, m_2 + c\, m_2^2 \quad (6)$$

where

$$\alpha_{12} = b/2\mathrm{k},\ \beta_{12} = c/2\mathrm{k},\ \text{and}\ \mathrm{k} = (\mathrm{RT} \ln 10)F.$$

The standard deviations of the fit shown in Table I, as well as the smooth plot of α_{12} as a function of m in Figure 1, were used as criteria for testing the validity of Harned's Rule. Harned's Rule is a good description of the data for $m = 0.05$, 0.1, 0.25, 0.5, and perhaps (as a borderline case, based on the value of α_{12}) at $m = 1.0$ mol kg^{-1}. The nonlinear form of Equation 6 was used at $m = 1.0$ and 1.5 mol kg^{-1}, where the greatest deviations from linearity were thought to occur.

Two interpretations can be made from Figure 1. In the first place, the value of α_{12} increases at molalities below 0.1 mol kg^{-1}, which contradicts the prediction of the specific interaction theory. Next, the effect of addition of $(Bu)_4NBr$, $(Pr)_4NBr$, and $(Et)_4NBr$ (as compared with NH_4Br) on HBr is about two and a half to three times greater at a given molality (e.g., for $m = 0.25$ mol kg^{-1}, α_{12} for $NH_4Br = 0.0999$, whereas α_{12} for $(Pr)_4NBr$ is equal to 0.3079). This variation of α_{12} with m is consistent with results obtained in previous studies (*13, 14, 15, 16*).

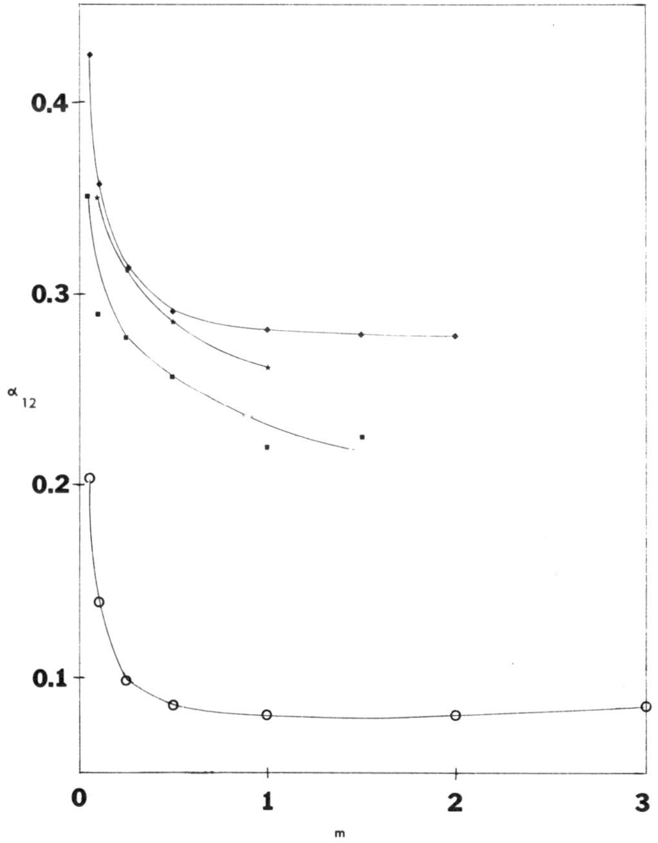

*Figure 1. Harned interaction coefficient vs. total molality of Br^- for the HBr + $(Pr)_4NBr$ + H_2O, HBr + $(Bu)_4NBr$ + H_2O, HBr + $(Et)_4NBr$ + H_2O, and HBr + NH_4Br + H_2O systems at 25°C: ◆, HBr + Pr_4NBr + H_2O; *, HBr + Bu_4NBr + H_2O; ■, HBr + Et_4NBr + H_2O; and ○, HBr + NH_4Br + H_2O*

Bronsted–Guggenheim Equation. If Bronsted's principle of specific interaction is applicable at $m = 0.1$ mol kg^{-1}, then for mixtures of HBr-(MX) and $(Et)_4NBr$(NX), the equation for $\log \gamma_{HBr}$ (*21*) is

$$\log \gamma_{HBr} = -0.5108\, I^{1/2}\,(1 + I^{1/2}) + \tfrac{1}{2} B_{H,Br} \cdot m_{Br} + \tfrac{1}{2} B_{H,Br} \cdot m_{H} + \tfrac{1}{2} B_{(Et)_4N,Br} \cdot m_{(Et)_4N} \quad (7)$$

where $I = m_1 + m_2 = m$, and $B_{(Et)_4N,Br} = 2\beta/2.3026$, indicating the specific interaction coefficient arising from the interaction between the $(Et)_4N^+$ and Br^- ions.

Table II. Values of the Parameters $f(\Phi°)$, α_{12}, Θ, α_{21}, ΔG^E,

m	$f(\Phi°)$	$2.303\ \alpha_{12}$	$-\Theta$[a]
0.05	0.6923	0.8068	0.1145
0.10	0.6114	0.6664	0.0550
0.25	0.4877	0.6431	0.1554
0.50	0.3923	0.5936	0.2013
1.00	0.3077	0.4964	0.1887
1.50	0.2673	0.5232	0.2559

[a] The weighted average of $-\Theta = 0.21$.
[b] Present study.

The Harned coefficient, α_{12}, is related to the Bronsted–Guggenheim parameters by means of the following equation:

$$\alpha_{12} = \tfrac{1}{2}\ [B_{H,Br} - B_{(Et)_4N,\,Br}] \tag{8}$$

where the value of $B_{H,Br}$ is 0.287 kg mol^{-1} (*22*) and that for α_{12} (at m = 0.1 mol kg^{-1}) as taken from Table I is equal to 0.2894. Thus, the value of $B_{(Et)_4N,Br}$ (at m = 0.1 mol kg^{-1}) is − 0.292, which can be compared with $B_{NH_4,Br}$ = + 0.0122 (*14*), $B_{(Pr)_4N,Br}$ = − 0.427 (*15*), and $B_{(Bu)_4N,Br}$ = − 0.374 (*16*), implying more interaction than between NH_4^+ and Br^-.

The following simplified form of Equation 9 (true only for a 1:1 electrolyte, such as $(Et)_4NBr$) can be used to calculate the activity coefficient of $(Et)_4NBr$ at a given molality (for example, at m = 0.1 mol kg^{-1}).

$$\log \gamma_{(Et)_4NBr} = -0.5108\ m^{1/2}/(1 + m^{1/2}) + B_{(Et)_4N,\,Br} \cdot m_{(Et)_4NBr} \tag{9}$$

The value of $\gamma(Et)_4NBr$ at m = 0.1 mol kg^{-1}, as calculated by the use of the specific interaction coefficient, is 0.706. This can be compared qualitatively with that determined from isopiestic methods, $\gamma_{(Et)_4NBr}$ = 0.716 (*9*), and also with the value obtained using Pitzer's Equations (*3*), as computed by Leyendekkers and Hunter [$\gamma_{(Et)_4NBr}$ = 0.704, Table II (*12*)]. Quantitative significance should not be placed on the data obtained for a system at such dilute concentrations (e.g., m = 0.1 mol kg^{-1}) from isopiestic methods. However, the results obtained by the three various aforementioned methods are satisfactory in their agreement, considering the fact that the Bronsted–Guggenheim Theory deals with specific interaction between ions of unlike charges only.

Formalism According to Pitzer. The most common method for the evaluation of the activity and osmotic coefficients of an electrolyte in a binary mixture of strong electrolytes with a common ion is by Scatchard's Equations (*23*), the McKay–Perring treatment (*24*), Mayer's Equations

and γ^{tr}, as Deduced from an Application of Pitzer's Equations

$-\alpha_{21}$	$\Delta G^{E\,b}$ *(cal kg^{-1})*	$\Delta G^{E\,c}$ *(cal kg^{-1})*	γ^{tr}_{HBr}
0.2157	−0.15	−0.014	1.160
0.1804	−0.58	−0.06	1.162
0.1266	−3.63	−0.36	1.092
0.0852	−14.52	−1.4	0.942
0.0518	−58.06	−5.8	0.699
0.0310	−130.6	—	0.189

[c] Ref. *14* (the HBr + NH_4Br + H_2O system).

as extended by Friedman (*25, 26*), and the more recent equations of Pitzer (*1, 2, 3, 4, 5*). In the present investigation, we have used the more simplified treatment of Pitzer for the evaluation of α_{21} and the activity coefficients of HBr(MX) in mixtures of HBr + H_2O + $(Et)_4NBr$(NX), as has been successfully done by Pitzer himself (*4*) and during prior work in this laboratory (*14, 15, 16*). This is caused by the fact that each Pitzer parameter has physical interpretations and the treatment of formulas and computations are of a less complex nature as compared with those of Scatchard. The equation for the activity coefficient of HBr in HBr + $(Et)_4NBr$ + H_2O mixtures can be expressed as

$$\ln \gamma_1 = f^{\gamma} + m[B^{\gamma}_{MX} + y_2(B^{\Phi}_{NX} - B^{\Phi}_{MX} + \Theta_{MN}) + y_1y_2m\Theta'_{MN}] + m^2[C^{\gamma}_{MX} + y_2(C^{\Phi}_{NX} - C^{\Phi}_{MX} + \tfrac{1}{2}\Psi_{MNX}) + \tfrac{1}{2}\, y_1y_2\Psi_{MNX}] \qquad (10)$$

The symbols have the same physical significance as those in Pitzer's previous papers (*1, 2, 3, 4, 5*), which are based on a different theoretical framework from that of Scatchard and use the ions of the mixed electrolytes as components. The Θ_{MN} (the doublet cation–cation interaction) represents the interactions between H^+ and $(Et)_4N^+$, whereas Ψ_{MNX} (the triplet ion interaction) indicates the interactions between H^+, Br^-, and $(Et)_4N^+$. Thus, the quantities Θ, Θ', and Ψ are properties characteristic of the mixture, whereas B^{γ}, B^{Φ}, C^{γ}, and C^{Φ} are the properties of the single electrolyte solution, and are functions of the ionic strength. Equation 10 can be further reduced after imposing the conditions that $\Psi_{MNX} = 0$, $\Theta'_{MN} = 0$, and y_2 (at the limit) $= 0$:

$$\ln (\gamma_1/\gamma^{\circ}_1) = my_2\,(B^{\Phi}_{NX} - B^{\Phi}_{MX} + \Theta) + m^2y_2\,(C^{\Phi}_{NX} - C^{\Phi}_{MX}) \qquad (11)$$

and

$$\Theta = -2.3026\,\alpha_{12} + f(\Phi^{\circ}) \qquad (12)$$

which is obtained after combination and subsequent simplification of Equation 1 (the linear form) and Equation 11. The expressions for $f(\Phi°)$ and B_{MX}^{Φ} are:

$$f(\Phi°) = (\beta°_{MX} - \beta°_{NX}) + (\beta^1_{MX} - \beta^1_{NX}) [\exp(-2m^{1/2})] + m(C^{\Phi}_{MX} - C^{\Phi}_{NX}) \quad (13)$$

and

$$B^{\Phi}_{MX} = \beta°_{MX} + \beta^1_{MX} \exp(-2m^{1/2}) \quad (14)$$

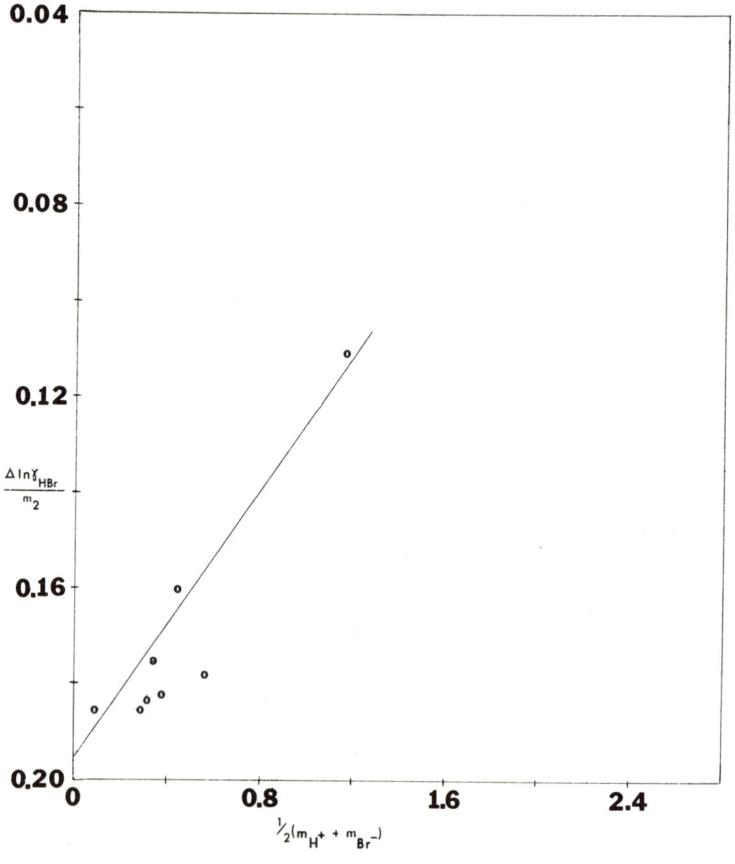

Figure 2. The term $\Delta ln\gamma/m_2$ vs. the parameter $\frac{1}{2}(m_{H^+} + m_{Br^-})$

The values of the parameters (i.e., the interaction coefficients for HBr and $(Et)_4NBr$) to be used in Equations 13 and 14 have been given by Pitzer and Mayorga (*3*), and by Pitzer (*27*),

$$\beta^{\circ}_{MX} = 0.1960 \qquad \beta^{\circ}_{NX} = -0.01755 \qquad C^{\Phi}_{MX} = 0.00827$$

$$\beta^{1}_{MX} = 0.3564 \qquad \beta^{1}_{NX} = -0.3938 \qquad C^{\Phi}_{NX} = 0.01561$$

A fixed value of C_{MX}^{Φ} or C_{NX}^{Φ} usually applies up to quite high molalities of a given electrolyte. After computation, Equation 13 becomes

$$f(\Phi^{\circ}) = 0.21355 + 0.7502 \exp(-2m^{1/2}) - 0.00734m \tag{15}$$

The results of α_{12} (from Table I, with the first five values determined from the linear plots, and the remaining one from the nonlinear plot), Θ, and $f(\Phi^{\circ})$ are presented in Table II. The emf of Cell I is a function of Θm rather than Θ. Hence, the weighted average value of Θ (which equals -0.21 (from Table II)) must be known and has more significance at higher molalities than the unweighted average. It is evident that Θ varies with *m*, but at this stage, the appropriate form of the dependence of Θ_{MN} on *m* (or I) is not known.

It is of interest to note from Table II that the values of Θ decrease (with the exception of $m = 0.10$ mol kg^{-1}) with an increase in *m*, whereas those values for $HCl + NH_4Cl + H_2O$ (*13*) and $HBr + NH_4Br + H_2O$ (*14*) systems illustrate that this trend is reversed in these cases. This decrease in the values of Θ at higher molalities may be attributed to the existence of less hard-core contact between the pairs of ions (H^+ and $(Et)_4N^+$).

All the calculations explained so far were based on the assumptions that $\Psi_{H^+, (Et)_4N^+, Br^-} = 0$. It seems very appropriate to compute the values of $\Theta_{H^+, (Et)_4N^+}$ and $\Psi_{H^+, (Et)_4N^+, Br^-}$ by means of a simplified equation, as described by Pitzer (*28*):

$$\Delta \ln \gamma_{HBr}/m_{(Et)_4N^+} = \Theta_{H^+, (Et)_4N^+} + \tfrac{1}{2}(m_{H^+} + m_{Br^-})\,\Psi_{H^+, (Et)_4N^+, Br^-} \tag{16}$$

The quantity on the left was plotted against the coefficient of Ψ on the right (as shown in Figure 2). A linear plot was obtained, with the intercept $\Theta = -0.20$ (as compared with the weighted average value of $\Theta = -0.21$, cf. Table II), and the slope was $\Psi = -0.19$. In Equation 16, $\Delta \ln \gamma_{HBr}$ is the difference between the experimental value of $\ln \gamma_{HBr}$ and the value calculated with the appropriate parameters for the pure single electrolyte terms, but with a value of $\Theta = \Psi = 0$. The following pertinent equations were used to calculate $\ln \gamma_{HBr}$ (theoretical):

$$\ln \gamma_{HBr} = f^{\gamma} + m[B^{\gamma}_{MX} + y_2(B^{\Phi}_{NX} - B^{\Phi}_{MX})] + m^2[C^{\gamma}_{MX} + y_2(C^{\Phi}_{NX} - C^{\Phi}_{MX})] \quad (17)$$

where

$$f^{\gamma} = -A_{\Phi}\,[m^{1/2}/(1 + 1.2m^{1/2}) + (2/1.2)\ \ln\ (1 + 1.2m^{1/2})] \quad (18)$$

and

$$A_{\Phi} = 1/3\,A_{\gamma} = 0.392 \text{ for } H_2O \text{ at } 298.15\text{ K}; \quad C^{\gamma}_{MX} = 3/2\ C^{\Phi}_{MX} \quad (19)$$

where C^{Φ}_{MX} is known for the pure electrolyte; the expressions for B^{Φ}_{MX} and $B_{NX}{}^{\Phi}$ are given in Equation 14; and

$$B^{\gamma}_{MX} = 2\beta^{\circ}_{MX} + (2\,\beta^{1}_{MX}/4m)\,[1 - \exp\,(-2m^{1/2})\ (1 + 2m^{1/2} - 2m)] \quad (20)$$

The values of $\Delta \ln\ \gamma_{HBr}$ are usually very small (i.e., negligible) in the molality range of $m = 0.05$–0.5 mol kg^{-1}.

In Equation 16, the Pitzer parameter Θ is truly independent of the common ion, and is equal to ½ $(b_{A,B}{}^{0,1})$ in the Scatchard treatment, or $g_{M,N}$ according to Friedman. Also, the Pitzer and Scatchard equations for uni-univalent, three-ion systems are of comparable forms, with $b_{A,B}{}^{0,2} = 2\ \Psi_{H^+,\ (Et)_4N^+,\ Br^-}$.

It is interesting and of some significance to evaluate α_{21} (the Harned interaction coefficient in Equation 2) and $\gamma_1{}^{tr}$, the trace activity coefficient, from the following pertinent expressions (*15*) (using the value of $\Theta = -0.20$ from Figure 2):

$$-2.3026\ \alpha_{21} = (B^{\Phi}_{MX} - B^{\Phi}_{NX}) + m(C^{\Phi}_{MX} - C^{\Phi}_{NX}) + \Theta \quad (21)$$

and

$$\log\ (\gamma_1{}^{tr}/\gamma_1{}^{\circ}) = -\ m\alpha_{12} \quad (22)$$

Specific effects on ΔG^E, the excess Gibbs free energy, are expected in tetraethyl ammonium bromide (as compared with ammonium bromide), because of the presence of alkyl chains as substituents. The equation needed to calculate ΔG^E is given by

$$\Delta G^{E} = 2\ y_1 y_2\ RT\ m^2\ \Theta \quad (23)$$

where $y_1 = y_2 = 0.5$ and $\Theta = -0.20$. The values of α_{21}, $\gamma_1{}^{tr}$, and ΔG^E are listed in Table II.

Discussion

In this investigation, Pitzer's less complex, two-parameter treatment has been used successfully. It is evident from Table III that the values of Θ are usually small for unsubstituted ammonium halides (i.e., $HCl + NH_4Cl + H_2O$ and $HBr + NH_4Br + H_2O$). This observation was confirmed by Pitzer for several systems (*1, 4*). According to Pitzer, there should be only a single value of Θ_{H^+, NH_4^+}, but different values of Ψ_{H^+, NH_4^+, Cl^-} and Ψ_{H^+, NH_4^+, Br^-}. This small negligible difference (0.003) in the values of Θ_{H^+, NH_4^+} can be largely attributed to errors in experimental measurements. The physical significance of the parameter, Θ, is a measure of the difference between the virial coefficients of the mutual interactions of H^+ and $(Et)_4N^+$ and the average of the interactions of the pairs of like ions. The interactions of like ions (such as H^+–H^+, Br^-–Br^-, and $(Et)_4N^+$–$(Et)_4N^+$) are usually too low to be significant.

The value of Θ as shown in Figure 2 is -0.20, which is in good agreement with the weighted average ($\Theta = -0.21$), where the individual values of Θ at each constant total molality, *m*, were solved by assuming $\Theta' = 0$, with the values of Θ being weighted according to molality, a feature usually used with the Pitzer equations. As evident from Table II, the value of $\Theta_{H^+, (Et)_4N^+}$ decreases from a higher value at low molality to a relatively constant value at high molality, and reflects the increasing intensity of the interactions (caused by interpenetration and entangling by the ethyl chains) between H^+ and $(Et)_4N^+$. It is interesting to note that for Θ_{H^+, NH_4^+}, the trend with an increase in *m* is reversed (*14*).

The triplet-ion interactions (such as Ψ_{H^+, H^+, H^+}, Ψ_{Br^-, Br^-, Br^-}, and $\Psi_{(Et)_4N^+, (Et)_4N^+, (Et)_4N^+}$) are negligibly small, but $\Psi_{H^+, Br^-, (Et)_4N^+}$ is of considerable importance at higher concentrations. Frank and Evans (*29*),

Table III. Comparison of the Values of the Pitzer Interaction Parameters Θ and Ψ for $HCl + NH_4Cl$, $HBr + NH_4Br$, and HBr + Various Substituted Tetraalkylammonium Bromides in Water at 25°C

System	m_{max}[a]	Θ	Ψ
$HCl + NH_4Cl + H_2O$[b]	3.0	-0.016_5	0.0 (assumed)
$HBr + NH_4Br + H_2O$[c]	3.0	-0.019_5	0.0 (assumed)
$HBr + (C_2H_5)_4NBr + H_2O$[d]	1.5	−0.20	−0.19
$HBr + (C_3H_7)_4NBr + H_2O$[e]	2.0	−0.17	−0.15
$HBr + (C_4H_9)_4NBr + H_2O$[f]	1.0	−0.22	0.0 (assumed)

[a] Maximum molality studied.
[b] Ref. *13*.
[c] Ref. *14*.
[d] Present investigation.
[e] Ref. *15*.
[f] Ref. *16*.

Rasaiah (*30*), and Wood and Anderson (*31*) have studied the effects of large-sized tetraalkyl hydrophobic cations on mixed strong electrolyte systems. Their observations show that this kind of salt tightens the structure of water around them in a way similar to some aliphatic hydrocarbons. Hence, the structural effect (or entropy effect) of $(Et)_4N^+$ in comparison with NH_4^+ will be large. The more negative values of ΔG^E would suggest the formation of more $H^+(Et)_4N^+$ pairs than would be expected from statistical encounters. Similar studies on HBr + *N*-decyltriethylammonium bromide (a surfactant) are at present in progress to gain more insight into the nature of complex interactions of mixed aqueous electrolyte solutions (*32*).

Acknowledgment

The authors are grateful to T. White, who performed some of the preliminary measurements, and to the Computer Center at Southwest Missouri State University in Springfield, on whose IBM 360 the calculations were done. Acknowledgment is made to the Donors of the Petroleum Research Fund (Grant 9725-B5), administered by the American Chemical Society, for complete support of this research.

Literature Cited

1. Pitzer, K. S., *J. Phys. Chem.* (1973) **77**, 268.
2. Pitzer, K. S., Mayorga, G., *J. Solution Chem.* (1974) **3**, 539.
3. Pitzer, K. S., Mayorga, G., *J. Phys. Chem.* (1973) **77**, 2300.
4. Pitzer, K. S., Kim, J. J., *J. Am. Chem. Soc.* (1974) **96**, 5701.
5. Silvester, L. F., Pitzer, K. S., *J. Solution Chem.*, submitted for publication.
6. Kessler, Y. M., Grouba, V. D., Kiryanov, V. A., Yemelin, V. P., *J. Solution Chem.* (1977) **6**, 231.
7. Boyd, G. E., *J. Solution Chem.* (1977) **6**, 135.
8. Harned, H. S., Robinson, R. A., "Multicomponent Electrolyte Solutions," Pergamon, Oxford, 1968.
9. Lindenbaum, S., Boyd, G. E., *J. Phys. Chem.* (1964) **64**, 911.
10. Wen, W., Miyajima, K., Otsuka, A., *J. Phys. Chem.* (1971) **75**, 2148.
11. Ramanathan, P. S., Krishnan, C. V., Friedman, H. L., *J. Solution Chem.* (1972) **1**, 237.
12. Leyendekkers, J. V., Hunter, R. J., *J. Electroanal. Chem.* (1977) **81**, 123.
13. Robinson, R. A., Roy, R. N., Bates, R. G., *J. Solution Chem.* (1974) **3**, 837.
14. Roy, R. N., Swensson, E., *J. Solution Chem.* (1975) **4**, 431.
15. Roy, R. N., Gibbons, J. J., Snelling, R., Moeller, J., White, T., *J. Phys. Chem.* (1977) **81**, 391.
16. Roy, R. N., Gibbons, J. J., Krueger, C., White, T., *J. Chem. Soc., Faraday Trans.* I (1976) **72**, 2197.
17. Roy, R. N., Gibbons, J. J., Buechter, K., Faszholz, S., *J. Solution Chem.*, to be submitted soon.
18. Unni, A. K. R., Elias, L., Schiff, H. I., *J. Phys. Chem.* (1963) **67**, 1216.
19. Bates, R. G., "Determination of pH," 2nd ed., pp. 283, 331, Wiley, New York, 1973.

20. Hetzer, H. B., Robinson, R. A., Bates, R. G., *J. Phys. Chem.* (1962) **66,** 1423.
21. Lewis, G. N., Randall, M., "Thermodynamics," rev. by K. S. Pitzer, L. Brewer, Chap. 23, McGraw-Hill, New York, 1961.
22. Guggenheim, E. A., Turgeon, J. C., *Trans. Faraday Soc.* (1955) **51,** 747.
23. Scatchard, G., *J. Am. Chem. Soc.* (1961) **83**, 2636.
24. McKay, H. A. C., *Trans. Faraday Soc.* (1955) **51,** 903.
25. Mayer, J. E., *J .Chem. Phys.* (1950) **18,** 1426.
26. Friedman, H. L., "Ionic Solution Theory," Wiley, New York, 1962.
27. Pitzer, K. S., *J. Phys. Chem.* (1974) **78,** 2698.
28. Pitzer, K. S., *J. Solution Chem.* (1975) **4,** 249.
29. Frank, H. S., Evans, M. W., *J. Phys. Chem.* (1945) **13,** 507.
30. Rasaiah, J. C., *J. Chem. Phys.* (1970) **52,** 704.
31. Wood, R. H., Anderson, H. L., *J. Phys. Chem.* (1967) **71,** 1871.
32. Pitzer, K. S., Roy, R. N., Silvester, L. F., *J. Am. Chem. Soc.* (1977) **99,** 4930.

RECEIVED March 10, 1978.

Thermodynamic Study of Glycine in Different Tetrahydrofuran–Water Mixtures at Several Temperatures from 278.15 to 328.15 K

R. N. ROY, J. J. GIBBONS, J. L. PADRON,
K. BUECHTER, and S. FASZHOLZ

Department of Chemistry, Drury College, Springfield, MO 65802

The first and second dissociation constants of glycine have been determined by precise emf methods in 10, 30, and 50 mass % tetrahydrofuran–water mixtures at eleven different temperatures (ranging from 278.15 to 328.15 K at intervals of 5 K). The thermodynamic quantities ($\Delta G°$, $\Delta S°$, $\Delta H°$, and $\Delta C_p°$) were derived from the variation of these pK's with temperature. For glycine in 10 mass % THF–H_2O, pK_1 = 2715.7/T − 15.031 + 0.02805T, whereas that for pK_2 = 4622.3/T − 13.521 + 0.02607T. For glycine in 30 mass % THF–H_2O, pK_1 = 2084.1/T − 10.677 + 0.02126T, while that for pK_2 = 3561.7/T − 6.9091 + 0.01577T. For glycine in 50 mass % THF–H_2O, pK_1 = 703.6/T − 1.2781 + 0.006203T, whereas that for pK_2 = 3880.8/T − 9.3273 + 0.020258T. The results have been discussed in terms of the solute–solvent interactions and were compared with those in water, as well as in 50 mass % methanol–water and in 50 mass % monoglyme–water.

Considerable attention has been devoted to the nature of the solvent effects (as determined in water and in various mixed solvents) on the ionic dissociations (and related thermodynamic quantities) and other acid–base properties of aliphatic zwitterionic compounds. Such investigations include studies of tricine in 50 mass % methanol–water (*1*), "Bes" in pure water and in 50 mass % methanol–water (*2, 3*), glycine in 50 mass % monoglyme–water (*4*), and glycine in pure water and in 50 mass % methanol–water (*5, 6, 7*). The numerous factors (*8, 9, 10*) which

0-8412-0428-4/79/33-177-277$05.25/1

describe the preferential (or selective or specific) solute–solvent interactions, and are useful in accounting for the equilibrium behavior of electrolytes in solution, include the hydrogen-bonding capabilities of the solute and solvent species (*11*) and the dielectric properties of the medium and the ionic charge (*12*). The changes in pK as functions of the properties of the solvent (which allow for a wide range in the dielectric constant) may provide useful information in regard to the identification of the nature of these solute–solvent interaction patterns in binary solvent systems. However, direct measurements of this kind of interaction are difficult to perform.

As a part of the continuing studies of the effect of dipolar aprotic solvent plus water mixtures on these specific solute–solvent interactions, we have examined the first and second dissociation steps of glycine and computed their associated thermodynamic quantities in 10 mass % tetrahydrofuran–water (THF–H_2O) solvents (dielectric constant, ϵ = 71.8 at 298.15 K), 30 mass % THF–H_2O ($\epsilon = 56.6$ at 298.15 K), and 50 mass % THF–H_2O ($\epsilon = 40.0$ at 298.15 K) from 278.15 to 328.15 K.

These specific mixed solvents were chosen for this study, since solvent systems of this type have been used in similar studies such as the study of glucose mutarotation kinetics and equilibria in 50 mass % THF–H_2O (*13*), the investigation of ionization constants of weak acids (such as phenol), the determination of solubility product constants, K_{sp}, of slightly soluble ionogens (such as AgCl) in 10, 30, 50, and 70 mass % THF–H_2O (*14*), and the study of the standard Gibbs energy of transfer of *n*-$(Bu)_4$-NBr from water to THF–H_2O mixtures (*15*). Other studies have included the determination of the dissociation constant of *m*-nitroanilinium ion in various THF–H_2O mixtures (*16*), the solvation study of KCl and $KClO_4$ in THF–H_2O mixtures (*17*), and the conductance study of CsBr in 50 mass % THF–H_2O (*18*).

Measurements were made of the emf of cells of the type (without liquid junction): for the first dissociation step,

$$\text{Pt}; H_2(\text{g, 1 atm}) \mid \text{GHBr}(m_1), \; G^{\pm}(m_2) \text{ in } x \text{ mass \% THF–}H_2O \mid \text{AgBr, Ag} \quad \text{(I)}$$

and for the second dissociation stage:

$$\text{Pt}; H_2(\text{g, 1 atm}) \mid G^{\pm}(m_1), \text{NaG}(m_2), \; \text{KBr}(m_3) \text{ in } x \text{ mass \% THF–}H_2O \mid \text{AgBr, Ag} \quad \text{(II)}$$

where x = 10, 30, and 50 mass % THF–H_2O, and m represents the respective molality.

Experimental Procedure and Preparation of Solutions

The precise emf method originally introduced by Harned and Ehlers (*19*), and modified by Bates and Robinson, was used in the present study. The values of the emf were corrected to a partial pressure of one atmosphere (101.325 kp_a) of hydrogen. The vapor pressure data required for emf correction of 10, 30, and 50 mass % THF–H_2O at different temperatures are given elsewhere (*20*). The corrected emf data at the respective solvent compositions are given in Tables I, II, and III (for Cell I), and in Tables IV, V, and VI (for Cell II), respectively. The emf data were stable, as is evident from the data at 298.15 K, which were recorded at the beginning, in the middle, and at the end of each run. The mean difference between the first and the last readings was never more than 0.08 mV. The values of the emf were measured at five-degree intervals from 273.15 to 298.15 K, using a precision potentiometer (Leeds and Northrup Type K-5), which was standardized against an Eppley standard cell, and equipped with a Leeds and Northrup D.C. null detector (model 9829), using a sensitivity of 25 μV. The temperature of the constant temperature bath was known to within 0.02 K.

The design of the cells, the preparations of the silver + silver bromide electrodes (of the thermal type), and the hydrogen electrodes, have been described elsewhere (*21, 22*). The bias potential of the silver–silver bromide electrodes was always within 0.05 mV.

The cell solutions for the determination of the pK_1 were prepared by dilution with THF–H_2O solvent of a stock solution made by dissolving glycine in a portion of the hydrobromic acid solution and adding the proper amounts of water and pure THF. For the determination of the pK_2, stock solutions were prepared by mixing accurately weighed portions of glycine, sodium hydroxide solutions, and the required amounts of water and THF. These stock solutions then were brought to the proper molalities by dilution with distilled water and THF. Molalities are estimated to be accurate to within 0.05%. Dissolved air was removed by bubbling purified hydrogen gas through the solutions before the emf cells were filled. Vacuum corrections were applied to all weighings.

A commercial sample of glycine (Sigma Chemical Co.) was recrystallized twice from low-conductivity water ($\kappa = 1 \times 10^{-6}$ ohm^{-1} cm^{-1} at 298.15 K, as determined using a YSI Model 31 conductivity bridge). It assayed at 99.96% (standard deviation = 0.12) when titrated with a standard solution of NaOH to the phenolphthalein end point. The detailed procedure to check this purity has been given elsewhere (*7*). Recrystallized KBr (free from contamination by chloride) was used. Tetrahydrofuran was refluxed over anhydrous sodium sulfate for 3 days and then vacuum-distilled (*23, 24*).

Table I. Electromotive Force (E) of the Cell: Pt/H_2(g)/GHBr(m_1), Temperatures (T) from

m_1/mol kg^{-1}	m_2/mol kg^{-1}	T/K − 273.15			
		5	10	15	20
0.097950	0.302699	0.31383	0.31509	0.31623	0.31722
0.077198	0.324519	0.32609	0.32745	0.32881	0.33001
0.040002	0.120139	0.33109	0.33257	0.33388	0.33522
0.068372	0.328423	0.33121	0.33265	0.33413	0.33555
0.030001	0.089629	0.33686	0.33734	0.33936	0.34109
0.019992	0.060077	0.34551	0.34820	0.34983	0.35143
0.009997	0.029635	0.36129	0.36346	0.36564	0.36767

Table II. Electromotive Force (E) of the Cell: Pt/H_2(g)/GHBr(m_1), Temperatures (T) from

m_1/mol kg^{-1}	m_2/mol kg^{-1}	T/K − 273.15			
		5	10	15	20
0.100010	0.301070	0.32387	0.32433	0.32464	0.32489
0.086069	0.350453	0.33403	0.33461	0.33419	0.33483
0.040004	0.121078	0.34242	0.34478	0.34405	0.34459
0.069656	0.330622	0.34190	0.34275	0.34421	0.34473
0.030001	0.091001	0.34847	0.34946	0.34976	0.35060
0.019991	0.060511	0.35670	0.35785	0.35951	0.36042
0.010000	0.029677	0.37110	0.37250	0.37378	0.37498

Table III. Electromotive Force (E) of the Cell: Pt/H_2(g)/GHBr(m_1), Temperatures (T) from

m_1/mol kg^{-1}	m_2/mol kg^{-1}	T/K − 273.15			
		5	10	15	20
0.098864	0.148651	0.31565	0.31412	0.31253	0.30995
0.080142	0.171518	0.32778	0.32662	0.32499	0.32326
0.070149	0.182134	0.33574	0.33441	0.33308	0.33121
0.060113	0.191735	0.34333	0.34198	0.34101	0.33974
0.039814	0.119902	0.35053	0.34996	0.34931	0.34797
0.029997	0.090539	0.35573	0.35523	0.35458	0.35390
0.020000	0.060533	0.36359	0.36325	0.36286	0.36233
0.009998	0.030410	0.37704	0.37695	0.37684	0.37656

$G^{\pm}(m_2)$ in Water + 10 Mass Percent Tetrahydrofuran/AgBr/Ag at 278.15 to 328.15 K

T/K − 273.15						
25	*30*	*35*	*40*	*45*	*50*	*55*
0.31837	0.31922	0.32012	0.32090	0.32190	0.32312	0.32644
0.33130	0.33255	0.33300	0.33487	0.33604	0.33727	0.34067
0.33630	0.33809	0.33882	0.34074	0.34214	0.34367	0.34539
0.33697	0.33841	0.33934	0.34091	0.34222	0.35053	0.34743
0.34268	0.34425	0.34580	0.34731	0.34888	0.36135	0.35272
0.35306	0.35514	0.35521	0.35824	0.36042	0.36343	0.36309
0.36978	0.37192	0.37300	0.37562	0.37756	0.37962	0.38221

$G^{\pm}(m_2)$ in Water + 30 Mass Percent Tetrahydrofuran/AgBr/Ag at 278.15 to 328.15 K

T/K − 273.15						
25	*30*	*35*	*40*	*45*	*50*	*55*
0.32575	0.32547	0.32633	0.32650	0.32606	0.32651	0.32476
0.33675	0.33676	0.33772	0.33916	0.33946	0.33980	0.33795
0.34516	0.34560	0.34635	0.34690	0.34681	0.34709	0.34704
0.34517	0.34586	0.34680	0.34711	0.34749	0.34772	0.34733
0.35123	0.35204	0.35295	0.35362	0.35479	0.35581	0.35504
0.36040	0.36185	0.36262	0.36327	0.36391	0.36430	0.36442
0.37592	0.37710	0.37783	0.37886	0.37998	0.38059	0.38131

$G^{\pm}(m_2)$ in Water + 50 Mass Percent Tetrahydrofuran/AgBr/Ag at 278.15 to 328.15 K

T/K − 273.15						
25	*30*	*35*	*40*	*45*	*50*	*55*
0.30747	0.30764	0.30651	0.30476	0.30203	0.29846	0.29607
0.32194	0.32109	0.32038	0.31874	0.31627	0.31383	0.31102
0.33096	0.32983	0.32895	0.32732	0.32529	0.32312	0.33096
0.33917	0.33839	0.33770	0.33630	0.33466	0.33283	0.33978
0.34713	0.34651	0.34567	0.34489	0.34379	0.34212	0.34218
0.35296	0.35267	0.35185	0.35117	0.35040	0.34935	0.34004
0.36169	0.36123	0.36057	0.36023	0.35945	0.35840	0.34753
0.37617	0.37606	0.37575	0.37548	0.37505	0.37485	0.37401

Table IV. Electromotive Force (E) of the Cell: $Pt/H_2(g)/G^{\pm}(m_1)$, AgBr/Ag at Temperatures

m_1/mol kg^{-1}	m_2/mol kg^{-1}	m_3/mol kg^{-1}	T/K − 273.15			
			5	10	15	20
0.100440	0.099993	0.100250	0.70652	0.70731	0.70809	0.70872
0.081639	0.079977	0.079431	0.71100	0.71195	0.71282	0.71350
0.069554	0.069978	0.068937	0.71501	0.71599	0.71679	0.71749
0.060209	0.059933	0.060679	0.71731	0.71838	0.71935	0.72012
0.031846	0.029965	0.030237	0.73103	0.73250	0.73378	0.73482
0.019601	0.019925	0.019662	0.74291	0.74453	0.74601	0.74729
0.010286	0.009993	0.011249	0.75455	0.75638	0.75807	0.75955

Table V. Electromotive Force (E) of the Cell: $Pt/H_2(g)/G^{\pm}(m_1)$, AgBr/Ag at Temperatures

m_1/mol kg^{-1}	m_2/mol kg^{-1}	m_3/mol kg^{-1}	T/K − 273.15			
			5	10	15	20
0.099301	0.099977	0.101190	0.70447	0.70466	0.70467	0.70460
0.079570	0.080009	0.079156	0.70891	0.70900	0.70911	0.70906
0.070927	0.069994	0.070684	0.71037	0.71070	0.71090	0.71099
0.060574	0.059995	0.060940	0.71355	0.71391	0.71411	0.71420
0.040763	0.039976	0.041685	0.72099	0.72158	0.72211	0.72240
0.031108	0.029932	0.031125	0.72660	0.72722	0.72768	0.72792
0.022737	0.019997	0.020702	0.73420	0.73495	0.73556	0.73598
0.011232	0.009282	0.009836	0.74927	0.75046	0.75144	0.75224

Table VI. Electromotive Force (E) of the Cell: $Pt/H_2(g)/G^{\pm}(m_1)$, AgBr/Ag at Temperatures

m_1/mol kg^{-1}	m_2/mol kg^{-1}	m_3/mol kg^{-1}	T/K − 273.15		
			5	10	15
0.10810	0.10020	0.10200	0.70082	0.70034	0.69909
0.07232	0.07978	0.08058	0.70675	0.70602	0.70518
0.06850	0.07001	0.07464	0.70496	0.70436	0.70359
0.04090	0.06000	0.04860	0.72055	0.72055	0.71983
0.03285	0.05000	0.04112	0.72370	0.72340	0.72313
0.02432	0.04000	0.02999	0.73132	0.73096	0.73083
0.00935	0.03000	0.02190	0.75340	0.75359	0.75363
0.00437	0.01000	0.01006	0.76091	0.76123	0.76181

NaG(m_2), KBr(m_3) in Water + 10 Mass Percent Tetrahydrofuran/ (T) from 278.15 to 328.15 K

T/K − 273.15						
25	30	35	40	45	50	55
0.70927	0.70964	0.70996	0.71027	0.71061	0.71084	—
0.71422	0.71472	0.71520	0.71570	0.71621	0.71684	0.71767
0.71821	0.71880	0.71928	0.71968	0.72029	0.72106	0.72277
0.72099	0.72167	0.72226	0.72286	0.72345	0.72409	0.72526
0.73600	0.73693	0.73786	0.73873	0.73964	0.74055	0.74186
0.74863	0.74980	0.75086	0.75192	0.75307	0.75442	0.75689
0.76116	0.76256	0.76396	0.76533	0.76684	0.76875	0.77074

NaG(m_2), KBr(m_3) in Water + 30 Mass Percent Tetrahydrofuran/ (T) from 278.15 to 328.15 K

T/K − 273.15						
25	30	35	40	45	50	55
0.70449	0.70413	0.70370	0.70323	0.70242	0.70154	—
0.70878	0.70897	0.70864	0.70822	0.70755	0.70663	0.70548
0.71096	0.71093	0.71068	0.71030	0.70990	0.70944	0.70789
0.71411	0.71436	0.71421	0.71404	0.71269	0.71222	0.71141
0.72257	0.72279	0.72273	0.72271	0.72228	0.72208	0.72130
0.72827	0.72904	0.72917	0.72920	0.72907	0.72884	0.72830
0.73598	0.73746	0.73705	0.73720	0.73712	0.73717	0.73703
0.75246	0.75354	0.75387	0.75406	0.75424	0.75452	0.75561

NaG(m_2), KBr(m_3) in Water + 50 Mass Percent Tetrahydrofuran/ (T) from 278.15 to 323.15 K

T/K − 273.15						
20	25	30	35	40	45	50
0.69769	0.69707	0.69467	0.69422	0.69360	0.69412	0.69627
0.70368	0.70271	0.70061	0.70035	0.69959	0.69981	0.70159
0.70199	0.70091	0.69888	0.69865	0.69796	0.69813	0.69945
0.71833	0.71701	0.71571	0.71504	0.71499	0.71525	0.71605
0.72161	0.72027	0.71911	0.71909	0.71845	0.71872	0.71932
0.72933	0.72805	0.72652	0.72742	0.72656	0.72679	0.72693
0.75253	0.75119	0.74999	0.75100	0.75051	0.75071	0.75157
0.76045	0.75883	0.75809	0.75913	0.75857	0.75885	0.75884

Methods and Results

Glycine is normally written in the form $\overset{+}{N}H_3–CH_2–COO^-$. If this zwitterion is represented by $G^\pm$, then the first dissociation process is the deprotonation of the carboxylate group of the species HG^+:

$$HG^+ \overset{K_1}{\rightleftarrows} H^+ + G^\pm \tag{1}$$

while the second dissociation step is the equilibrium given by

$$G^\pm \overset{K_2}{\rightleftarrows} H^+ + G^- \tag{2}$$

where K_1 and K_2 are the thermodynamic equilibrium constants, which were determined in 10, 30, and 50 mass % THF–H_2O at eleven temperatures from 278.15 to 328.15 K.

The pK_1 was determined by extrapolating the values of the pK_1' function (the apparent dissociation constant) to infinite dilution (i.e., zero ionic strength, I).

The following two equations are important:

$$pK_1 = \frac{E - E^\circ}{(RT \ln 10)/F} + \log \frac{m_{HG^+}\, m_{Br^-}}{m_{G^\pm}} + \log \frac{\gamma_{HG^+}\, \gamma_{Br^-}}{\gamma_{G^\pm}} \tag{3}$$

and

$$pK_1' \equiv pK_1 - \beta I = \frac{E - E^\circ}{(RT \ln 10)/F} + \log \frac{m_1(m_1 - m_H')}{(m_2 + m_H')} - \frac{2AI^{\frac{1}{2}}}{1 + Ba_oI^{\frac{1}{2}}} \tag{4}$$

Equation 3 was obtained by combining the Nernst equation for the emf of Cell I with the equilibrium constant of the acidic dissociation of glycine. In Equations 2 and 3, E° is the standard emf of the cell in the respective solvent composition and these values were obtained from an earlier work (*20*). In Equation 4, β is the linear slope parameter for the plot of pK_1' vs. I, a_o is the ion-size parameter, A and B are the Debye–Huckel constants on the molal scale (*20*) for the respective mixed solvent systems, and I is the ionic strength given by m_1.

Because of the appreciable dissociation of the species, HG^+, at low pH values of the buffer solutions, $m_{HG^+} = m_1 - m_H'$, and $m_{G^\pm} = m_2 + m_H'$. This apparent hydrogen ion molality, m_H', was estimated according to the equation,

$$-\log m_H' = \frac{E - E^\circ}{(RT \ln 10)/F} + \log m_1 - \frac{2AI^{\frac{1}{2}}}{1 + Ba_oI^{\frac{1}{2}}} \quad (5)$$

The least-squares estimate of the intercept for the linear regression of pK_1' vs. I from Equation 4 (using the proper value of a_o) was termed the actual pK_1, the values of which are presented in Tables VII, VIII, and IX, for 10, 30, and 50 mass % THF–H_2O, respectively. In order to obtain the best linear fit of the data, several values of a_o (from 0.1 to 0.8 nm) were inserted into this equation. The true values of pK_1 were those which gave the smallest standard deviation of regression.

These experimental values of pK_1 were fitted into an equation selected by Harned and Robinson (25), by the least-squares method, with the following results:

$$pK_1 \text{ (in 10 mass \% THF–H}_2\text{O)} = 2715.7/T - 15.031 + 0.02805T \quad (6)$$

where T is the thermodynamic temperature in K.

$$pK_1 \text{ (in 30 mass \% THF–H}_2\text{O)} = 2084.1/T - 10.677 + 0.02126T \quad (7)$$

and

$$pK_1 \text{ (in 50 mass \% THF–H}_2\text{O)} = 703.6/T - 1.2781 + 0.006203T \quad (8)$$

In order to get a satisfactory fit of the data, it was necessary to use double precision arithmetic (16 digits). The average deviations of the fit for Equations 6, 7, and 8 were 0.007, 0.003, and 0.003 pK units, respectively.

The second dissociation constant, pK_2 (which involves the removal of the proton from the nitrogen atom in the process $G^\pm \rightleftarrows H^+ + G^-$), can be obtained by extrapolation of pK_2' values (the apparent second dissociation constants) to zero ionic strength I, where $I = m_2 + m_3$. The pertinent equation is

$$pK_2' = \frac{E - E^\circ}{(RT \ln 10)/F} + \log \frac{m_3(m_1 + m_{OH^-}')}{(m_2 - m_{OH^-}')} + \log \frac{\gamma_{Br^-}\gamma_{G^\pm}}{\gamma_{G^-}} \quad (9)$$

In Equation 9, $\gamma_{G^\pm}$ is assumed to be equal to unity; hence, the last term is directly proportional to the ionic strength I. Since the anion G^- undergoes hydrolysis at high pH values, the hydroxyl ion concentration, m_{OH^-}', which is very small in comparison with m_1 and m_2, can be evaluated with sufficient accuracy by the equation

$$-\log m_{OH^-}' = pK_w + \log m_H\gamma_H \quad (10)$$

The contribution of m_{OH^-}' to the value of pK_2 (for example, at 298.15 K) lies within 0.0005 pK unit, which is well within the experimental error. Thus, the simplified version of Equation 9 (ignoring these negligible corrections) was used:

$$pK_2' = \frac{E - E°}{(RT \ln 10)/F} + \log \frac{m_1 m_3}{m_2} \qquad (11)$$

The values of pK_2 listed in Tables VII, VIII, and IX were obtained from the linear plots of pK_2' vs. I (where $I = m_2 + m_3$), when extrapolated to zero. The variation in the observed values of the pK_2 with temperature can be well represented by Harned and Robinson's equation (*25*). The final forms resulting from these equations are:

$$pK_2 \text{ (in 10 mass \% THF–H}_2\text{O)} = 4622.3/T - 13.521 + 0.02607T \qquad (12)$$

$$pK_2 \text{ (in 30 mass \% THF–H}_2\text{O)} = 3561.7/T - 6.9091 + 0.01577T \qquad (13)$$

$$pK_2 \text{ (in 50 mass \% THF–H}_2\text{O)} = 3880.8/T - 9.3273 + 0.020258T \qquad (14)$$

The average deviations between the experimental values and those calculated from Equations 12, 13, and 14 are 0.008, 0.003, and 0.007, respectively.

The thermodynamic functions for the first and second ionization processes were evaluated from Equations 6, 7, and 8 (for pK_1) and 12, 13, and 14 (for pK_2) in 10, 30, and 50 mass % THF–H_2O, respectively, using the customary thermodynamic formulae. The values of the standard Gibbs energy ($\Delta G°$), enthalpy ($\Delta H°$), entropy ($\Delta S°$), and heat capacity

Table VII. The pK_1, pK_2, σ, and a_0 of Protonated Glycine (GH^+ and $G^\pm$) in 10 Mass Percent Tetrahydrofuran–Water from 278.15 to 328.15 K on the Molal Scale

T/K − 273.15	*pK_1 (obs)*	*pK_1 (calc)*	*σ (pK_1)*	a_0/*nm*	*pK_2 (obs)*	*pK_2 (calc)*	*σ (pK_2)*
5	2.527	2.535	0.002	0.25	10.338	10.348	0.003
10	2.502	2.503	0.008	1.50	10.186	10.185	0.003
15	2.481	2.476	0.006	1.50	10.038	10.032	0.003
20	2.463	2.456	0.005	1.50	9.895	9.889	0.002
25	2.447	2.441	0.005	1.50	9.762	9.755	0.002
30	2.440	2.431	0.005	1.50	9.634	9.629	0.002
35	2.411	2.426	0.002	0.50	9.512	9.512	0.002
40	2.423	2.425	0.004	0.25	9.396	9.403	0.002
45	2.424	2.429	0.006	0.50	9.288	9.302	0.002
50	2.429	2.437	0.008	0.25	9.195	9.207	0.003
55	2.462	2.450	0.006	1.50	9.138	9.119	0.005

Table VIII. The pK_1, pK_2, σ, and a_0 of Protonated Glycine (GH^+ and $G^{\pm}$) in 30 Mass Percent Tetrahydrofuran–Water from 278.15 to 318.15 K on the Molal Scale

$T/K - 273.15$	pK_1 (obs)	pK_1 (calc)	σ (pK_1)	a_0/nm	pK_2 (obs)	pK_2 (calc)	σ (pK_2)
5	2.723	2.727	0.002	0.25	10.282	10.282	0.003
10	2.705	2.701	0.008	0.25	10.136	10.135	0.003
15	2.681	2.680	0.008	0.25	9.996	9.995	0.003
20	2.663	2.663	0.007	0.25	9.860	9.863	0.003
25	2.651	2.650	0.002	0.25	9.731	9.738	0.003
30	2.641	2.641	0.003	0.25	9.628	9.620	0.004
35	2.632	2.635	0.004	0.25	9.510	9.508	0.002
40	2.629	2.634	0.006	0.25	9.402	9.402	0.004
45	2.640	2.635	0.008	0.25	9.301	9.303	0.004

Table IX. The pK_1, pK_2, σ, and a_0 of Protonated Glycine (GH^+ and $G^{\pm}$) in 50 Mass Percent Tetrahydrofuran–Water from 278.15 to 328.15 K on the Molal Scale

$T/K - 273.15$	pK_1 (obs)	pK_1 (calc)	σ (pK_1)	a_0/nm	pK_2 (obs)	pK_2 (calc)	σ (pK_2)
5	2.972	2.977	0.004	2.00	10.246	10.260	0.002
10	2.969	2.963	0.005	4.00	10.119	10.115	0.003
15	2.955	2.951	0.006	4.00	10.002	9.978	0.002
20	2.940	2.941	0.005	4.00	9.854	9.850	0.001
25	2.929	2.931	0.004	2.00	9.716	9.729	0.001
30	2.922	2.924	0.004	4.00	9.599	9.616	0.003
35	2.913	2.917	0.003	4.00	9.516	9.509	0.003
40	2.912	2.911	0.004	6.00	9.409	9.409	0.002
45	2.909	2.907	0.005	4.00	9.320	9.316	0.003
50	2.909	2.904	0.003	4.00	9.228	9.228	0.005
55	2.898	2.902	0.007	6.00	—	—	—

ΔC_p°) changes for both the processes are summarized in Tables X, XI, and XII (on the molal scale), and in Tables XIII, XIV, and XV (on the mole-fraction basis), together with the respective estimates of the errors. The standard deviations of these quantities were estimated by the application of the method of propagation of errors as used by Please (*26*).

Discussion

The pK_1 of glycine in water is 2.350 at 298.15 K (*6*), as compared with 2.447 (Table VII) in 10 mass % THF–H_2O, 2.651 (Table VIII) in 30 mass %, and 2.929 (Table IX) in 50 mass %, respectively. The dielectric constants for each of the above solvent mixtures (including pure water) at 298.15 K are 78.3 (for water), 71.8 (for 10 mass % THF–H_2O),

Table X. Thermodynamic Functions (on the Molal Scale) for the at Temperatures (*T*)

T / *K − 273.15*	ΔG° / *cal mol⁻¹*	σ(ΔG°)	ΔH° / *cal mol⁻¹*	σ(ΔH°)
		First Dissociation Process		
5	3226	8	2496	201
15	3265	4	1769	130
25	3330	4	1017	68
35	3420	4	239	72
45	3536	4	−565	141
		Second Dissociation Process		
5	13170	4	11923	107
15	13227	2	11248	69
25	13308	2	10548	36
35	13412	2	9825	38
45	13541	2	9078	75

56.6 (for 30 mass %), and 40.0 (for 50 mass % THF–H_2O), respectively (*20*). It is evident from the above data that the HG^+ species (that is, NH_3–CH_2–COO^-) decreases in acidic strength when THF is added to a solvent that is initially pure water, which follows the same general trend as that found in other dipolar aprotic solvent–water media, and also in amphiprotic solvent–water media. For example, the pK_1 of glycine in 50 mass % monoglyme–water at 298.15 K is 2.806 (*4*), whereas the pK_1 of glycine in 50 mass % methanol–water at 298.15 K is 2.961 (*7*). The deprotonation of the carboxylate group in the species HG^+ mimics the

Table XI. Thermodynamic Functions (on the Molal Scale) for the at Temperatures (*T*)

T / *K − 273.15*	ΔG° / *cal mol⁻¹*	σ(ΔG°)	ΔH° / *cal mol⁻¹*	σ(ΔH°)
		First Dissociation Process		
5	3471	8	2011	274
15	3533	4	1460	153
25	3615	5	890	78
35	3716	4	301	164
45	3836	8	−308	299
		Second Dissociation Process		
5	13086	5	10715	162
15	13178	3	10307	91
25	13285	3	9884	46
35	13406	3	9446	97
45	13542	5	8994	177

Dissociations of Glycine in 10 Mass Percent Tetrahydrofuran–Water from 278.15 to 318.15 K

$\Delta S°$ cal K^{-1} mol^{-1}	$\sigma(\Delta S°)$	$\Delta C_p°$ cal K^{-1} mol^{-1}	$\sigma(\Delta C_p°)$
	First Dissociation Process		
−2.6	0.7	−71.4	7.5
−5.2	0.4	−74.0	7.8
−7.8	0.2	−76.5	8.0
−10.3	0.2	−79.1	8.3
−12.9	0.4	−81.7	8.6
	Second Dissociation Process		
−4.5	0.4	−66.3	4.0
−6.9	0.2	−68.7	4.1
−9.2	0.1	−71.1	4.3
−11.6	0.1	−73.5	4.4
−14.0	0.2	−75.9	4.6

behavior of similar model neutral acids, such as acetic acid, since the pK_a of acetic acid in water is 4.756 at 298.15 K, as compared with 5.660 in 50 mass % methanol–water (*27*) and 5.645 in 50 mass % ethylene carbonate–water (*28*).

It is evident from Tables VII, VIII, and IX that the pK_2 (involving the removal of the proton from the nitrogen atom) at 298.15 K is 9.716 in 50 mass % THF–H_2O, as compared with 9.780 in water (*6*) and 9.453 in 50 mass % monoglyme–water (*4*) at the same temperature. Thus, the acidic strength of the dissociation of $G^{\pm}$ increases when pure water is

Dissociations of Glycine in 30 Mass Percent Tetrahydrofuran–Water from 278.15 to 318.15 K

$\Delta S°$ cal K^{-1} mol^{-1}	$\sigma(\Delta S°)$	$\Delta C_p°$ cal K^{-1} mol^{-1}	$\sigma(\Delta C_p°)$
	First Dissociation Process		
−5.2	1.0	−54.1	12.9
−7.2	0.5	−56.0	13.3
−9.1	0.3	−58.0	13.8
−11.1	0.5	−60.0	14.3
−13.0	1.0	−61.9	14.7
	Second Dissociation Process		
−8.5	0.6	−40.1	7.6
−10.0	0.3	−41.6	7.9
−11.4	0.2	−43.0	8.2
−12.8	0.3	−44.5	8.4
−14.3	0.6	−45.9	8.7

Table XII. Thermodynamic Functions (on the Molal Scale) for the at Temperatures (*T*)

T / *K − 273.15*	ΔG° / *cal mol⁻¹*	σ(ΔG°)	ΔH° / *cal mol⁻¹*	σ(ΔH°)
		First Dissociation Process		
5	3789	7	1024	179
15	3891	4	863	116
25	3999	4	697	61
35	4113	4	524	65
45	4232	4	347	126
		Second Dissociation Process		
5	13058	3	10586	98
15	13156	2	10061	59
25	13272	2	9517	29
35	13408	2	8955	44
45	13561	2	8375	85

substituted by 10, 30, and 50 mass % THF–H_2O as the solvent. This observation conforms to the same trend as that of a similar model nitrogen acid, such as NH_4^+, whose pK_a in water is 9.245 at 298.15 K, as compared with 8.687 in 50 mass % methanol–water (*29*).

With regard to the first dissociation step, the transfer Gibbs energy of dissociation, $\Delta G_t°$(diss), pertains to the process

$$HG^+(s) + G^{\pm}(w) + H^+(w) = HG^+(w) + G^{\pm}(s) + H^+(s) \quad (15)$$

Table XIII. Thermodynamic Functions (on the Mole-Fraction Scale) for the Dissociations of Glycine in 10 Mass Percent Tetrahydrofuran–Water at Temperatures (*T*) from 278.15 to 318.15 K

T / *K − 273.15*	ΔG° / *cal mol⁻¹*	ΔH° / *cal mol⁻¹*	ΔS° / *cal K⁻¹ mol⁻¹*	$\Delta C_p°$ / *cal K⁻¹ mol⁻¹*
		First Dissociation Process		
5	3140	2496	−2.3	−71.4
15	3176	1769	−4.9	−74.0
25	3238	1017	−7.4	−76.5
35	3325	239	−10.0	−79.1
45	3437	−565	−12.6	−81.7
		Second Dissociation Process		
5	13084	11923	−4.2	−66.3
15	13138	11248	−6.6	−68.7
25	13216	10548	−8.9	−71.1
35	13317	9825	−11.3	−73.5
45	13442	9078	−13.7	−75.9

Dissociations of Glycine in 50 Mass Percent Tetrahydrofuran–Water from 278.15 to 318.15 K

ΔS° / *cal K⁻¹ mol⁻¹*	$\sigma(\Delta S^\circ)$	$\Delta C_p{}^\circ$ / *cal K⁻¹ mol⁻¹*	$\sigma(\Delta C_p{}^\circ)$
	First Dissociation Process		
−9.9	0.6	−15.8	6.7
−10.5	0.4	−16.4	6.9
−11.1	0.2	−16.9	7.2
−11.6	0.2	−17.5	7.4
−12.2	0.4	−18.1	7.6
	Second Dissociation Process		
−8.8	0.3	−51.6	4.1
−10.7	0.2	−53.4	4.2
−12.6	0.1	−55.3	4.4
−14.4	0.1	−57.1	4.5
−16.3	0.3	−59.0	4.6

while the transfer process associated with the second ionization step is

$$G^{\pm}(s) + G^{-}(w) + H^{+}(w) = G^{\pm}(w) + G^{-}(s) + H^{+}(s) \qquad (16)$$

where w indicates water and s represents the mixed solvent, respectively. The $\Delta G_t{}^\circ$ (diss) then can be mathematically expressed by the equation

$$\Delta G_t{}^\circ(\text{diss}) = (RT \ln 10)[\text{pK}_s(m) - \text{pK}_w(m)] \qquad (17)$$

Table XIV. Thermodynamic Functions (on the Mole-Fraction Scale) for the Dissociations of Glycine in 30 Mass Percent Tetrahydrofuran–Water at Temperatures (*T*) from 278.15 to 318.15 K

T / *K − 273.15*	ΔG° / *cal mol⁻¹*	ΔH° / *cal mol⁻¹*	ΔS° / *cal K⁻¹ mol⁻¹*	$\Delta C_p{}^\circ$ / *cal K⁻¹ mol⁻¹*
		First Dissociation Process		
5	3189	2011	−4.2	−54.1
15	3241	1460	−6.2	−56.0
25	3313	890	−8.1	−58.0
35	3404	301	−10.1	−60.0
45	3514	−308	−12.0	−61.9
		Second Dissociation Process		
5	12804	10715	−7.5	−40.1
15	12886	10307	−9.0	−41.6
25	12983	9884	−10.4	−43.0
35	13094	9446	−11.8	−44.5
45	13220	8994	−13.3	−45.9

Table XV. Thermodynamic Functions (on the Mole-Fraction Scale) for the Dissociations of Glycine in 50 Mass Percent Tetrahydrofuran–Water at Temperatures (T) from 278.15 to 318.15 K

$\frac{T}{K - 273.15}$	$\frac{\Delta G^\circ}{cal\ mol^{-1}}$	$\frac{\Delta H^\circ}{cal\ mol^{-1}}$	$\frac{\Delta S^\circ}{cal\ K^{-1}\ mol^{-1}}$	$\frac{\Delta C_p^\circ}{cal\ K^{-1}\ mol^{-1}}$
		First Dissociation Process		
5	3269	1024	−8.1	−15.8
15	3353	863	−8.6	−16.4
25	3442	697	−9.2	−16.9
35	3537	524	−9.8	−17.5
45	3638	347	−10.3	−18.1
		Second Dissociation Process		
5	12538	10586	−7.0	−51.6
15	12618	10061	−8.9	−53.4
25	12715	9517	−10.7	−55.3
35	12832	8955	−12.6	−57.1
45	12967	8375	−14.4	−59.0

where m indicates the value of pK on the molal scale. The values of ΔG_t°(diss) on this scale, as determined from Equation 17, can easily be calculated from the data given in Tables VII, VIII, and IX as well as from Reference 6, where data for $pK_w(m)$ are available.

Since it is easier to make a direct comparison of values for equal numbers of solvent molecules, all the data listed in Tables VII, VIII, and

Table XVI. Comparison of the Transfer Dissociation Energies (on Tetrahydrofuran, from Water to 50 Mass Percent Methanol,

	First Transfer Process		
	50 Mass % THF	*50 Mass % MeOH*[a]	*50 Mass % Monoglyme*[b]
$\frac{\Delta G_t^\circ (diss)}{cal\ mol^{-1}}$	790	541	16
$\frac{\Delta H_t^\circ (diss)}{cal\ mol^{-1}}$	696	550	−2572
$\frac{\Delta S_t^\circ (diss)}{cal\ K^{-1}\ mol^{-1}}$	−1.6	0.1	−8.7
$\frac{\Delta C_{pt} (diss)}{cal\ K^{-1}\ mol^{-1}}$	17.1	−19	86

[a] Ref. *1*.
[b] Ref. *4*.

IX have been converted to the mole-fraction scale. The following standard equation has been used for this conversion:

$$\mathrm{pK_s}(N) = \mathrm{pK_s}(m) + 2 \log (1000/M_s) \tag{18}$$

where M_s is the mean molar mass of the mixed solvent, and N denotes a value on the mole-fraction scale. The standard Gibbs energy of transfer $\Delta G_t{}^\circ$ is given by the equation

$$\Delta G_t{}^\circ(N) = 2.3026\, RT\, [\mathrm{pK_s}(N) - \mathrm{pK_w}(N)] \tag{19}$$

The values of $\Delta G_t{}^\circ$(diss) and $\Delta S_t{}^\circ$(diss) are different based on different scales, whereas those of $\Delta H_t{}^\circ$(diss) and $\Delta C_{pt}{}^\circ$(diss) are the same on each scale.

As can be seen from Table XVI, the value of $\Delta G_t{}^\circ$(diss) at 298.15 K for the first transfer process is + 790 cal mol^{-1}. This positive value indicates the preferential stabilization of HG^+ by the mixed solvent and the species $G^\pm$ and H^+ by water.

The value of $\Delta G_t{}^\circ$(diss) for the second transfer process at 298.15 K is − 87 cal mol^{-1}. This negative value strongly suggests (*a*) the stabilization of the zwitterionic species, $G^\pm$, by water through hydrogen bonding and (*b*) the interactions of the centers of the ionic charges G^- and H^+ with the mixed solvent. As evident from Table XVI, a similar trend was observed for the transfer dissociation energies in the behavior of glycine in 50 mass % methanol–water (*1*) and in 50 mass % monoglyme–water

the Mole-Fraction Scale) for Glycine from Water to 50 Mass Percent and from Water to 50 Mass Percent Monoglyme, at 25°C

Second Transfer Process		
50 Mass % THF	*50 Mass % MeOH*[a]	*50 Mass % Monoglyme*[b]
−87	−569	−1052
9506	−290	−3312
−1.3	1.0	−7.6
−43.3	−32	24

mixed solvent (*4*). According to Parker (*30*), the anion solvation (for example, solvation of G^-, the glycinate ion) by dipolar aprotic solvents (such as THF) is linked closely with delocalization of the negative charge.

The values (listed in Tables X–XV) of the standard Gibbs free energies of ionization, $\Delta G°$, for both processes increase with the increasing organic content of the solvent mixture. This trend is consistent with the effect of the lowered dielectric constant of the mixed solvent and it is caused by the increase of the electrostatic free energies of the ions produced in the dissociation process.

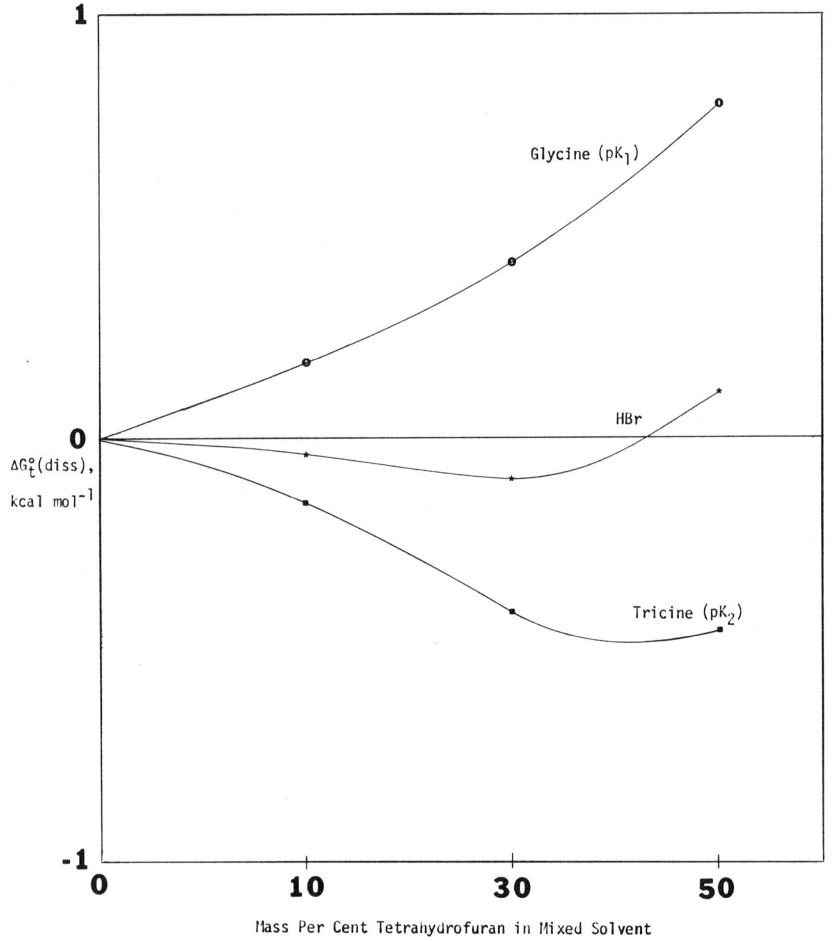

Figure 1. Transfer free energies (on the mole-fraction scale) for hydrobromic acid, glycine (pK_1), and tricine (pK_2) as a function of the composition of tetrahydrofuran–water solvent

It is interesting to note that the standard entropy of dissociation of glycine (for both processes) becomes steadily more negative (cf. Tables X–XV) with increasing THF content in the solvent mixtures, as opposed to pure water (*6*). The qualitative explanations (*31*) may be offered on this point. The THF–H_2O solvent mixtures are less structured than pure water; hence, the species (such as $G^{\pm}$, H^+, G^-, etc.) will orient solvent molecules to a greater extent in THF–H_2O media than in water. Consequently, this greater degree of orientation results in a more negative standard entropy of dissociation in the THF–H_2O solvent, which is produced by a decrease in the disorder of the system.

Electrostatic interactions between ions such as $G^{\pm}$, H^+, G^-, etc., and dipolar solvent molecules such as THF, monoglyme, etc., would likely alter the values of $\Delta S°$ and $\Delta C_p°$ of the system in the same direction. More explicitly, the orientation of solvent molecules by the ions in close proximity should decrease the values of these quantities. Indeed, this is the trend that is observed in this study (cf. Tables X–XV).

In Figure 1, it may be concluded that the Gibbs energy and activity of hydrobromic acid in water-rich THF–H_2O solvents are much lower than those of glycine · H^+ (pK_1). It also indicates that the dipolar aprotic solvent–water mixture has a greater affinity for the protons of HBr than those of the large-sized protonated glycine. The curve for the protonated glycine (HG^+) is thus seen to resemble that of acetic acid in methanol–water, whereas the graph of the second ionization constant of tricine is similar in nature to that of the ammonium ion (NH_4^+) in methanol–water. Figure 1 also points out that the solvent effects on the changes in pK (namely $\Delta G_t° = RT \ln 10\,[pK_s(N) - pK_w(N)]$), as THF is added to the aqueous solvent, is strongly dependent on the charge type (*32*).

Acknowledgment

The authors are grateful to T. White and R. Snelling who assisted in performing such preliminary work as recrystallization, standardization of solutions, etc., and also to the computer center of Southwest Missouri State University for providing the computational facilities.

Acknowledgment is made to the Donors of the Petroleum Research Fund (Grant 9274-B5), administered by the American Chemical Society, for partial support of this research.

Glossary of Symbols

pK_1 or $pK_1(m)$ = first thermodynamic dissociation constant on the molal scale [(N) denotes mole-fraction scale for any quantity]

pK_2 or $pK_2(m)$ = second thermodynamic dissociation constant on the molal scale
ΔG° = standard Gibbs free energy of dissociation on the molal scale (subscript t denotes transfer function)
ΔH° = standard enthalpy of dissociation on the molal scale
ΔS° = standard entropy of dissociation on the molal scale
ΔC_p° = standard heat capacity of dissociation on the molal scale
THF = tetrahydrofuran
Bes = *N,N*-Bis-(2-hydroxyethyl)-2-amino ethane sulfonic acid
tricine = *N*-Tris(hydroxymethyl)methyl glycine
Cell I m_1 = molality of the ampholyte in protonated form
Cell I m_2 = molality of the ampholyte in zwitterionic form
Cell II m_1 = molality of the ampholyte in zwitterionic form
Cell II m_2 = molality of the ampholyte in anionic form
1 atm = 101.325 kp_a
HG^+ = protonated glycine
$G^\pm$ = zwitterionic glycine
G^- = anionic glycine
pK_1' = apparent first thermodynamic dissociation constant
pK_2' = apparent second thermodynamic dissociation constant
β = linear slope parameter
E = corrected emf (experimental)
E° = standard electrode potential
R = 8.314 joule K^{-1} mol^{-1}
T = temperature, K
F = 96,487 coulomb g eq^{-1}
m_H' = apparent hydrogen ion molality
I = ionic strength
a_o = ion-size parameter
A = Debye–Hückel limiting law constant
B = Debye–Hückel limiting law constant
γ_x = molal activity coefficient of species X
m_{OH}' = apparent hydroxyl ion molality
pK_w = dissociation constant in water on the molal scale
pK_s = dissociation constant in the mixed solvent on the molal scale
M_w = mean molar mass of water (18.016 g)
M_s = mean molar mass of the mixed solvent

Literature Cited

1. Bates, R. G., Roy, R. N., Robinson, R. A., *J. Solution Chem.* (1974) **3**, 905.
2. Roy, R. N., Swensson, E. E., LaCross, G., Jr., Krueger, C. W., *Anal. Chem.* (1975) **47**, 1407.
3. Roy, R. N., Gibbons, J. J., Krueger, C. W., LaCross, G., Jr., *J. Chem. Thermodyn.* (1977) **9**, 4697.
4. Roy, R. N., Gibbons, J. J., Snelling, R., *J. Solution Chem.* (1977) **6**, 475.
5. Owen, B. B., *J. Am. Chem. Soc.* (1934) **56**, 24.
6. King, E. J., *J. Am. Chem. Soc.* (1951) **73**, 155.
7. Roy, R. N., Gibbons, J. J., LaCross, G., Jr., Krueger, C. W., *J. Solution Chem.* (1976) **5**, 333.
8. Bates, R. G., *J. Electroanal. Chem.* (1971) 29, 1.
9. Bates, R. G., "Hydrogen-Bonded Solvent Systems," A. K. Covington, P. Jones, Eds., Taylor Francis Ltd., 1968.
10. Sager, E. E., Robinson, R. A., Bates, R. G., *J. Res. Natl. Bur. Stand.* (1964) **68A**, 305.
11. Clare, B. W., Cook, D., Ko, E. C. F., Mac, Y. C., Parker, A. J., *J. Am. Chem. Soc.* (1966) **88**, 1911.
12. Ohtaki, H., *Bull. Chem. Soc. Jpn.* (1969) **42**, 1573.
13. Livingstone, G., Franks, F., Aspinall, L. J., *J. Solution Chem.* (1977) **6**, 203.
14. Panichajakul, C. C., Woolley, E. M., ADV. CHEM. SER. (1976) **155**, 263.
15. Treiner, C., Tzias, P., ADV. CHEM. SER. (1976) **155**, 203.
16. Robinson, R. A., *J. Res. Natl. Bur. Stand.* (1970) **74A**, 495.
17. Bury, R., Treiner, C., *J. Chem. Phys.* (1968) **64** 94.
18. Justice, J. C., Bury, R., Treiner, C., *J. Chem. Phys.* (1968) **65**, 1708.
19. Harned, H. S., Ehlers, R. W., *J. Am. Chem. Soc.* (1932) **54**, 1350.
20. Roy, R. N., Swensson, E. E., LaCross, G., Jr., *J. Chem. Thermodyn.* (1975) **7**, 1015.
21. Bates, R. G., "Determination of pH," 2nd ed., Chap. 10, Wiley, New York, 1973.
22. Roy, R. N., Robinson, R. A., Bates, R. G., *J. Chem. Thermodyn.* (1973) **5**, 559.
23. Paul, D. E., Lippin, D. J., *J. Am. Chem. Soc.* (1956) **78**, 116.
24. Roy, R. N., Sen, B., *J. Chem. Eng. Data* (1967) **12**, 584.
25. Harned, H. S., Robinson, R. A., *Trans. Faraday Soc.* (1940) **36**, 973.
26. Please, N. W., *Biochem. J.* (1954) **56**, 196.
27. Schindler, P., Robinson, R. A., Bates, R. G., *J. Res. Natl. Bur. Stand.* (1968) **72A**, 141.
28. Halle, J. C., Bates, R. G., *J. Solution Chem.* (1975) **4**, 1033.
29. Paabo, M., Bates, R. G., Robinson, R. A., *J. Phys. Chem.* (1966) **70**, 247.
30. Parker, A. J., *Chem. Rev.* (1969) **69**, 1.
31. Latimer, W. M., Slansky, C. M., *J. Am. Chem. Soc.* (1940) **62**, 2019.
32. Bates, R. G., Robinson, R. A., "Acid–Base Behavior in Methanol–Water Solutions," *in* "Chemical Physics of Ionic Solutions," B. E. Conway, R. G. Barradas, Eds., Wiley, New York, 1966.

RECEIVED February 23, 1978.

19

The Excess Gibbs Free Energy and the Activity Coefficients of the $Nd(NO_3)_3$-HNO_3-H_2O System at 25°C

WILLIAM G. O'BRIEN[1] and RENATO G. BAUTISTA

Ames Laboratory USDOE and Department of Chemical Engineering, Iowa State University, Ames, IA 50011

The Gibbs–Duhem equation was used to determine the activity and the activity coefficient of the free neodymium (III) ion. The vapor pressures of binary electrolyte solutions consisting of neodymium nitrate and nitric acid were measured. Harned's activity coefficient equations for a binary electrolyte were used to model the data. The excess Gibbs free energy of the electrolyte mixtures also was calculated. The hydrogen ion activity in the neodymium nitrate–nitric acid binary electrolyte solutions was determined using the pH electrode. The nitrate ion electrode was used to measure the nitrate ion activity of the neodymium nitrate–nitric acid binary electrolyte solutions. The strong liquid junction potential generated at the reference electrode by the presence of the H^+ ions was eliminated by careful calibration of the nitrate electrode with nitric acid.

The activity behavior of a metallic ion species in aqueous solution is of primary importance in understanding any hydrometallurgical operation. A knowledge of the activity behavior of the components in the system are generally more significant in answering questions relating to a rate of chemical reaction or the position of chemical equilibrium.

[1] Presently Senior Research Engineer, Photo Products Department, DuPont, Towanda, PA 18848

0-8412-0428-4/79/33-177-299$05.25/1

In the past the prominent investigators (*1–6*) of electrolyte solutions were able to measure only the mean activity coefficient of the salt. The development of ion-selective electrodes has made the measurement of the activity of many ions possible. The development, reliability, and general theory of many different ion-selective electrodes is discussed extensively in a National Bureau of Standards (NBS) publication edited by Durst (*7*).

In this chapter the thermodynamic analysis of the $Nd(NO_3)_3$–HNO_3–H_2O system is presented. The determination of the activity and the activity coefficient of the free neodymium (III) ion and the development of an accurate model for the activity behavior of aqueous binary electrolyte solutions consisting of the neodymium nitrate and nitric acid are presented.

Theory

The Debye–Hückel theory forms the basis for the modern interpretation of the behavior of electrolyte solutions. Guggenheim (*3*) proposed the first modern theory dealing both with single electrolyte and mixtures of electrolytes over a wide concentration range. Essential to his theory is a series of constants analogous to the second virial coefficients which represent the net effect of various short-range forces between the cation and anion and exclude terms relating to the interaction of ions of like sign which leads to some difficulties at higher concentrations. Improvements of Guggenheim's theory have been made through the years by Scatchard (*4*) and his co-workers (*5*). Provision was made for different distances of closest approach for the solute components and terms were included for the short range interaction of ions of like sign. Arrays of third virial coefficients were added also. These complicated Guggenheim–Scatchard equations allowed Lietzke and Stoughton (*8*) to represent accurately the osmotic coefficients of twenty pure electrolytes and several mixed-electrolyte systems.

Pitzer (*9*) recently has developed a system of equations for electrolyte thermodynamics which yields comparable results to the Guggenheim–Scatchard equations for both single and mixed electrolytes. The main feature of the Pitzer equations is that they include an ionic-strength dependence on the short-range forces in binary interactions.

Meissner et al. (*10, 11, 12*) have presented a method of estimating the activity coefficient of an electrolyte in a single component or in a solution. The technique is based on graphical analysis of the reduced mean activity coefficient for a broad class of electrolytes. The method has good estimates of activity coefficients when compared with the published values for single components and mixtures of electrolytes.

Lanthanide Complexes in Aqueous Solution

Owing to its + 3 charge the lanthanide ion forms stable complexes with many anions such as chloride, nitrate, and sulfate. The degree of complexation is measured by the inorganic stability constant of the complex which is analogous to an acid dissociation constant. A frequently used technique to estimate complex stability constants involves the solvent extraction of the uncomplexed species into an organic phase. This distribution coefficient method has been reviewed thoroughly by Zozula and Peskova (*13*). Peppard, Mason, and Hucher (*14*) used this method to determine stability constants for the chloride and nitrate single-ligand complexes of some lanthanides and actinides. The nitrate ion formed a stronger complex with the lanthanide ion than with the chloride ion. Choppin et al. (*15, 16*) improved the estimates for some lanthanide chloride and nitrate stability constants by using radioactive tracers with the distribution method. The lanthanide chloride and nitrate complexes were primarily of the outer-sphere type.

The nitrate ion-selective electrode and IR and Raman spectroscopy were used by Knoeck (*17*) to investigate the lanthanum nitrate complexes in aqueous solution. Krumholtz (*18*), using IR spectroscopy, devised a method for determining the stability constants of weak complexes in the neodymium nitrate system. The fractions of the total neodymium concentration existing as the complexes $Nd(NO_3)^{+2}$ and $Nd(NO_3)_2^{+1}$ together with the stability constants of the first complex over a wide range of total ionic strengths were reported. Goto and Smutz (*19*) determined the stability constants of some lanthanide chloride complexes by using a potentiometric method to measure the change in the free chloride ion concentration. The values obtained by this method for the stability constants of the MCl^{+2} type agree well with the values reported by Peppard et al. (*14*) and Choppin and Unrein (*15*). The lanthanide ions (because of their high charge) have a tendency to form complexes with various inorganic anions in aqueous solution. Generally the degree to which these complexes form is dependent only on the total ionic strength of the solution. The complexing reactions for lanthanide nitrates are:

$$M^{+3} + NO_3^{-} \xrightarrow{K_1} MNO_3^{+2} \qquad (1)$$

$$MNO_3^{+2} + NO_3^{-} \xrightarrow{K_2} M(NO_3)_2^{+1} \qquad (2)$$

$$M(NO_3)_2^{+1} + NO_3^{-} \xrightarrow{K_3} M(NO_3)_3 \qquad (3)$$

Choppin and Strazik (*16*) have shown that lanthanide nitrate complexes are outer-sphere in nature, meaning that there is a monolayer of water molecules separating the nitrate and lanthanide. This layer of water acts as a dielectric and reduces the strength of the ionic bond that forms the complex. For this reason it is assumed that the neutral species ($M(NO_3)_3$) is present only at lanthanide nitrate concentrations approaching the solubility limit. Since all of the solutions studied in this work were below the solubility limit, the existence of the neutral species ($M(NO_3)_3$) is neglected.

Results and Discussion

The complex formation stability constants (K_1, K_2, K_3) are

$$K_1 = \frac{[MNO_3^{+2}]}{[M^{+3}][NO_3^{-}]} \tag{4}$$

$$K_2 = \frac{[M(NO_3)_2^{21}]}{[MNO_3^{+2}][NO_3^{-}]} \tag{5}$$

$$K_3 \simeq 0 \tag{6}$$

where the brackets denote chemical activity rather than concentration. For a system consisting of the lanthanide nitrate dissolved in water the Gibbs–Duhem equation can be written:

$$\begin{aligned} &55.1\, d(\ln[H_2O]) + [m_{NO_3}{}^{T} - m_M{}^{T}(\alpha_1 + 2\alpha_2)]\, d\ln([NO_3^{-}]) \\ &+ m_M{}^{T}(1 - \alpha_1 - \alpha_2)\, d(\ln[M^{+3}]) + m_M{}^{T}\alpha_1 d(\ln K_1[M^{+3}][NO_3^{-}]) \\ &+ m_M{}^{T}\alpha_2 d(\ln K_1K_2[M^{+3}][NO_3^{-}]^2) = 0 \end{aligned} \tag{7}$$

where α_1 is the fraction of the total metal species existing as MNO_3^{+2}, α_2 is the fraction of the total metallic species existing as $M(NO_3)_2^{+1}$, $1 - \alpha_1 - \alpha_2$ is the fraction existing as M^{+3}, $m_M{}^T$ is the total metallic ion concentration, and $m_{NO_3}{}^T$ is the nitrate ion concentrations. By writing the log of a product as the sum of logs and using $m_{NO_3}{}^T = 3m_M{}^T$ Equation 7 becomes:

$$\begin{aligned} 0 = &\ 55.51\, d(\ln[H_2O]) + 3\, m_M{}^T\, d(\ln[NO_3^{-}]) + m_M{}^T\, d(\ln[M^{+3}]) \\ &+ m_M{}^T\, (\alpha_1 + \alpha_2)\, d(\ln K_1) + m_M{}^T\alpha_2\, d(\ln K_2) \end{aligned} \tag{8}$$

The values of $[NO_3^{-}]$ are a function of $m_M{}^T$ by regression analysis of nitrate electrode data. Similarly, regression analysis of the data of Krumholtz (*18*) gives α_1, α_2, and K_1 as functions of m_M^T. Values of K_2 can be estimated from the relation

$$K_2 = \frac{[M(NO_3)_2^{+1}]}{[MNO_3^{+2}][NO_3^-]} \simeq \frac{\alpha_2}{\alpha_1[m_{NO_3}^T - m_M^T(\alpha_1 + 2\alpha_2)]} \quad (9)$$

Integration of Equation 8 from 0 to m_M^T yields the folowing:

$$-\ln[M^{+3}] = 55.51 \int_0^{m_M^T} \frac{1}{m_M^T}\, d\ln[H_2O] + \int_0^{m_M^T} 3\, d\ln[NO_3^-] + \int_0^{m_M^T} (\alpha_1 + \alpha_2)\, d\ln K_1 + \int_0^{m_M^T} \alpha_2\, d\ln K_2 \quad (10)$$

Thus the activity of the free lanthanide ion $[M^{+3}]$ can be obtained.

For the binary electrolyte system $M(NO_3)_3$–HNO_3–H_2O the Gibbs–Duhem equation can be written:

$$-m_M^T\, d(\ln[M^{+3}]) = 55.51\, d(\ln[H_2O]) + m_H^T\, d(\ln[H^+]) + [m_{NO_3}^T - m_M^T(\alpha_1 + 2\alpha_2)]\, d(\ln[NO_3^-]) + m_M^T(\alpha_1 + \alpha_2)\, d(\ln K_1) + m_M^T\alpha_2\, d(\ln K_2) \quad (11)$$

Solving this problem at constant total ionic strength and varying composition percentage (y_B) of electrolyte $M(NO_3)_3$ eliminates the terms involving K_1 and K_2 from Equation 11. The values of K_1 and K_2 were assumed to be constant at constant total ionic strength. The integrated form of Equation 11 becomes

$$-\ln[M^{+3}] = 55.51 \int_0^{y_B} \frac{1}{m_M^T}\, d(\ln[H_2O]) + \int_0^{y_B} \frac{m_H^T}{m_M^T}\, d(\ln[H^+]) + \int_0^{y_B} \left[\frac{m_{NO_3}^T}{m_M^T} - (\alpha_1 + 2\alpha_2)\right] d(\ln[NO_3^-]) \quad (12)$$

The ionic concentrations in Equation 12 can be eliminated by the following:

$$m_M^T = \frac{y_B I}{6}$$

$$m_H^T = (1 - y_B)I$$

$$m_{NO_3}^T = m_H^T + 3m_M^T = (1 - y_B/2)I$$

Through these relationships, Equation 12 becomes:

$$-\ln[M^{+3}] = \frac{333.06}{I}\int_0^{y_B} \frac{1}{\bar{y}_B}\, d(\ln[H_2O]) + 6\int_0^{y_B}\left(\frac{1-\bar{y}_B}{\bar{y}_B}\right) d(\ln[H^+])$$
$$+ 6\int_0^{y_B}\left(\left(\frac{1-\bar{y}_B/2}{\bar{y}_B}\right) - (\alpha_1 + 2\alpha_2)\right) d(\ln[NO_3^-]) \qquad (13)$$

$\bar{y}_B$ denotes the dummy variable of integration. The values of α_1 and α_2 are assumed constant at constant total ionic strength. Similarly, as in the solution of Equation 10 for the single-component electrolyte, the data obtained by the vapor pressure measurements and the nitrate and hydrogen ion electrode measurements on the binary electrolyte mixtures are analyzed by linear regression. Thus the three integrands of Equation 13 can be estimated as linear functions by y_B, the integration performed, and solving for $[M^{+3}]$.

Regression analysis can be used to estimate the various quantities in Equation 10 as linear polynominals of $m_M{}^T$. By letting $m_M{}^T$ equal x the following equations can be generated from the data:

$$\ln[H_2O] = a_0 + a_1x + a_2x^2 + a_3x^3 \qquad (14)$$

$$\ln[NO_3] = b_0 + b_1x + b_2x^2 + b_3x^3 \qquad (15)$$

$$\ln[K_1] = c_0 + c_1x + c_2x^2 + c_3x^3 \qquad (16)$$

$$\ln[K_2] = d_0 + d_1x + d_2x^2 + d_3x^3 \qquad (17)$$

$$\alpha_1 = e_0 + e_1x + e_2x^2 + e_3x^3 \qquad (18)$$

$$\alpha_2 = f_0 + f_1x + f_2x^2 + f_3x^3 \qquad (19)$$

where the various a's, b's, c's, d's, e's, and f's are the calculated regression coefficients from the data. The results of the regression analysis for the solution vapor pressure, specific nitrate electrode activity measurements, the stability constants K_1 and K_2, and the degree of formation of the complexes, α_1 and α_2, are summarized in Table I. These parameter estimates were used to calculate the activity of the free neodymium ion from 0.05 to 2.0 *M* and are shown in Figure 1 and Table II. By using the values of α_T to determine the free concentration of neodymium ion $(m_{Nd} + 3 = (1 - \alpha_T)m_{Nd}{}^T)$, the activity coefficient of the free neodymium ion was calculated also. The results are shown in Table II and Figure 2. The failure of the activity coefficient to go to unity smoothly as the concentration approaches zero is ascribed to the inaccuracy of the regression parameter estimates at these extremely low concentrations. The main sources of this error are the estimates for the d coefficients for the second stability constant, K_2, especially at the low concentrations. This can be seen in Figure 3.

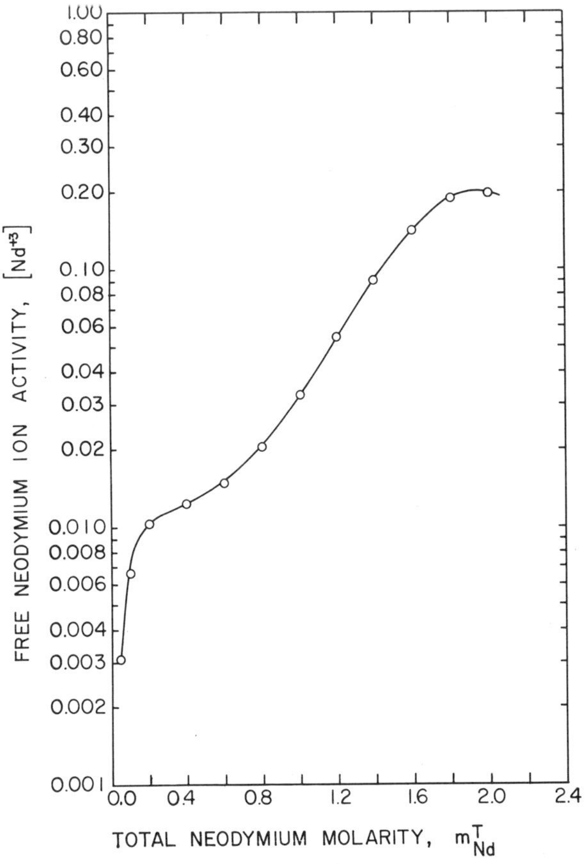

Figure 1. The activity of the free neodymium ion as a function of total neodymium concentration in aqueous nitrate solution

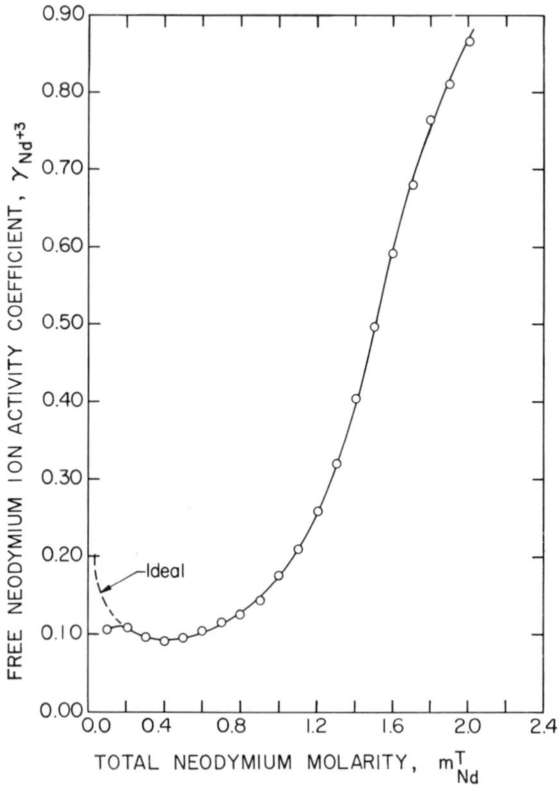

Figure 2. The activity coefficient of the free neodymium ion in aqueous nitrate solution as a function of total neodymium concentration

Table I. Results of Regression Analysis for Various Parameters in Equation 10

$a_0 = 0.00078$	$b_0 = -2.5402$
$a_1 = -0.03155$	$b_1 = 4.4132$
$a_2 = -0.01372$	$b_2 = -3.0549$
$a_3 = -0.00871$	$b_3 = 0.76930$
(R^2) of 0.999	(R^2) is 0.988
$c_0 = 0.63469$	$d_0 = -0.75684$
$c_1 = -3.6772$	$d_1 = -0.25626$
$c_2 = 2.8944$	$d_2 = 0.83916$
$c_3 = -0.72912$	$d_3 = -0.05297$
$R^2 = 0.955$	$R^2 = 0.790$
$e_0 = 0.14608$	$f_0 = -0.00312$
$e_1 = 1.4319$	$f_1 = 0.42777$
$e_2 = -1.6355$	$f_2 = 0.07158$
$e_3 = 0.44981$	$f_3 = -0.03897$
$R^2 = 0.932$	$R^2 = 0.999$

Table II. Activities and Activity Coefficients of the Free Neodymium Ion in Neodymium Nitrate Aqueous Solutions

m_{Nd}^{T}	a_{Nd}^{+3}	α_T	m_{Nd}^{+3}	γ_{Nd}^{+3}
0.1	0.0066	0.370	0.0630	0.1049
0.2	0.0105	0.510	0.0980	0.1071
0.3	0.0116	0.595	0.1215	0.0956
0.4	0.0124	0.670	0.1320	0.0937
0.5	0.0134	0.725	0.1375	0.0972
0.6	0.0149	0.762	0.1428	0.1043
0.7	0.0173	0.784	0.1512	0.1141
0.8	0.0207	0.796	0.1632	0.1268
0.9	0.0256	0.805	0.1755	0.1459
1.0	0.0325	0.815	0.1850	0.1775
1.1	0.0419	0.820	0.1980	0.2114
1.2	0.0545	0.825	0.2100	0.2593
1.3	0.0709	0.830	0.2210	0.3207
1.4	0.0914	0.839	0.2254	0.4054
1.5	0.1156	0.845	0.2325	0.4974
1.6	0.1421	0.850	0.2400	0.5922
1.7	0.1679	0.855	0.2465	0.6813
1.8	0.1889	0.863	0.2466	0.7661
1.9	0.2009	0.870	0.2480	0.8103
2.0	0.1990	0.875	0.2500	0.8652

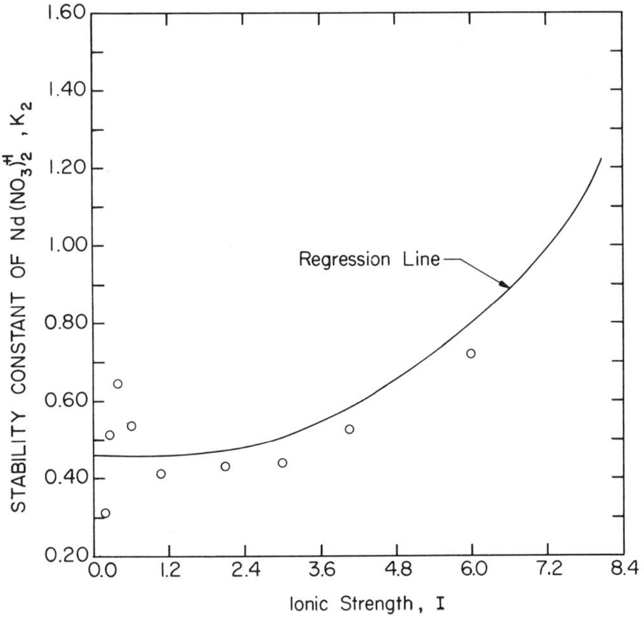

Figure 3. The stability constant of the $Nd(NO_3)_2^{+1}$ complex as a function of the total ionic strength: ○, calculated data; $\ln K_2 = 0.75684 - 0.04211\ I + 0.02331\ I^2 + 0.00025\ I^3$; $R^2 = 0.790$.

Stability Constant Data

The values of the stability constant for the complex $NdNO_3^{+2}$ (K_1) and the relative percentages of complexed species present in solution as determined by Krumholtz (*18*) are given in Table III. The values of K_2 were estimated by using this data in Equation 9. The values of K_1, α_T, and α_2 were extrapolated to regions of higher ionic strength. These extrapolated values were used to calculate values for α_1 and K_2 at higher ionic strengths. The values of K_2, K_1, and α_T, α_1, and α_2 are shown in Figures 3, 4, and 5, respectively. These figures show why K_1, α_T, and α_2 were chosen to be extrapolated and K_2 and α_1 to be calculated.

Linear regression analysis of the stability constants and the complex fractions yielded the following results:

$$\ln K_1 = 0.634694 - 0.61287I + 0.08040I^2 - 0.003381^3 \qquad R^2 = 0.955 \tag{20}$$

$$\ln K_2 = -0.75684 - 0.04211I + 0.02331I^2 - 0.00025I^3 \qquad R^2 = 0.790 \tag{21}$$

$$\alpha_1 = 0.14608 + 0.23866I - 0.04543I^2 - 0.00208I^3 \qquad R^2 = 0.932 \tag{22}$$

$$\alpha_2 = -0.00312 + 0.07129I + 0.00199I^2 - 0.00018I^3 \qquad R^2 = 0.999 \tag{23}$$

Table III. Stability Constant and Complex Fraction Data

Ionic Strength I	K_1	K_2 [a]	α_T [b]	α_1	α_2
0.18	2.20	0.310	0.115	0.112	0.003
0.24	1.80	0.515	0.183	0.173	0.010
0.36	1.40	0.641	0.265	0.240	0.025
0.60	1.10	0.536	0.370	0.325	0.045
1.08	0.90	0.413	0.500	0.420	0.080
2.12	0.70	0.429	0.640	0.480	0.160
3.00	0.60	0.439	0.725	0.500	0.225
4.12	0.50	0.762	0.780	0.473	0.308
6.00	0.44	0.711	0.815	0.365	0.450
9.00	0.40	1.789	0.845	0.170	0.675
12.00	0.39	—	0.885	0.055	0.825

[a] Calculated by Equation 9.
[b] $\alpha_T = \alpha_1 + \alpha_2$.

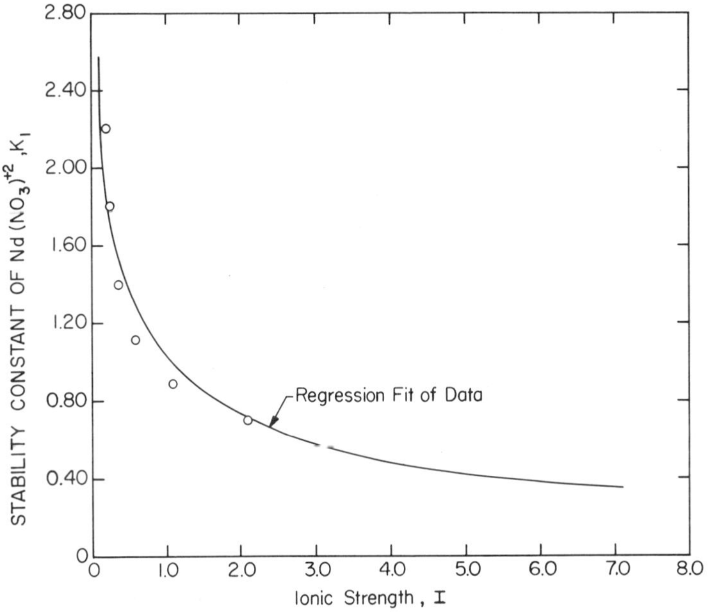

Figure 4. The stability constant of the $Nd(NO_3)^{+2}$ complex as a function of total ionic strength: $\ln K_1 = 0.634694 - 0.61287\ I + 0.08040\ I^2 - 0.00338\ I^3$; $R^2 = 0.955$

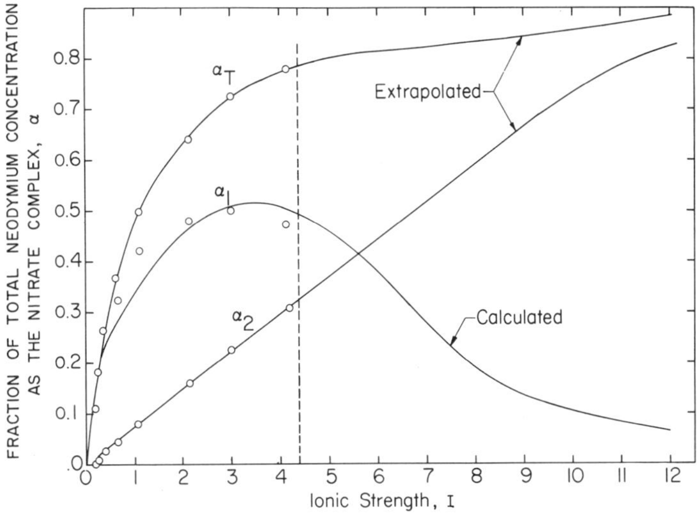

Figure 5. The fraction of the total neodymium concentration existing as the nitrate complexes as a function of ionic strength: $\alpha_1 = 0.14608 + 0.23866\ I - 0.04543\ I^2 - 0.00208\ I^3$, $R^2 = .932$; $\alpha_2 = -0.00312 + 0.07129\ I + 0.0019\ I^2 - 0.00018\ I^3$, $R^2 = 0.999$

The values of the two stability constants (K_1 and K_2) for the two neodymium–nitrate complexes existing in the $Nd(NO_3)_3$–HNO_3 aqueous phase were determined from the total ionic strengths of the aqueous phase using Equations 20 and 21. The total ionic strength was obtained from the relationship:

$$I = m_H + 6m_{Nd} \tag{24}$$

where I is the total ionic strength, m_H is the total hydrogen ion concentration and m_{Nd} is the total metal concentration of the aqueous phase.

Binary Electrolyte Solutions

The results of static vapor pressure measurements of aqueous mixtures of $Nd(NO_3)_3$–HNO_3 at various constant total ionic strengths are shown in Figures 6 and 7 and in Table IV. The vapor pressures at $y_B = 1.0$ came from the single-component vapor pressure data for $Nd(NO_3)_3$. The vapor pressures at $y_B = 0$ (pure HNO_3 solutions) were taken from Robinson and Stokes (2). The logarithm of the water activity referenced to pure nitric acid solutions is illustrated in Figure 8. The common intercept at zero is caused by the referencing of the water activity to the pure nitric acid solutions.

Table IV. $Nd(NO_3)_3$–HNO_3–H_2O Data

I	y_B	P	$\ln\left(\frac{P}{P_{HNO_3}}\right)$	$(6000/18\ y_B)$-$\ln(P/P_{HNO_3})$
3.0	0.20	10.911	0.02986	49.86
3.0	0.20	10.907	0.02949	49.25
3.0	0.40	11.172	0.05350	44.55
3.0	0.60	11.409	0.07449	41.37
3.0	0.80	11.502	0.08261	34.44
3.0	0.80	11.544	0.08626	35.93
3.0	1.00	11.640	0.09454	31.50
1.50	0.20	11.5326	0.01596	27.03
1.50	0.40	11.6642	0.02731	22.76
1.50	0.60	11.7606	0.03554	19.72
1.50	0.80	11.8102	0.03975	16.53
1.50	1.00	11.8976	0.04712	15.71
0.6	0.20	11.8162	0.00562	9.360
0.6	0.20	11.8186	0.00582	9.703
0.6	0.40	11.8706	0.01021	8.510
0.6	0.40	11.8796	0.01097	9.140
0.6	0.60	11.9276	0.01517	8.330
0.6	0.80	11.9436	0.01634	6.680
0.6	0.80	11.9556	0.01735	7.227
0.6	1.00	11.9476	0.01668	5.560

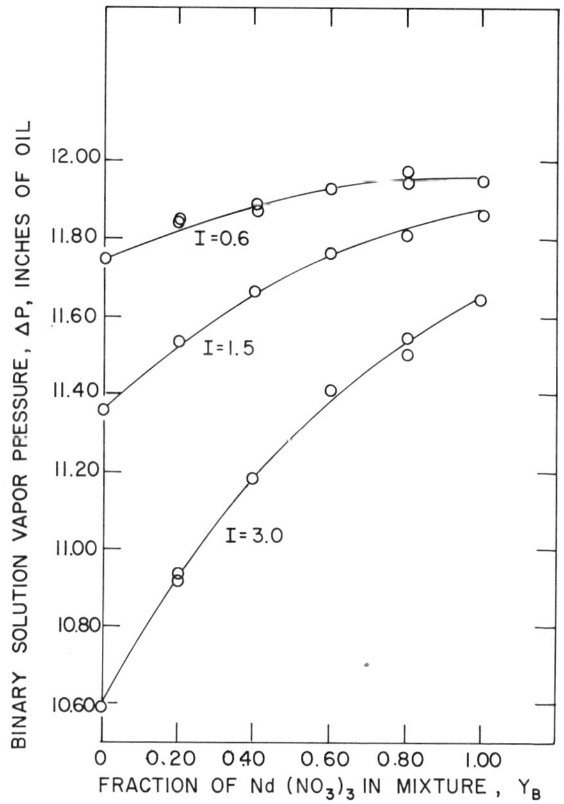

Figure 6. The vapor pressure of the $Nd(NO_3)_3$–HNO_3–H_2O binary electrolyte solution

The method of Harned and Robinson (*1*) analyzes a solution of two electrolytes (B and C) at constant total ionic strength I with the following model for mean ionic activity behavior in the mixture.

$$\begin{aligned} \ln \gamma_{\pm B} &= \ln \gamma_{\pm B}{}^0 + Q_B y_C I + R_B y_C{}^2 I^2 \\ \text{in } \gamma_{\pm C} &= \ln \gamma_{\pm C}{}^0 + Q_C y_B I + R_C y_B{}^2 I^2 \end{aligned} \qquad (25)$$

where $\gamma_{\pm B}$ and $\gamma_{\pm C}$ are the mean ionic activity coefficients of the B and C salt, respectively, in the solution mixture, and $\gamma_{\pm B}{}^0$ and $\gamma_{\pm C}{}^0$ are the mean activity coefficients of pure solutions of B and C, respectively, at ionic strength I. Q's and R's represent Harned's first and second coefficients for the electrolyte and y represents the fraction of the respective electrolyte in the mixture.

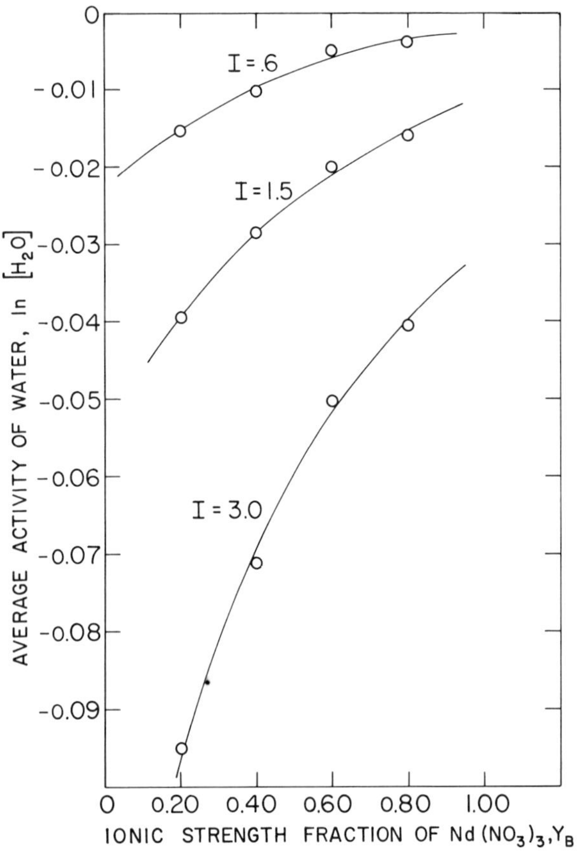

Figure 7. Average water activity as a function of ionic strength fraction of neodymium nitrate in the $Nd(NO_3)_3$–HNO_3–H_2O binary electrolyte solution

The constant total ionic strength I is made up to two linearly varying components, I_B and I_C, the individual ionic strengths of the B and C electrolyte in the mixture.

$$y_b = \frac{I_B}{I}, y_C = \frac{I_C}{I} \tag{26}$$

The individual ionic strengths are related to their respective electrolyte molarities by the equations

$$I_B = \frac{1}{2} \nu_B Z_{B^+} + Z_{B^-} m_B = k_B m_B$$

$$I_C = \frac{1}{2} \nu_C Z_{C^+} + Z_{C^-} m_C = k_C m_C$$

where ν refers to the total number of ions produced by the specific electrolyte and Z_+ and Z_- refer to the charge of the cation and anion, respectively, produced by the specific electrolyte.

Application of the Gibbs–Duhem equation to the $Nd(NO_3)_3$–HNO_3–H_2O system gives

$$-\frac{1000\ k_B k_C}{W_A y_B} \ln P_A/P_A(C) = (\nu_B k_C - \nu_C k_B + \nu_C k_B Q_C I) I$$

$$- \frac{1}{2} (\nu_B y_C Q_C + \nu_C k_B Q_C) I^2 y_B \tag{27}$$

$$- (\nu_B k_C R_B - \nu_C k_B R_C) I^3 \left(y_B - \frac{2}{3} y_B^2 \right)$$

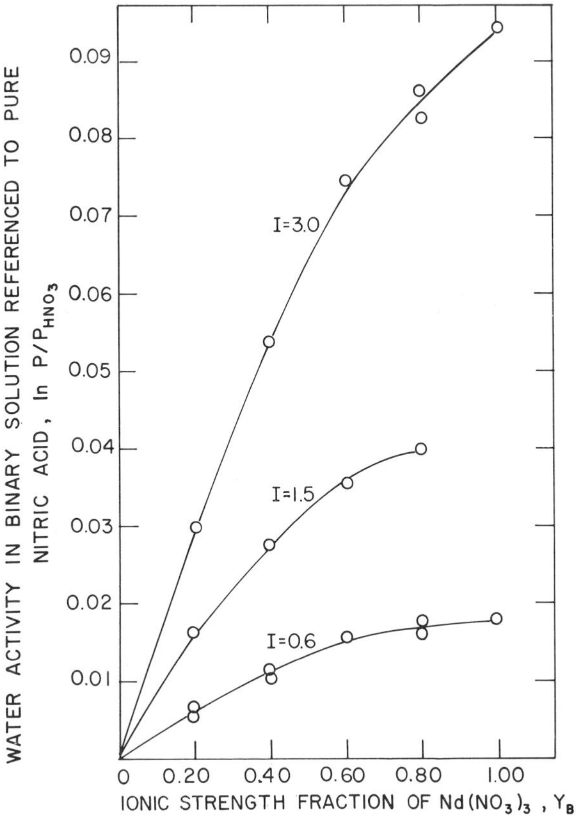

Figure 8. The water activity in the $Nd(NO_3)_3$–HNO_3–H_2O binary electrolyte solution to the water activity referenced to pure HNO_3 as a function of the ionic strength fraction of $Nd(NO_3)_3$

Table V. HNO_3–H_2O Data[a]

m	Φ	$\gamma_{\pm}$	P[b]
0.1	0.940	0.791	11.960
0.2	0.935	0.754	11.920
0.3	0.936	0.735	11.875
0.4	0.940	0.725	11.838
0.5	0.944	0.720	11.795
0.6	0.950	0.717	11.750
0.7	0.957	0.717	11.710
0.8	0.964	0.718	11.670
0.9	0.971	0.721	11.625
1.0	0.979	0.724	11.580
1.2	0.994	0.734	11.495
1.4	1.009	0.745	11.405
1.6	1.025	0.758	11.308
1.8	1.042	0.775	11.214
2.0	1.060	0.793	11.150
2.5	1.106	0.846	10.860
3.0	1.154	0.909	10.590

[a] Taken from Ref. *2*.
[b] Measured in in. of 702 pump oil.

in which $P_A(C)$ is the solvent vapor pressure of a pure C electrolyte solution of total ionic strength I (Figure 8 and Table V). It is possible to obtain estimates of the Q coefficients by measuring the vapor pressures of electrolyte mixtures at constant ionic strengths and fitting the data to Equation 27 by regression analysis. The cross differentiation relation

$$\nu_B \left(\frac{\partial \ln \gamma_B}{\partial m_C}\right)_{m_B} = \nu_C \left(\frac{\partial \ln \gamma_C}{\partial m_B}\right)_{m_C} \qquad (28)$$

is a general result of the property of a chemical potential, being a partial differential coefficient of the total free energy with respect to the concentration. Harned and Robinson (*1*) have used this to determine an independent relationship between the Q and R coefficients:

$$(\nu_B k_C Q_B + \nu_C k_B Q_C) = \text{constant} - 2(\nu_B k_C R_B + \nu_C k_B R_C) I \qquad (29)$$

Experience with other electrolyte mixtures has shown that the values of the R coefficients are very small and can be taken as being independent of the total ionic strength. By making this assumption, Equation 29 can be evaluated at two different values of ionic strength and the constant evaluated. Equation 29 can be used now with Equation 27 to determine the values of the R coefficients.

Harned's plot for a binary electrolyte solution at constant total ionic strength is presented in Figure 9. The utility of this plot can be seen from Equation 27. Because of the linearity of the data, the slope and intercept of each line determines the value of the Q coefficients at each total ionic strength.

Through linear regression analysis of the data the slope and intercept of the line at $I = 3.0$ were -23.024 and 54.142, respectively. The correlation coefficient was 0.9949 and the mean deviation about the regression line was 0.6818. Similarly for the line at $I = 1.5$ the slope and the intercept were -17.27 and 30.145, respectively, while the correlation coefficient and the mean deviation about the regression line were 0.9968 and 0.3120, respectively. At $I = 0.6$ the slope was -4.775 while the intercept was 10.678. The correlation coefficient was 0.9664 and the mean deviation

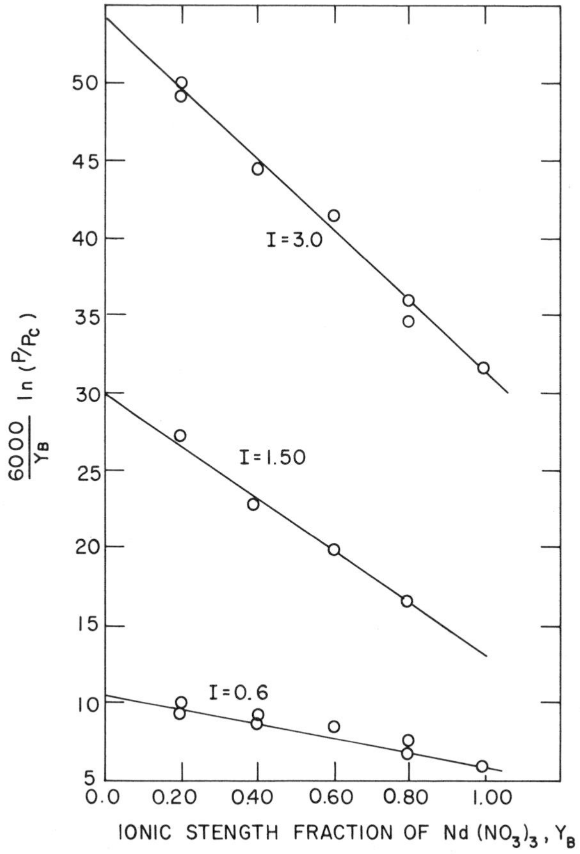

Figure 9. Harned's plot of binary $Nd(NO_3)_3$–HNO_3–H_2O electrolyte solution vapor pressure data as a function of ionic strength fraction of $Nd(NO_3)_3$

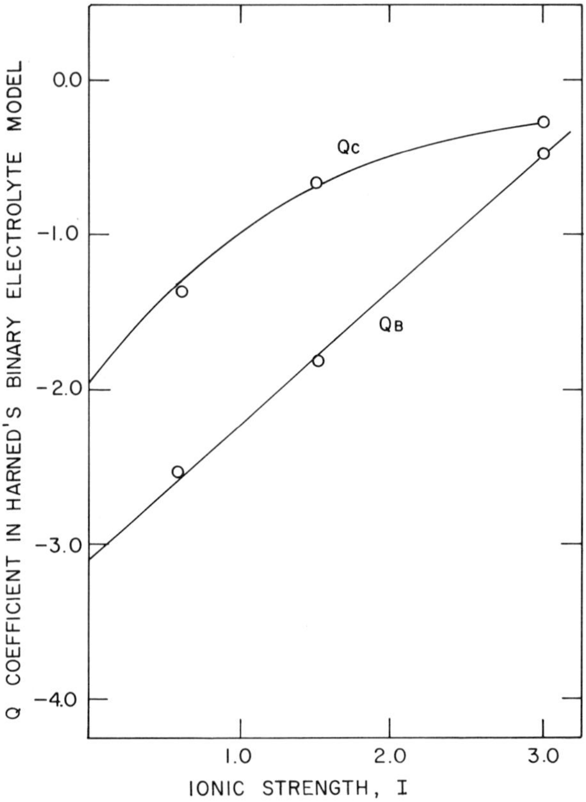

Figure 10. Q coefficients for Harned's binary electrolyte model as a function of ionic strength

0.3537. The Q coefficients were calculated from these slopes and intercepts by using Equation 27. They are illustrated in Figure 10 and Table VI, and appear to be strongly dependent on total ionic strength. The R coefficients can be calculated by using Equation 29 from the values of the Q coefficients at different total ionic strengths. The following values were obtained: $R_B = -0.5900$ and $R_C = -0.1967$. They are assumed to be independent of the total ionic strength of the mixture.

Table VI. Q Coefficients

I	Q_B	Q_C
3.00	−0.4418	−0.2791
1.50	−1.8217	−0.6720
0.60	−2.5502	−1.3606

The concept of excess Gibbs free energy is useful in describing mixtures of electrolyte solutions. The excess free energy is the free energy of the mixed solution over and above that possessed by the single-electrolyte solutions which comprise the mixture.

The total excess free energy of the electrolyte mixture can be evaluated from the equation

$$\frac{\Delta G^E}{RT} = 1/2\, I^2 y_B y_C \left[\left(\frac{\nu_B}{k_B} Q_B + \frac{\nu_C}{k_C} Q_C\right) + I\left(\frac{\nu_B}{k_B} R_B + \frac{\nu_C}{k_C} R_C\right)\right.$$
$$\left. + 1/3\, I\left(\frac{\nu_C}{k_C} R_C - \frac{\nu_B}{k_B} R_B\right)(y_B - y_C)\right] \tag{30}$$
$$+ \nu_B y_B \frac{I}{k_B} \ln y_B + \nu_C y_C \frac{I}{k_C} \ln y_C$$

The last two terms on the right-hand side of Equation 30 denote the ideal excess free energy of the mixture. These terms are caused by the mixing alone and in no way represent ionic interaction. Equation 30 therefore can be broken up into two portions, one dealing with the excess free energy owing to ionic interaction ΔG_I^E, the other owing to mixing ΔG_M^E. Using these definitions:

$$\Delta G^E = \Delta G_I^E + \Delta G_M^E \tag{31}$$

where

$$\frac{\Delta G_M^E}{RT} = \nu_B y_B \frac{I}{k_B} \ln y_B + \nu_C y_C \frac{I}{k_C} \ln y_C \tag{32}$$

With the Q and R coefficients fully determined the excess Gibbs free energy of the mixture can be evaluated using Equation 30. These results are shown in Figure 28 and illustrate the deviation of the mixture from ideality. The data for the excess Gibbs free energy are summarized in Table VII and Figure 11.

The logarithm of the water activity, the hydrogen ion activity, and the nitrate ion activity for $Nd(NO_3)_3$–HNO_3–H_2O solutions vs. the ionic strength fraction of neodymium nitrate (y_B) are shown in Figures 7, 12, and 13 respectively. In Figure 7 the logarithm of the water activity approaches a finite value at $y_B = 0$ and $y_B = 1.0$. This corresponds to a pure nitric acid solution ($y_B = 0$) and a pure $Nd(NO_3)_3$ solution ($y_B = 1.0$). In Figure 12 the log of the hydrogen ion activity goes to minus infinity as y_B nears 1.0. This corresponds to a zero hydrogen ion concentration or a pure $Nd(NO_3)_3$ solution as indicated by $y_B = 1.0$. Thus the pH electrode behaves correctly at high $Nd(NO_3)_3$, low H^+ concentrations with little appreciable response to the polyvalent metal ion.

Table VII. Excess Gibbs Free Energy

	$-\Delta G^E/RT$		
y_B	I = *3.0*	I = *1.5*	I = *0.6*
0.20	2.313	0.3279	0.1409
0.40	3.470	0.4918	0.2114
0.50	3.615	0.5124	0.2202
0.60	3.410	0.4918	0.2114
0.80	2.313	0.3279	0.1409
0.10	1.301	0.1845	0.0792
0.05	0.687	0.0973	—
0.02	0.283	—	—

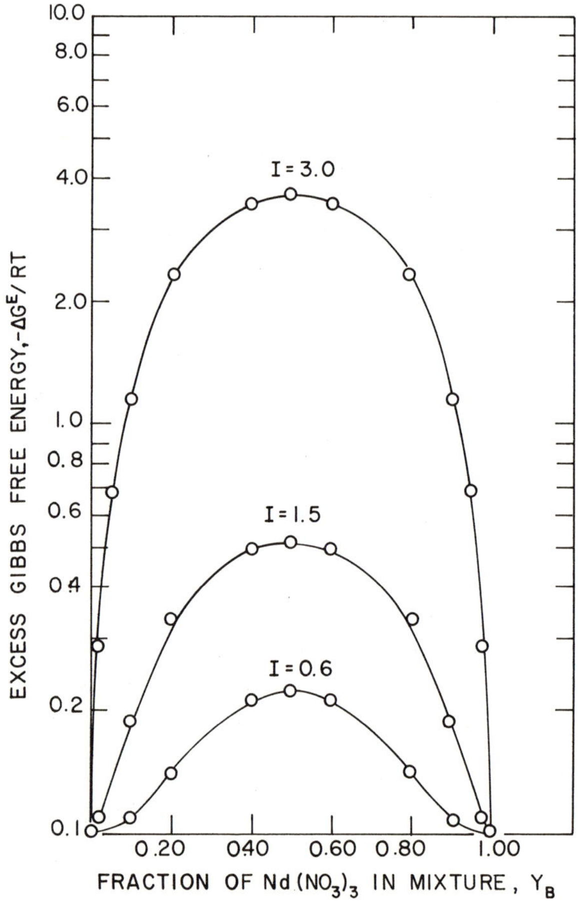

Figure 11. Excess Gibbs free energy of binary $Nd(NO_3)_3$–HNO_3–H_2O electrolyte mixture as a function of the fraction of $Nd(NO_3)_3$ in mixture

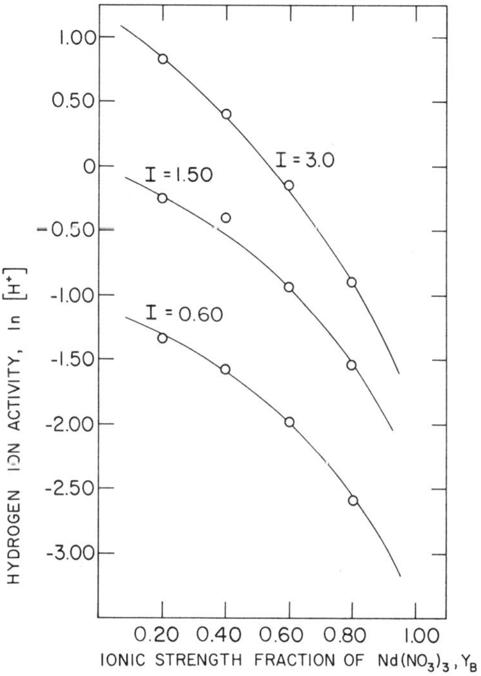

Figure 12. The hydrogen ion activity in the $Nd(NO_3)_3$–HNO_3–H_2O electrolyte solution as a function of ionic strength fraction of $Nd(NO_3)_3$

Table VIII. Nitrate Electrode Measurements on Neodymium Nitrate–Nitiric Acid Solutions at Constant Total Ionic Strength (*I*)

I	y_B [a]	(H^+) [b]	E_a *(mv)* [c]	E *(mv)*	$[NO_3^-]$ [d]
0.6	0.80	0.12	−108.1	−55.3	0.1281
0.6	0.60	0.24	−113.6	−63.9	0.1444
0.6	0.40	0.36	−119.1	−71.9	0.1596
0.6	0.20	0.48	−124.4	−79.6	0.1746
1.5	0.80	0.30	−116.4	−85.6	0.3012
1.5	0.60	0.60	−129.7	−102.6	0.3489
1.5	0.40	0.90	−142.3	−119.0	0.4031
1.5	0.20	1.20	−154.4	−134.0	0.4518
3.0	0.80	0.60	−129.7	−124.8	0.8278
3.0	0.60	1.20	−154.4	−155.3	1.0352
3.0	0.40	1.80	−176.7	−181.2	1.1910
3.0	0.20	2.40	−196.6	−212.6	1.8674

[a] Fraction of neodymium nitrate making up the total ionic strength.
[b] Hydrogen ion concentration of the aqueous phase.
[c] Calculated by Equation 33.
[d] Calculated by the Nernst Equation.

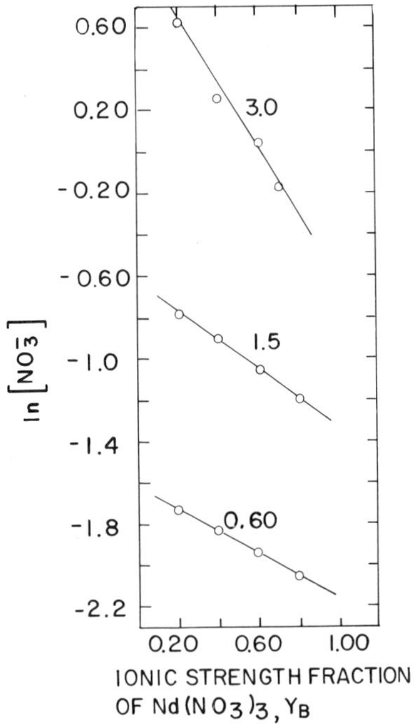

Figure 13. The nitrate ion activity in the $Nd(NO_3)_3$–HNO_3–H_2O electrolyte solution as a function of ionic strength fraction of $Nd(NO_3)_3$

The nitrate ion activities shown in Figure 13 were measured using a special nitrate electrode standardization technique for samples with a high hydrogen ion concentration or acidity. Pure samples of known HNO_3 concentration and nitrate activity were analyzed with the nitrate

Table IX. pH Electrode Measurements on Neodymium Nitrate–Nitric Acid Solutions at Constant Total Ionic Strength (*I*)

I	y_B[a]	E *(mv)*	$[H^+]$[b]
0.6	0.80	346.7	0.0745
0.6	0.60	363.4	0.1378
0.6	0.40	373.8	0.2066
0.6	0.20	380.0	0.2630
1.5	0.80	375.0	0.2116
1.5	0.60	390.6	0.3972
1.5	0.40	404.1	0.6718
1.5	0.20	407.7	0.7729
3.0	0.80	390.4	0.4107
3.0	0.60	410.5	0.8607
3.0	0.40	424.6	1.4918
3.0	0.20	435.6	2.2819

[a] Fraction of neodymium nitrate making up the total ionic strength.
[b] Calculated by $E = E_a + 59.16 \log [H^+]$ where E_a was determined by buffer solutions to be 414.32 mv.

electrode. From the data the reference voltage as a function of the hydrogen ion concentration could be calculated. Thus for a sample with a known concentration of H^+, the reference voltage could be calculated and the nitrate ion activity thus measured. All of the calibrations were consolidated to produce one equation to predict the reference voltage, E_a, as a function of sample acidity.

Through regression analysis of the data the following equation was estimated:

$$E_a = -102.449 - 47.394\, m_{HNO_3} + 3.410\, m^2_{HNO_3} \qquad (33)$$

where m_{HNO_3} is the sample molarity of nitric acid. The remaining data for the determination of $[NO_3^-]$ and $[H^+]$ is presented in Table VIII and IX.

Acknowledgment

This work was supported by the U.S. Department of Energy, Office of Energy Research, Division of Basic Energy Sciences. It was prepared for the U.S. Department of Energy under Contract No. W-7405-eng-82.

Literature Cited

1. Harned, H. S., Robinson, R. A., "Multicomponent Electrolyte Solutions," The International Encyclopedia of Physical Chemistry and Chemical Physics, Pergamon, Glasgow, Scotland, 1968.
2. Robinson, R. A., Stokes, R. H., "Electrolyte Solutions," Academic, London, 1955.
3. Guggenheim, E. A., *Trans. Faraday Soc.* (1955) **51**, 747.
4. Scatchard, G., *J. Am. Chem. Soc.* (1961) **83**, 2636.
5. Scatchard, G., Rush, R. M., Johnson, J. S., *J. Phys. Chem.* (1970) **74**, 3786.
6. Bechthold, M. F., Newton, R. F., *J. Am. Chem. Soc.* (1940) **62**, 1390.
7. Durst, R., "Ion Selective Electrodes," N.B.S. Special Publication **314**, 1969.
8. Lietzke, M. H., Stoughton, R. W., *J. Phys. Chem.* (1962) **66**, 508.
9. Pitzer, K. S., *J. Phys. Chem.* (1973) **77**, 268.
10. Meissner, H. W., Kusik, C. L., *Am. Inst. Chem. Eng. J.* (1972) **18**, 294.
11. Meissner, H. W., Tester, J. W., *Ind. Eng. Chem.* (1972) **11**, 128.
12. Meissner, H. W., Tester, J. W., Kusik, C. L., *Am. Inst. Chem. Eng. J.* (1972) **18**, 661.
13. Zozula, A. P., Peskova, V. M., *Russ. Chem. Rev. (Engl. Transl.)* (1960) **2**, 101.
14. Peppard, D. F., Mason, G. W., Hucher, I., *J. Inorg. Nucl. Chem.* (1962) **24**, 881.
15. Choppin, G., Unrein, P. J., *J. Inorg. Nucl. Chem.* (1963) **25**, 387.
16. Choppin, G., Strazik, W. F., *Inorg. Chem.* (1965) **9**, 1250.
17. Knoeck, J., *J. Anal. Chem.* (1969) **41**, 2069.
18. Krumholtz, P., in "Advances in the Chemistry of Coordination Compounds," S. Kirshner, Ed., Macmillan, New York, 1961.
19. Goto, J., Smutz, M., *J. Inorg. Nucl. Chem.* (1965) **27**, 663.

RECEIVED December 22, 1977.

20

The Activity Behavior of the $NdNO_3$-HNO_3-H_2O-HDEHP-AMSCO System

WILLIAM G. O'BRIEN[1] and RENATO G. BAUTISTA

Ames Laboratory USDOE and Department of Chemical Engineering, Iowa State University, Ames, IA 50011

The thermodynamic behavior of the two-phase $NdNO_3$–HNO_3–H_2O–HDEHP (Di-2-ethylhexyl phosphoric acid)-AMSCO Odorless Mineral Spirits system has been studied to provide experimental data for use with a model to predict the distribution coefficients of neodymium. The presence of high hydrogen ion concentration was taken into account in determining the nitrate ion activities in the aqueous phase using the nitrate ion selective electrode. The nitrate ion concentration in the organic phase was back-extracted with sulfuric acid before analysis with the nitrate and sulfate concentrations. The activity coefficients of the HDEHP dimer were estimated from the data of Baes who made isopiestic comparisons with triphenylmethane in n-*octane as the reference solution assuming that the activity coefficient of triphenylmethane is unity.*

Liquid–liquid extraction of the lanthanides from an aqueous salt solution to a chelate in an organic phase has been somewhat difficult to analyze thermodynamically for a number of reasons. First, very little data exists on the activity behavior of a single-component lanthanide salt in the aqueous solution. These data, generally more common for 1:1 electrolytes such as NaCl, KNO_3, etc., are in the form of the mean activity coefficient of the salts. Second, this value of the mean activity coefficient for a particular salt is a measure of the combined effect of both

[1] Present address: Photo Products Department, Du Pont, Towanda, PA 18848.

0-8412-0428-4/79/33-177-323$05.25/1

the cation and the anion on the solvent. Thus the mean activity coefficient reveals nothing about the nature of the behavior of the individual ions in solution. Third, owing to the nature of the extraction mechanism considerable amounts of H^+ ions are liberated into the aqueous phase especially where organophosphoric acids such as di-2-ethylhexyl phosphoric acid (HDEHP) are used as the extractant. Thus for the extraction of a lanthanide salt the final equilibrated aqueous phase will consist of an electrolyte solution, containing the remaining lanthanide ions, the liberated H^+ ions and the common anion. Fourth, at relatively high concentrations the lanthanide ion forms a series of complexes with the anion. Thus, the lanthanide can exist in the following forms in the aqueous phase: M^{+3}, MA^{+2}, MA_2^{+1} and MA_3 where A represents the anion species. Fifth, until recently the extraction mechanism at higher concentrations was not well defined. The uncertainty was caused by the presence of the various complexes of the lanthanide in the aqueous phase and their relative extractability. With recent advances in the liquid–liquid extraction of lanthanide–HDEHP chemistry and chemistry of the lanthanide–anion complexes, a more exact thermodynamic analysis of the liquid–liquid extraction reaction becomes possible (*2, 3*). The thermodynamic behavior of the extraction of neodymium from the system $Nd(NO_3)_3$–HNO_3–H_2O–HDEHP–AMSCO Odorless Mineral Spirits is reported in this chapter.

Lanthanide Extraction by Organophosphorous Acids

Organophosphorus compounds used in the liquid–liquid extraction of rare earths can be categorized as neutral or acidic depending on the general chelation mechanism particular to either class of organic. The neutral compounds, the primary example being tri-*n*-butyl phosphate (TBP), extract the neutral salt species while the organic species remain unchanged by the complexing. Both Peppard (*5*) and Eyring (*6*) have presented thorough reviews of the studies of TBP and other neutral extractants. Of the acidic organophosphoric extractants, HDEHP is the primary example. Upon complexing with a positive ion, the acidic extractant liberates a proton to the aqueous phase. The number of extractant molecules per complex is equal to the charge of the metallic ion or the extractable complex in the aqueous phase. The strength of this complex is attributable to its ionic characteristic and also to hydrogen bonding among the phosphate groups.

The mechanism of extraction in tracer rare earth concentrations using HDEHP was investigated by Peppard et al. (*7, 8, 9*). Freezing-point depression measurements of toluene by the addition of HDEHP showed that HDEHP is dimeric in solution (*9*). On the basis of all their findings at tracer concentrations Peppard et al. proposed the following ion exchange mechanism for the extraction of lanthanide ions by HDEHP:

$$M^{+3} + 3(HG)_2 \rightleftharpoons M(HG_2)_3 + 3H^+ \quad (1)$$

where $(HG)_2$ represents the dimeric HDEHP and $M(HG_2)_3$ the organometallic complex. This stoichiometry was valid (*7, 8*) at tracer concentrations of trivalent lanthanides in aqueous chloride, nitrate, and perchlorate solutions at low acidities. This HDEHP dimer was shown by Lenz and Smutz (*10*) to be the dominant species in solutions of HDEHP. Bauer (*11*) studied the extraction of cerium (III) by HDEHP from acidic aqueous nitrate solutions and found that the cerium extraction decreases as the acidity increases. It reaches a minimum value around acid concentrations of 5*M* and increases again at higher acidities. This would indicate that at the lower acidities the mechanism of Peppard et al. (*7, 8*) was obeyed, while at higher acidities cerium nitrate complexes are extracted.

Kosinski and Bostian (*12*) reported the extraction of lanthanum by HDEHP from aqueous nitrate solutions over a wide concentration range. Using mass balance data and IR analysis of the organic phase, they proposed three extraction reactions which occur simultaneously:

$$La^{+3} + 3(HG)_2 \cdot H_2O \rightleftharpoons La(HG_2)_3 + 3H_2O + 3H^+ \quad (2)$$

$$LaNO_3^{+2} + 2(HG)_2H_2O \rightleftharpoons LaNO_3(HG_2)_2 + 2H_2O + 2H^+ \quad (3)$$

$$La(NO_3)_2^{+1} + (HG)_2 \cdot H_2O \rightleftharpoons La(NO_3)_2 \cdot HG_2 + H_2O + H^+ \quad (4)$$

Extensive studies involving the separation of two or more rare earth species by HDEHP liquid–liquid extraction from aqueous nitrate solution is found in the work of Smutz et al. (*13, 14, 15, 16, 17*). The reviews of Peppard (*5*), Erying (*6*), and more recently by Golinski (*18*) give additional information on extraction by organophosphoric acids other than HDEHP on aqueous systems having common anions other than nitrate such as chloride, sulfate, and perchlorate. A further complication of the organic phase is the possible formation of HDEHP trimers in the organic phase by the following reaction:

$$3(HG)_2 \rightleftharpoons 2(HG)_3 \quad (5)$$

Baes et al. (*19*) have investigated this trimer formation in *n*-hexane through isopiestic measurements and have found little indication of the trimer species. In subsequent investigations Baes (*20*) and Baes and Baker (*21*) suggested HDEHP partial trimerization as an explanation for

HDEHP–octane nonideal solution behavior. However, they state quite plainly that without more direct evidence for Equation 5 this interpretation should be considered no more than a good indication of the nature of HDEHP-*n*-octane nonideal solution behavior. Peppard et al. (*9*) through freezing-point depression and IR spectroscopy determined that HDEHP was strongly dimerized in benzene and naphthalene while being monomeric in solution with high molecular weight alcohols and monocarboxylic acids. Other work by Baes (*20*) has included activity coefficient determination of the HDEHP dimer in *n*-octane through isopiestic measurements. The structure of these hydrogen-bonded complexes has been reported by Peppard et al. (*9*) and Baes (*20*).

Experimental Procedures

The neodymium oxide used in this work had a purity of greater than 99.9% by emission spectroscopy and was prepared at the Ames Laboratory, USDOE. The neodymium nitrate stock solutions were prepared by dissolving the neodymium oxide in concentrated reagent grade HNO_3. The amount of acid used was 50% in excess of the required stoichiometric amount. The excess HNO_3 was removed by boiling down the solution on a hot plate and subsequent dilution with distilled water. To eliminate hydrolysis of the highly charged neodymium ions, the stock solution was titrated with dilute HNO_3 to the equivalence point where the ratio of neodymium to nitrate is 1:3. This was determined by the point of discontinuity of a plot of pH vs. milliliters of acid added. The equivalence point of neodymium nitrate solutions occurs at about pH = 2.0. Thus all stock solutions were adjusted to this final value of pH to insure negligible hydrolysis effects. The concentrations of these stock solutions were determined accurately by the total precipitation of neodymium by the addition of oxalic acid. The precipitate was converted to the oxide by roasting and subsequently weighed to determine the concentration of the neodymium stock solutions.

The HDEHP obtained from the Union Carbide Corporation had a purity of 99.1%. Solutions of dimerized HDEHP were prepared by diluting pure HDEHP with AMSCO Odorless Mineral Spirits, a high molecular weight hydrocarbon diluent. The molarity of the HDEHP was 1*M* on a monomer basis, thus the dimerized HDEHP was 0.5*M* on a dimer basis.

The liquid–liquid extraction feed solutions were prepared by dilution of the neodymium nitrate stock solution with distilled water and nitric acid. These then were contacted with equal volumes of 1*M* HDEHP in separatory funnels, agitated for .5 hr by a mechanical shaker, allowed to separate for .5 hr, shaken another .5 hr, then allowed to settle for 12 hr before the phases were separated carefully. A volumetric sample of the organic phase was back-extracted four times with an equal volume of 6*M* HNO_3. This solution was evaporated to dryness in order to remove the excess acid and then diluted to an appropriate concentration. The pH of this solution was adjusted to 3.0 to eliminate any hydrolysis effects and then titrated with a standardized EDTA solution using arsenazo as the indicator and pyridine as the buffer to determine the neodymium concentration in the organic phase. EDTA titration also was used to calculate the neodymium concentration in the aqueous phase.

The hydrogen ion concentration of the aqueous phase cannot be determined directly by titration with NaOH because of the hydrolysis with the neodymium ion. The cation exchange resin, Dowex 50 x 8, is used to adsorb the neodymium ion (*11, 22*)., A volumetric sample of the aqueous phase is passed through an ion exchange column containing the Dowex resin. For every neodymium ion attached to the resin, three H^+ ions are liberated to the aqueous phase. This aqueous phase, after complete neodymium removal by passage through the column, is collected and analyzed for total acidity by titration with NaOH using phenolphtalein as the indicator. The equilibrium acidity is the total acidity minus three times the aqueous neodymium ion concentration.

The hydrogen ion activity of the equilibrated aqueous phase was determined with a Beckman 39301 glass pH electrode and a Beckman saturated calomel electrode. The electrode potential was measured by an Orion 801 Ionanalyzer. The reference voltage of the system was calculated through electrode measurement of standard buffer solutions. From the Nernst Equation the reference voltage becomes

$$E_a = E + \frac{\mathrm{RT}}{\mathrm{F}} \ln[\mathrm{H}^+] \qquad (6)$$

or since pH $= -\log\,[\mathrm{H}^+]$, Equation 6 becomes

$$E_a = E - 59.16\,\mathrm{pH} \quad (T = 25°\mathrm{C}) \qquad (7)$$

where E is measured in millivolts.

The polyvalent neodymium ions are assumed to have a slight effect on the glass pH electrode measurements. It also should be mentioned that the pH electrode was calibrated before and during use with buffer solutions of 4.01 and 2.27 pH values.

The following general procedures were followed for electrode measurements to increase the experimental accuracy. (1) Both electrodes were rinsed with deionized water and patted dry with absorbent tissue after each measurement. An Orion microsample dish was used to hold all solutions. (2) The same electrode equilibration time was used for both the unknown and the standard. An equilibration time of 200 sec gave good reproducible results over the concentration range studied in this work. (3) It was observed that the electrode exhibits quicker response times when going from lower to higher concentration solutions than vice versa. (4) When working with high concentration solutions, the organic ion exchange liquid was changed frequently to prevent electrode failure. One sign of electrode failure for the nitrate system is the formation of a red crystal in the organic phase. Under these conditions the electrode response is very erratic until fresh organic is used. (5) Standardization with several standard solutions bracketing the ion activity of the unknown solutions is beneficial. This procedure diminishes errors introduced by non-Nernstian electrode response and by the residual liquid junction potential.

The analysis of the organic phase for nitrate content was carried out by back-extraction of the organic phase with sulfuric acid, dilution, and measurement of the nitrate concentration with the nitrate ion electrode. Ten standards were prepared having varying concentrations of nitrate ion and a constant background concentration of sulfuric acid (0.3*M*). Equal volumes of 3*M* H_2SO_4 were contacted three times with the organic phase. The resulting aqueous solution was diluted and analyzed by the nitrate electrode. Suitable standards were made up containing sulfuric acid and sodium nitrate to determine exact nitrate ion concentrations of the samples.

The percentage of sulfate ion interference for the nitrate electrode was calculated to be negligible using the empirical equation

$$E = \text{constant} + \frac{2.303\ \mathrm{RT}}{\mathrm{Z_A F}} \log\left[a + \Sigma \mathrm{k_i} b_i^{\mathrm{Z_B/Z_A}}\right] \qquad (8)$$

where a and b represent the activities of the sought after ion, A, and the interfering ion, B, respectively. The activity of each interfering ion in the sample, raised to a power equal to the ratio of the charge Z, is multiplied by a weighting factor, the selectivity constant k_i. Thus with a knowledge of the concentration of interfering ion in the sample and its selectivity constant, the electrode response caused by the interfering ion can be predicted. For the specific nitrate ion electrode, the manufacturer's k_i value is 3×10^{-5} for SO_4^{2-}. The voltage of the nitrate electrode in these various standards was measured and is used to construct the nitrate–sulfate calibration curve. The unknown samples were diluted appropriately so that the background H_2SO_4 concentration is 0.3*M*. The nitrate content of the organic phase samples was obtained using the electrode voltage, the dilution factor, and the calibration curve.

Results and Discussions

Thermodynamics can offer no unique guide or formula in separating the well-defined properties of an electrolyte into those for the individual ions. In this work, the presence of H^+ ions in the sample caused serious interference to the nitrate electrode operation. The emf readings were too highly negative for the approximate nitrate concentrations of the samples. The varying background concentration of hydrogen ions was assumed responsible for this. A possible explanation for this H^+ electrode interference is the greater ability of the smaller, more mobile hydrogen ion to diffuse into the liquid junction of the reference electrode than the larger, more bulky NO_3^- ion. This liquid junction potential could be the cause of the high negative readings.

This liquid junction potential or diffusion potential is caused by slight separation of ionic charges that results from the tendencies of the various ions to diffuse at unequal rates across the boundary of the solutions. The effects of the liquid junction potential can be minimized by using a solution of approximately the same ionic strength in the reference electrode and having similar values for the ionic mobilities as that for the samples to be tested. This reduces the degree of ionic diffusion and hence the liquid junction potential. Another method of reducing the liquid junction potential effect is by using several reference solutions to standardize the electrodes throughout the anticipated concentration range.

One way of correcting for this H^+ diffusion effect is to evaluate the reference voltage for each nitrate activity and measured electrode potential and make the reference voltage functionally dependent on the concentration of the H^+ ions in the solution. Thus, through careful calibration with pure HNO_3 solutions and a knowledge of the H^+ concentration of the sample, the reference voltage can be calculated and nitrate ion electrode can be used to determine the nitrate ion activity of the sample. The individual calibration corresponds to a series of experimental nitrate

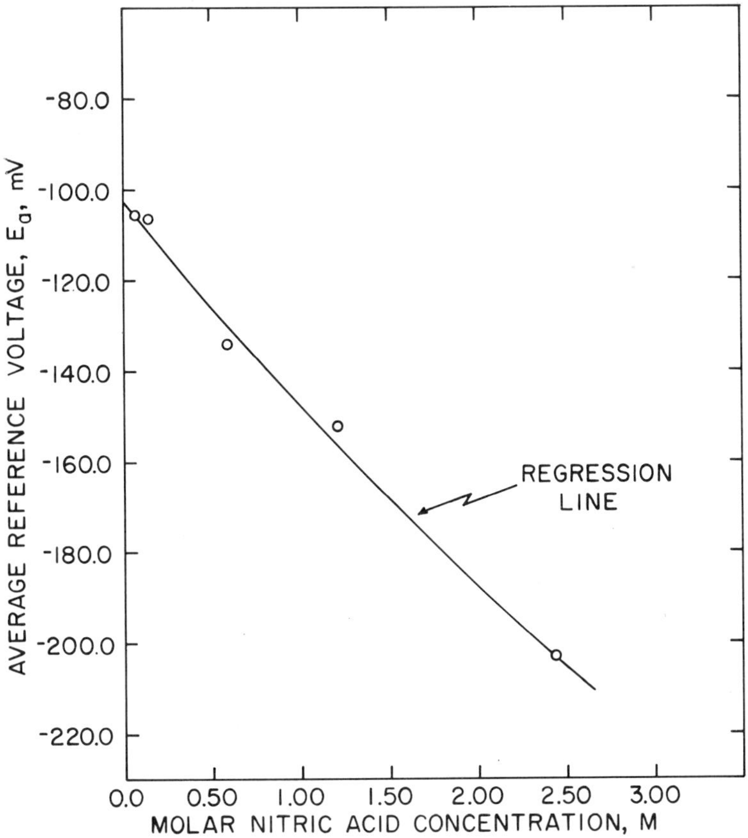

Figure 1. Average value of the reference voltage, E_a as a function of the molar nitric acid concentration

Table I. Nitrate Electrode Data for Equilibrated Aqueous Phase

Sample	(H^+)[a] M	E_a *(mV)*[b]	E *(mV)*	$[NO_3^-]$[c]
1	0.2210	−112.8	−74.8	0.2626
2	0.3224	−117.4	−89.2	0.3340
3	0.4550	−123.3	−103.6	0.4644
4	0.5560	127.7	−113.2	0.5677
5	0.6792	−133.1	−124.8	0.7249
6	0.7731	−137.1	−129.1	0.7338
7	1.2720	−157.2	−138.9	0.4902
8	1.7900	−176.3	137.5	0.6560
9	0.2710	−115.0	−85.5	0.3167
10	0.3625	−119.2	−87.7	0.2937
11	0.4325	−122.3	−93.5	0.3259
12	0.5311	−126.7	−100.7	0.3641
13	0.5049	−125.5	−107.6	0.4981
14	0.8527	−140.4	−131.1	0.6952
15	0.9836	−145.8	−139.7	0.7872
16	1.2158	−155.0	−145.9	0.7021
17	0.8121	−138.7	−127.4	0.6444
18	1.3068	−158.6	−134.7	0.3950
19	1.8486	−178.4	−181.1	1.1104
20	2.4456	−198.0	−193.6	0.8421
21	0.3779	−119.9	−86.7	0.2750
22	0.5181	−126.1	−111.7	0.5712
23	0.6796	−133.1	−113.4	0.4648
24	0.8032	−138.3	−131.5	0.7665
25	0.5516	−127.6	−105.8	0.4289
26	0.9341	−143.7	−126.1	0.5032
27	1.3427	−159.9	−144.4	0.5462
28	1.7466	−174.8	−164.4	0.6665
29	0.2665	−112.9	−86.4	0.3563
30	0.3379	−117.0	−95.3	0.4302
31	0.3843	−120.2	−111.2	0.7056
32	0.4654	−123.8	−123.6	0.9935
33	0.5401	−127.1	−124.2	0.8731
34	0.9270	−143.5	−140.7	0.8962
35	1.4360	−163.5	−154.5	0.7051

[a] Hydrogen ion concentration of the equilibrated aqueous phase.
[b] Reference voltage calculated by Equation 9.
[c] Calculated by Nernst Equation.

electrode measurements. All of the calibrations were lumped together to produce one equation to predict the reference voltage, E_a as a function of sample acidity. Figure 1 shows the averaged value of the reference voltage calculated from the Nernst Equation as a function of the HNO_3 concentration. Through regression analysis of the data the following equation was obtained.

$$E_a = -102.449 - 47.394\, M_{HNO_3} + 3.410\, M_{HNO_3}^{\;2} \qquad (9)$$

where M_{HNO_3} is the nitric acid molarity. The reference voltage was calculated using the hydrogen ion concentrations of the equilibrated aqueous phase and the nitrate ion activity. These values are presented in Table I.

Bates and Alfenar (23) proposed that the activity of the chloride ion in NaCl solutions be taken as the activity standard. Using this chloride convention together with the published values of mean activity coefficients, the activity of any simple ionic species can be estimated. The equation used to estimate values of $\gamma_{NO_3}^{\;-}$ and thus $a_{NO_3}^{\;-}$ for nitric acid solutions is

$$\log \gamma_{NO_3^-} = \log \gamma_{Cl^-} + 2 \log(\gamma_{NO_3^-}/\gamma_{HCl}) \qquad (10)$$

where:

$$-\log\gamma_{Cl^-} = \frac{0.509\, I^{\frac{1}{2}}}{1.0 + 1{:}5\, I^{\frac{1}{2}}} \qquad (11)$$

and I is the ionic strength of the solution. The calculated values of $\gamma_{NO_3}^{\;-}$ and γ_{HNO_3} are given in Table II and are shown as a function of molar nitric acid concentraiton in Figure 2. With this information nitrate elec-

Table II. Data for the Calculation of the Nitrate Ion Activity Coefficient in Aqueous Solutions of Nitric Acid

(HNO_3) M	γ_{HCl}	γ_{HNO_3}	γ_{Cl^-} [a]	$\gamma_{NO_3^-}$ [b]
0.1	0.796	0.791	0.778	0.768
0.2	0.767	0.754	0.731	0.706
0.3	0.756	0.735	0.703	0.665
0.4	0.755	0.725	0.684	0.630
0.5	0.757	0.720	0.669	0.605
0.6	0.763	0.717	0.657	0.580
0.7	0.772	0.717	0.647	0.558
0.8	0.783	0.718	0.639	0.537
0.9	0.795	0.721	0.632	0.520
1.0	0.809	0.724	0.626	0.501
1.2	0.840	0.734	0.615	0.470
1.4	0.896	0.745	0.607	0.449
1.6	0.916	0.758	0.600	0.411
1.8	0.960	0.775	0.593	0.387
2.0	1.009	0.793	0.588	0.363
2.5	1.147	0.846	0.577	0.314
3.0	1.316	0.909	0.569	0.271

[a] Calculated using Equation 11.
[b] Calculated using Equation 10.

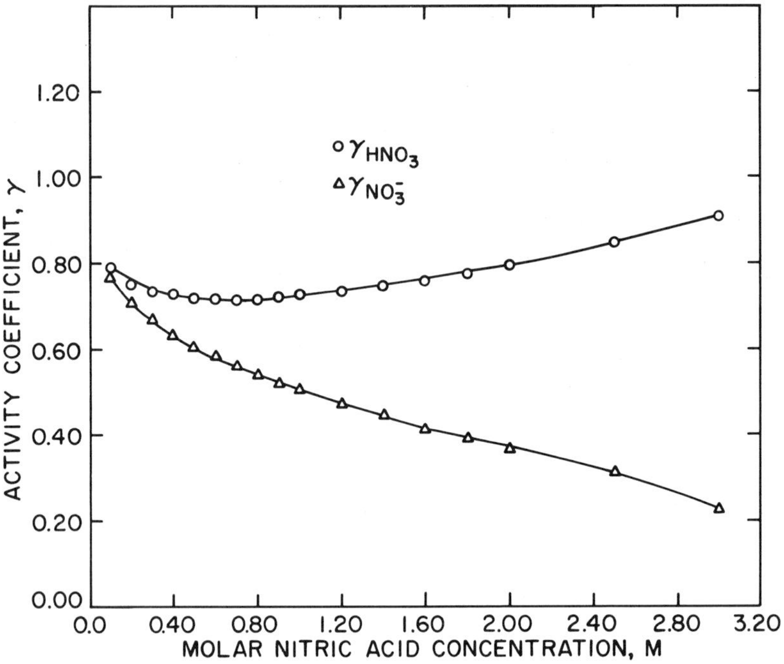

Figure 2. Activity coefficients of HNO_3 and NO_3^- as a function of the molar nitric acid concentration

trode measurements on various concentration HNO_3 solutions can be made and the Nernstian performance of the electrode evaluated. The data from nitrate electrode measurements of HNO_3 solutions deviates quite sharply from the Nernstian slope at the higher HNO_3 concentrations supporting the assumption that the interference is caused by H^+ diffusion into the liquid junction of the reference electrode.

The hydrogen ion activity of the equilibrium aqueous phase was measured with Beckman pH and saturated calomel electrodes. The reference voltage E_a of the system was calculated using the Nernst Equation and the electrode measurement of standard buffer solutions. For all pH electrode measurements E_a varied only slightly between 414 and 417 mV. The data are shown in Table III and the calculated hydrogen ion activity $[H^+]$ is based on the average of the two values of electrode voltage ($E1$ and $E2$).

The equilibrium metal concentrations in the aqueous and organic phases and the equilibrium acidity of the aqueous phase are presented along with the calculated values for the distribution coefficient in Table IV.

The total nitrate concentration in the organic phase owing to the extraction of neodymium complexes by HDEHP was calculated from nitrate electrode measurements after back-extraction of a volumetric sample of the organic phase. The nitrate content of the organic phase is given in Table V and indicates substantial amounts of NO_3^- in the organic phase. For high concentrations of $Nd(NO_3)_3$ in the aqueous phase the fraction of neodymium complexed with the nitrate ion increases sharply. Thus at higher feed concentrations more of the nitrate species

Table III. The pH Electrode Data for Equilibrated Aqueous Phase

Sample	E_a *(mV)*[a]	E*1 (mV)*	E*2 (mV)*	*[H^+]*[b]
1	416.16	356.8	363.1	0.1212
2	416.16	368.3	373.9	0.1871
3	416.16	377.9	384.0	0.2746
4	416.16	385.8	392.9	0.3807
5	416.16	394.4	400.7	0.5239
6	416.16	402.6	410.6	0.7451
7	416.16	412.3	419.4	1.0680
8	416.16	424.4	425.3	1.4638
9	416.27	372.9	370.7	0.1771
10	416.27	378.4	376.7	0.2224
11	416.27	383.3	380.7	0.2635
12	416.27	387.9	385.4	0.3157
13	416.27	384.1	385.1	0.2915
14	416.27	400.2	403.4	0.5700
15	416.27	405.5	409.6	0.7103
16	416.27	421.3	418.5	1.1510
17	416.27	405.1	404.3	0.6374
18	416.27	416.5	416.4	1.0070
19	416.27	426.5	424.5	1.4322
20	416.27	435.3	431.7	1.9560
21	414.60	382.9	380.5	0.2779
22	414.60	398.8	396.6	0.5181
23	414.60	405.9	403.5	0.6796
24	414.60	408.1	405.5	0.7391
25	414.60	395.7	392.7	0.4521
26	414.60	412.1	408.8	0.8510
27	414.60	424.0	420.3	1.3430
28	414.60	432.1	428.6	1.8466
29	416.07	369.2	367.3	0.1555
30	416.07	376.5	375.5	0.2102
31	416.07	383.6	382.2	0.2749
32	416.07	396.4	393.3	0.4420
33	416.07	405.9	403.5	0.6431
34	416.07	412.4	414.0	0.8942
35	416.07	424.7	422.9	1.3770

[a] Calculated from pH measurements of buffer solutions by Equation 6.
[b] Calculated from the averaged value of *E*1 and *E*2 using Nernst Equation.

Table IV. $Nd(NO_3)_3$–HNO_3–H_2O–HDEHP–AMSCO Extraction Data

Sample	$(Nd)_O$[a] M	$(Nd)_A$[b] M	(H^+) M	K_D[c]
1	0.0751	0.0144	0.2210	5.215
2	0.1031	0.1498	0.3224	0.6762
3	0.1110	0.3781	0.4550	0.2936
4	0.1166	0.6138	0.5560	0.1900
5	0.1221	0.8458	0.6792	0.1444
6	0.1313	1.1047	0.7731	0.1189
7	0.1350	1.3296	1.2720	0.1015
8	0.1365	1.5845	1.7900	0.0861
9	0.0567	0.0222	0.2710	2.5560
10	0.0422	0.0178	0.3625	2.3740
11	0.0240	0.0142	0.4325	1.6870
12	0.0085	0.0071	0.5311	1.1891
13	0.0556	0.1267	0.5049	0.4385
14	0.0268	0.1078	0.8527	0.2485
15	0.0132	0.0801	0.9836	0.1646
16	0.0038	0.0422	1.2158	0.0895
17	0.0434	0.3313	0.8121	0.1309
18	0.0150	0.2535	0.3068	0.0592
19	0.0271	0.1957	1.8486	0.1383
20	0.0049	0.0956	2.4456	0.0515
21	0.0612	0.0784	0.3779	0.7803
22	0.0097	0.0665	0.5181	0.1495
23	0.0061	0.0517	0.6796	0.1219
24	0.0071	0.0285	0.8032	0.2502
25	0.0534	0.2470	0.5516	0.2163
26	0.0238	0.1924	0.9341	0.1234
27	0.0095	0.1354	1.3427	0.0702
28	0.0033	0.0736	1.7466	0.0452
29	0.0548	0.0244	0.2665	2.2520
30	0.0695	0.0744	0.3379	0.9345
31	0.0744	0.1524	0.3843	0.4880
32	0.0724	0.2681	0.4654	0.2699
33	0.0689	0.3810	0.5401	0.1807
34	0.0771	0.6156	0.9270	0.1251
35	0.0652	0.8471	1.4360	0.0773

[a] Equilibrium organic neodymium concentration.
[b] Equilibrium aqueous neodymium concentration.
[c] Distribution coefficient.

is extracted into the organic phase along with the metal species as shown in Figure 3. The ratio of nitrate to metal increases quite sharply with increasing feed concentration as shown by Figure 4.

The equilibrated concentration of HDEHP dimer, $(HG)_2^E$, can be estimated from the initial HDEHP dimer concentration, $(HG)_2^I$, and the total concentration of neodymium, $(Nd)_0^T$, and nitrate ions, $(NO_3)_0^T$, in the organic phase by the equation

$$(HG)_2^E = (HG)_2^I - 3(Nd)_0^T + (NO_3^-)_0^T \tag{12}$$

The calculated free equilibrium HDEHP dimer concentration using Equation 12 is illustrated in Table VI. The relationship of the equilibrated HDEHP dimer molar concentration with the total neodymium concentration and the total nitrate concentration in the organic phase is shown in Figures 5 and 6, respectively.

The AMSCO Odorless Mineral Spirits is a mixture of high-purity aliphatic hydrocarbons. For this reason the degree of monomerization caused by polar solvent interaction is considered negligible but the estimation of accurate activity coefficients for the dimer is an extremely difficult task. Early attempts at the measurement of HDEHP dimer activity coefficient in AMSCO by considering the solvent to be a hypothetical pure component proved fruitless. Vapor pressure measurements on AMSCO–HDEHP solutions using averaged properties for the AMSCO, such as molecular weight and pure solvent vapor pressure, were nonreproducible. The lack of uniformity in sample outgassing caused different amounts of the most highly volatile portion of AMSCO to be lost even with identical samples.

Estimates of the activity coefficients of the HDEHP dimer in *n*-octane were made by Baes (*20*) on the basis of isopiestic comparisons with triphenylmethane in *n*-octane as the reference solution with the assumption that the activity coefficient of the triphenylmethane is unity. The results were formulated into the following empirical equation:

$$\log \gamma_{(HG)_2} = -0.5227 m_{(HG)_2}^{1/3} + 0.420 m_{(HG)_2} \tag{13}$$

where *m* refers to dimer molality. Through density measurements of HDEHP–*n*-octane solutions Baes (*20*) determined the relationship between dimer molality (*m*) and dimer molarity (*M*) as

$$M_{(HG)_2} = 0.6986 m_{(HG)_2} - 0.3207 m_{(HG)_2}^2 + 0.15 m_{(HG)_2}^3 \tag{14}$$

Baes and Baker (*21*) in the analysis of iron (III) extraction by HDEHP in *n*-octane showed that the dimer activity coefficient can be represented by

$$\log \gamma_{(HG)_2} = 0.6432 m_{(HG)_2}^{1/3} \tag{15}$$

Equations 13 and 15 were determined over a concentration range of $m_{(HG)_2} = 0.02 - 0.16$.

Table V. Nitrate Electrode Data for Organic

Sample	(SO_4^{2-}) M	*E (mV)*	(NO_3^-)[a] M
1	0.30	64.5	0.00030
2	0.30	51.5	0.00065
3	0.30	40.0	0.00120
4	0.30	32.3	0.00176
5	0.30	26.6	0.00240
6	0.30	21.9	0.00333
7	0.30	18.8	0.00345
8	0.30	15.4	0.00410
9	0.30	14.2	0.00435

[a] Taken from the nitrate–sulfate vs. emf calibration curve.
[b] Organic phase nitrate concentration calculated from the sample nitrate concentration and the dilution factor.
[c] Approximate initial feed molarity.

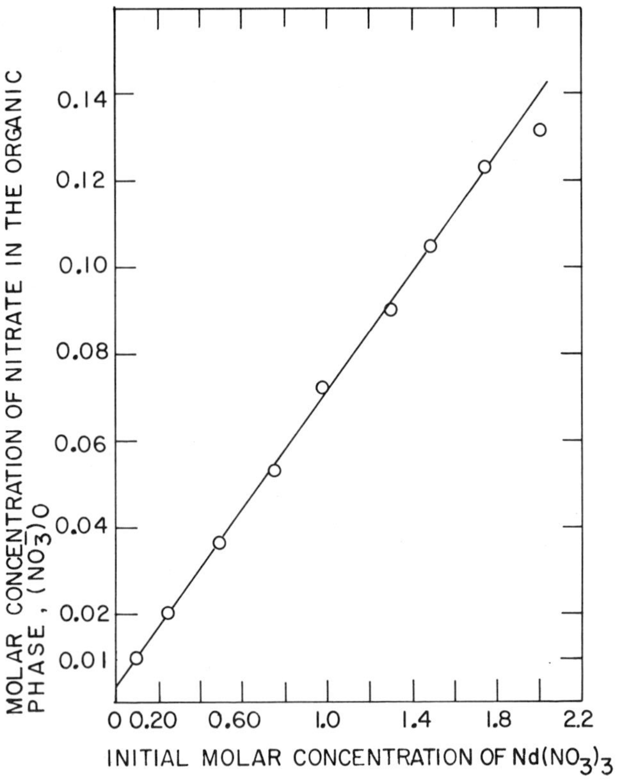

Figure 3. Nitrate concentration in the equilibrated organic phase as a function of the initial $Nd(NO_3)_3$ concentration

Phase Samples Back-Extracted with H_2SO_4

Dilution Factor (DF)	$(NO_3^-)_O$[b] M	$(Nd)_O$ M	$(Nd)_F$[c] M
30	0.0090	0.0751	0.10
30	0.0195	0.1013	0.25
30	0.0360	0.1100	0.50
30	0.0528	0.1166	0.75
30	0.0720	0.1211	1.00
30	0.1000	0.1311	1.25
30	0.1135	0.1350	1.50
30	0.1230	0.1365	1.75
30	0.1305	0.1434	2.00

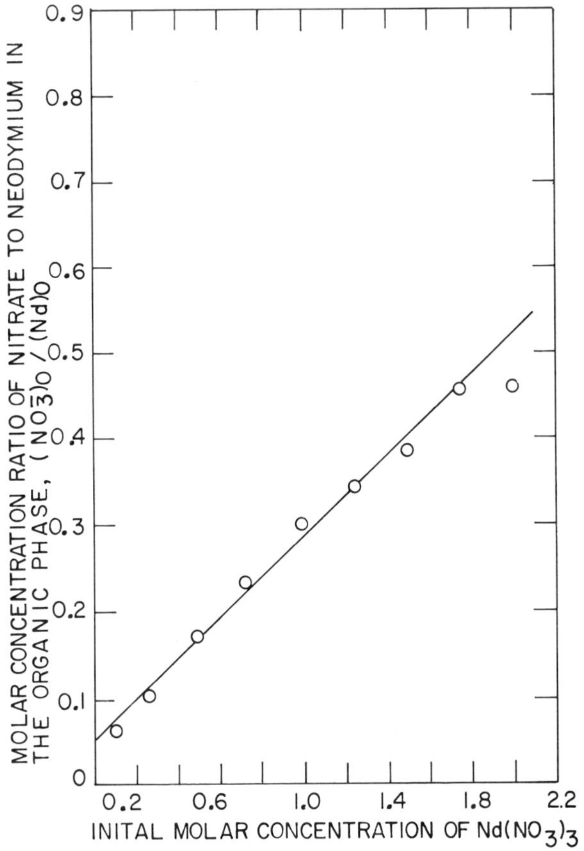

Figure 4. Nitrate-to-neodymium concentration ratio in the organic phase as a function of the initial $Nd(NO_3)_3$ concentration

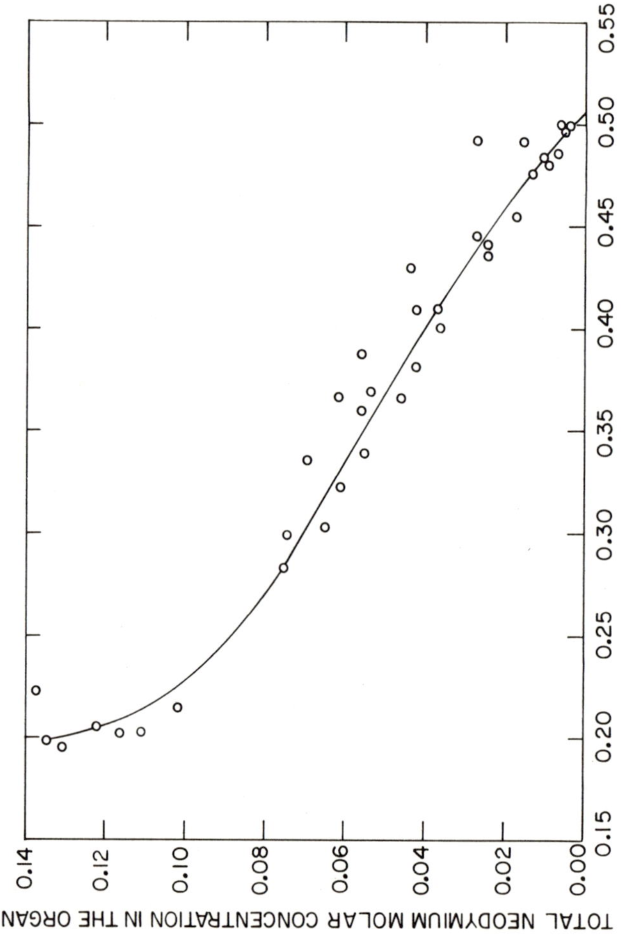

Figure 5. Total neodymium molar concentration in the organic phase as a function of the equilibrated HDEHP–dimer molar concentration

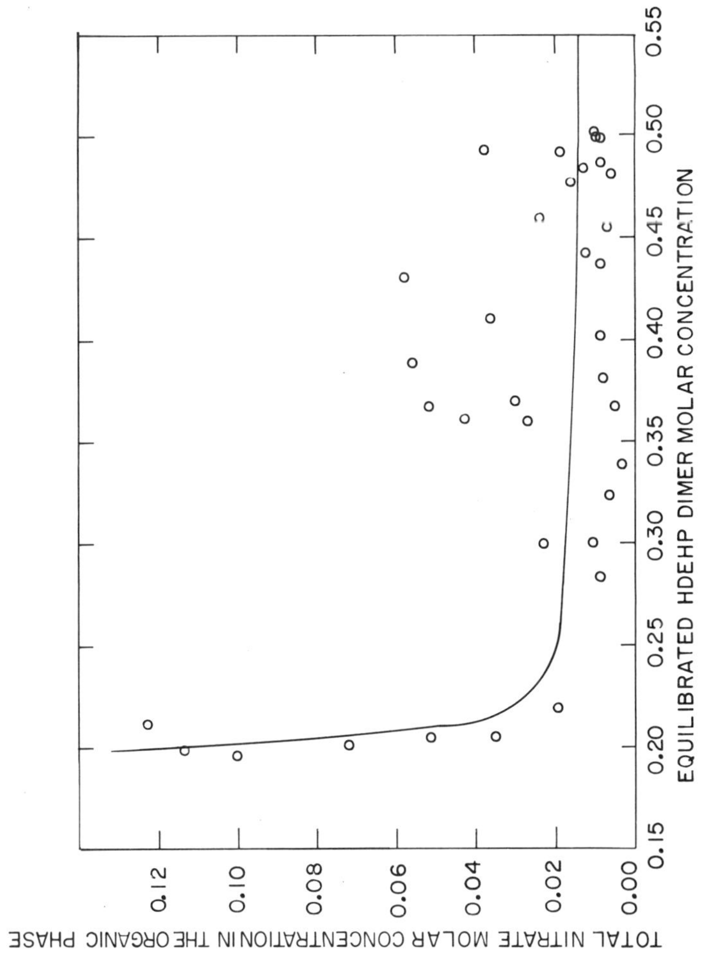

Figure 6. Total nitrate concentration in the organic phase as a function of the equilibrated HDEHP–dimer molar concentration

Table VI. Data for the Estimation of the Free HDEHP Dimer Concentration in the Equilibrated Organic Phase

Sample	$(HG)_2^{I\,a}$ M	$(Nd)_O^T$ M	$(NO_3^-)_O^T$ M	$(HG)_2^{E\,b}$ M
1	0.50	0.0751	0.0090	0.2837
2	0.50	0.1013	0.0195	0.2156
3	0.50	0.1110	0.0360	0.2030
4	0.50	0.1166	0.0528	0.2030
5	0.50	0.1221	0.0720	0.2057
6	0.50	0.1313	0.1000	0.1961
7	0.50	0.1350	0.1135	0.1985
8	0.50	0.1365	0.1230	0.2135
9	0.50	0.0456	0.0050	0.3683
10	0.50	0.0422	0.0080	0.3813
11	0.50	0.0240	0.0086	0.4366
12	0.50	0.0085	0.0057	0.4804
13	0.50	0.0556	0.0270	0.3603
14	0.50	0.0268	0.0236	0.4462
15	0.50	0.0132	0.0162	0.4767
16	0.50	0.0038	0.0090	0.4977
17	0.50	0.0434	0.0580	0.4297
18	0.50	0.0150	0.0380	0.4923
19	0.50	0.0271	0.0192	0.4922
20	0.50	0.0049	0.0096	0.4989
21	0.50	0.0612	0.0066	0.3225
22	0.50	0.0362	0.0096	0.4009
23	0.50	0.0173	0.0072	0.4552
24	0.50	0.0071	0.0086	0.4872
25	0.50	0.0534	0.0300	0.3697
26	0.50	0.0238	0.0128	0.4416
27	0.50	0.0095	0.0128	0.4843
28	0.50	0.0033	0.0096	0.4996
29	0.50	0.0548	0.0034	0.3389
30	0.50	0.0695	0.0110	0.3025
31	0.50	0.0744	0.0225	0.2994
32	0.50	0.0424	0.0370	0.4099
33	0.50	0.0689	0.0430	0.3364
34	0.50	0.0615	0.0520	0.3674
35	0.50	0.0561	0.0560	0.3878

[a] Initial HDEHP dimer concentration.
[b] Equilibrium dimer concentration (Equation 12).

Owing to the similarity of *n*-octane and AMSCO Odorless Mineral Spirits Equations 13 and 15 should provide a reasonable estimate for the activity coefficient to the HDEHP dimer in AMSCO. The similarities between the two organic diluents are as follows: *n*-octane is a nonpolar aliphatic hydrocarbon and AMSCO is a mixture of nonpolar aliphatic hydrocarbons; the molecular weight of *n*-octane is 114 while the mean molecular weight for AMSCO determined by freezing-point depression of benzene is 148.7; and the density of *n*-octane is 0.7061 gm/cm³ while that for AMSCO is 0.7428 gm/cm³.

The activity coefficient of the HDEHP dimer is estimated from Equations 14 and 16 as a function of molality using data from the iso-

Table VII. HDEHP Dimer Activity Coefficients Calculations

Point	$(HG)_2$ [a] M	$\gamma_{(HG)_2}$ [b]	$\gamma_{(HG)_2}$ [c]	I [d]
1	0.2837	0.6225	0.3102	0.3074
2	0.2156	0.6015	0.3489	1.2212
3	0.2030	0.5993	0.3573	2.7236
4	0.2030	0.5993	0.3573	4.2388
5	0.2057	0.5997	0.3555	5.7540
6	0.1961	0.5983	0.3622	7.4013
7	0.1985	0.5987	0.3605	9.2496
8	0.2135	0.6011	0.3502	11.2970
9	0.3683	0.6660	0.2740	0.4775
10	0.3813	0.6742	0.2693	0.4692
11	0.4366	0.7131	0.2512	0.5179
12	0.4804	0.7482	0.2389	0.5738
13	0.3603	0.6612	0.2770	1.2653
14	0.4462	0.7204	0.2484	1.4997
15	0.4767	0.7451	0.2398	1.4639
16	0.4977	0.7631	0.2344	1.4692
17	0.4279	0.7065	0.2539	2.8000
18	0.4930	0.7590	0.2356	2.8281
19	0.4922	0.7583	0.2358	3.0230
20	0.4989	0.7641	0.2341	3.0190
21	0.3225	0.6403	0.2923	0.8002
22	0.4009	0.6873	0.2626	0.9172
23	0.4552	0.7275	0.2458	0.9647
24	0.4872	0.7540	0.2371	0.9742
25	0.3697	0.6669	0.2735	1.7728
26	0.4416	0.7169	0.2498	2.0885
27	0.4843	0.7515	0.2379	2.1550
28	0.4996	0.7647	0.2339	2.1884
29	0.3389	0.6490	0.2854	0.4126
30	0.3025	0.6307	0.3012	0.7841
31	0.2994	0.6293	0.3027	1.2987
32	0.4099	0.6935	0.2596	2.0741
33	0.3364	0.6476	0.2865	2.8260
34	0.3674	0.6655	0.2744	4.6210
35	0.3878	0.6785	0.2670	6.5184

[a] Free concentration of HDEHP dimer.
[b] HDEHP dimer activity coefficient from isopiestic data.
[c] HDEHP activity coefficient from iron (III) extraction data of Baes and Baker (*21*).
[d] Total ionic strength of equilibrated aqueous phase.

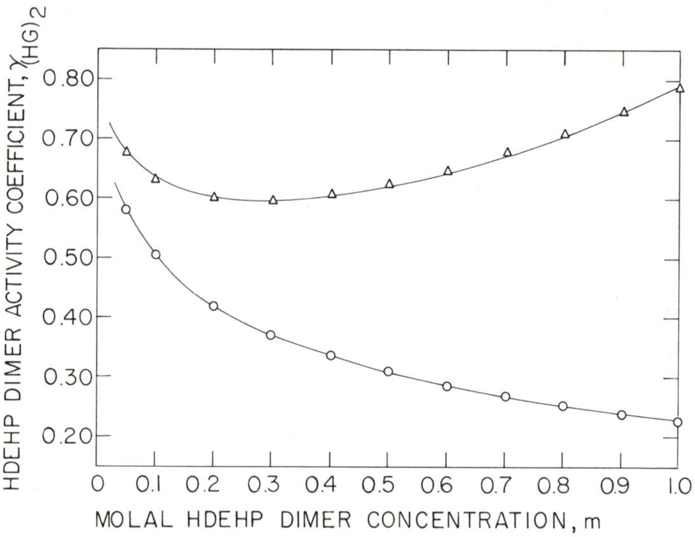

Figure 7. HDEHP dimer activity as a function of the molal HDEHP dimer concentration using the data of Baes (20) and Baes and Baker (21): △, isopiestic results (Baes); ○, iron(III) extraction results (Baes and Baker).

piestic experiments of Baes (*20*) and from the iron (*III*) extraction results of Baes and Baker (*21*). These are illustrated in Figure 7 and are summarized in Table VII.

Summary

The nitrate ion activities in the aqueous phase were measured with a nitrate ion selective electrode taking into account the presence of high hydrogen ion concentration by calibration of the nitrate electrode with nitric acid. The nitrate ion concentration in the organic phase owing to the extraction of neodymium complexes by HDEHP was determined by back-extraction of the organic phase with 3*M* sulfuric acid, dilution, and analysis with a nitrate ion electrode calibrated for different nitrate and sulfate concentrations. The amount of the nitrate species extracted into the organic phase increases as the initial neodymium nitrate concentration increases.

The neodymium ion concentration in the organic phase was measured by back-extraction with 6*M* nitric acid, removal of the neodymium by adsorption with the cation exchange resin, Dowex 50 x 8, and titration of the neodymium-free aqueous phase with NaOH using phenolphtalein as the indicator. The hydrogen ion activity in the equilibrated aqueous phase was determined by direct pH measurements since the polyvalent neodymium ions had negligible effect on the pH electrode.

Vapor pressure measurements to estimate the activity coefficients of the HDEHP dimer resulted in nonreproducible values because of the nonuniform outgassing of the most highly volatile portion of the HDEHP diluent, AMSCO. The isopiestic data of Baes with triphenylmethane in *n*-octane as the reference solution assuming an activity coefficient of unity for triphenylmethane were used in estimating the activity coefficients of the HDEHP dimer.

Literature Cited

1. Ioannou, T. K., Smutz, M., Bautista, R. G., *Proc. In. Solvent Extr. Conf., 1971* (1971) **II,** 957.
2. O'Brien, W. G., Bautista, R. G., *Proc. Int. Solvent Extr. Conf., 1974* (1974) **2,** 1399.
3. Hoh, Y. C., Nevarez, M., Bautista, R. G., *Ind. Eng. Chem., Process Des. Dev.* (1978) **17,** 88.
4. Hoh, Y. C., Bautista, R. G., *Ind. Eng. Chem., Process Des. Dev.* (1979) in press.
5. Peppard, D. F., "Fractionation of Rare Earths by Liquid–Liquid Extraction Using Phosphorous Based Extractants," U. S. Atomic Energy Report **TID-13573** (Technical Service Extension AEC), 1961.
6. Eyring, L., "Progress in the Science and Technology of the Rare Earths," Pergamon, New York, 1968.
7. Peppard, D. F., Mason, G. W., Driscoll, W. J., Sironen, R. J., *J. Inorg. Nucl. Chem.* (1958) **7,** 276.
8. Peppard, D. F., Mason, G. W., Maier, J. L., Driscoll, W. J., *J. Inorg. Nucl. Chem.* (1957) **4,** 334.
9. Peppard, D. F., Ferraro, J. R., Mason, G. W., *J. Inorg. Nucl. Chem.* (1958) **7,** 231.
10. Lenz, T., Smutz, M., *J. Inorg. Nucl. Chem.* (1966) **28,** 1119.
11. Shaw, V. E., Bauer, D. J., U. S. Bureau of Mines **RI 7691,** 1964.
12. Kosinski, F. E., Bostian, H., *J. Inorg. Nucl. Chem.* (1969) **31,** 3623.
13. Smutz, M., Gotto, T., *J. Inorg. Nucl. Chem.* (1965) **27,** 1369.
14. Smutz, M., Lenz, T., *J. Inorg. Nucl. Chem.* (1969) **30,** 621.
15. Smutz, M., Michaelson, O. B., *J. Inorg. Nucl. Chem.* (1971) **33,** 265.
16. Smutz, M., Bautista, R. G., Ioannou, T. K., *Metall. Trans.* (1972) **3,** 2639.
17. Smutz, M., Bautista, R. G., Harada, T., *Metall. Trans.* (1971) **2,** 195.
18. Golinski, M., *Wiad. Chem.* (1973) **27,** 1.
19. Baes, C. F., Zingaro, R. A., Coleman, C. F., *J. Phys. Chem.* (1958) **62,** 129.
20. Baes, C. F., *J. Phys. Chem.* (1962) **66,** 1629.
21. Baes, C. F., Baker, H. T., *J. Phys. Chem.* (1960) **64,** 89.
22. Sinclair, D. A., *J. Phys. Chem.* (1933) **37,** 495.
23. Bates, R. G., Alfenaar, M., "Ion Selective Electrodes," N.B.S. Publication, R. Durst, Ed. (1969) **314,** 191.

RECEIVED April 17, 1978. Prepared for the U.S. Department of Energy under Contract No. W-7405-eng-82.

21

Thermodynamics of Transfer of Hydrogen Iodide in Binary Mixtures of Ethylene Glycol + Diethylene Glycol from emf Measurements

C. KALIDAS[1] and P. SIVAPRASAD

Department of Chemistry, Indian Institute of Technology, Madras 600 036, India

Standard electrode potentials of the Ag–AgI electrode were determined in the temperature range 5°–35°C in 20–80 wt % ethylene glycol + diethylene glycol mixtures by emf measurements on the cell: Pt–H_2(g, 1 atm)/HOAC (m_1), NaOAC (m_2) KX (m_3)/AgX–Ag in the solvent. The standard molal potentials ${}_sE_m^\circ$, in the various solvent mixtures have been expressed as a function of temperature. The various thermodynamic parameters for the transfer of hydrogen iodide from ethylene glycol to these media at 25°C are reported, and their variation with solvent composition is discussed. The transfer free energies of the proton and the iodide at 25°C, on the basis of the ferrocene reference method with ethylene glycol as the reference solvent, are also reported in the mixtures.

Studies in isodielectric solvent mixtures have provided useful information recently (*1, 2, 3*) toward an understanding of the chemical effects of the solvent components on the thermodynamic behavior of electrolytes in these mixtures. This is mainly because the electrostatic effects arising out of the differences in their dielectric constants are expected to be small in such media. Very few emf studies have been reported in such media, especially in binary mixtures of glycols. In a continuation of our earlier investigations (*4, 5*) on the thermodynamic behavior of electrolytes and ions in glycolic solvents, the present work, therefore, deals with the

[1] To whom correspondence should be addressed.

0-8412-0428-4/79/33-177-345$05.00/1

determination of the standard potentials of the Ag–AgI electrode in the nearly isodielectric solvent mixtures containing 20–80 wt % ethylene glycol (EG) + diethylene glycol (DEG) in the temperature range 5°–35°C. The buffered cell (Cell 1):

$$\text{Pt–H}_2\text{(g, 1 atm) / HOAC } (m_1)\text{, NaOAC } (m_2)\text{, KX } (m_3) \text{ / AgX–Ag} \quad (1)$$
in the solvent

containing acetic acid (HOAC), sodium acetate (NaOAC), and the salt KX, where X = I^-, was used to determine the standard potentials of the Ag–AgI electrode in the different mixtures. The dissociation constants of acetic acid required to evaluate the standard potentials of the Ag–AgI electrode in these solvents were determined as described (*6*) previously.

Experimental

Ethylene glycol (BDH, LR) and diethylene glycol (BDH, LR) were purified by the methods described (*4, 7*) earlier. The Pt–H_2 and Ag–AgI electrodes were prepared according to the method of Ives and Janz (*8*). Glacial acetic acid (AR, BDH), purified and dried according to the method described previously (*9*), was used to prepare acetate buffers. Sodium acetate was prepared in situ by appropriately neutralizing the acid with sodium glycolate in the desired solvent. Other salts used in these measurements, such as potassium chloride and potassium iodide, were reagent grade and were suitably dried before use. The experimental setup and the general procedure used for emf measurements are identical to those described (*4*) earlier. All measurements were made with a pair of Pt–H_2 and four Ag–AgX (where X = a halogen) electrodes. The cells were thermostated at each temperature to an accuracy of ± 0.05°C. The Ag–AgI electrodes were stable over the entire temperature range, and the constancy of cell emf to ± 0.1 mV over a period of half an hour was considered as an adequate criterion of equilibrium in the emf measurements. The dielectric constants of the solvent mixtures at various temperatures were determined with a DKO3 dekameter (Wissenschaftlichetechnische Werkstatten, West Germany) as described (*4*) previously. The physical constants for the solvent mixtures are presented in Table I.

Results and Discussion

The determination of the standard potential of the Ag–AgI electrode in these mixtures is based on Owen's method (*10*), which involves the use of the buffered cell (Cell 1) to prevent the air oxidation of I^- to I_2. Using nearly identical molalities of HOAC, NaOAC, and KI and assuming that all electrolytes other than HOAC are completely dissociated, the emf of Cell 1 was measured in the various solvent mixtures, and the data

Table I. Physical Constants of Diethylene Glycol–Ethylene Glycol Mixtures[a]

Diethylene Glycol (wt %)	*Constant*	*5°C*	*15°C*	*25°C*	*35°C*	M_{xy}
0	ϵ	45.70	43.20	40.80	39.30	62.07
	d	1.1240	1.1160	1.1100	1.1030	
20	ϵ	43.40	41.30	39.05	37.20	67.69
	d	1.1279	1.1214	1.1143	1.1079	
40	ϵ	41.20	39.30	37.15	35.20	74.43
	d	1.1290	1.1222	1.1154	1.1088	
60	ϵ	39.05	37.20	35.10	33.00	82.65
	d	1.1291	1.1225	1.1161	1.1083	
80	ϵ	37.40	35.30	33.55	31.80	92.93
	d	1.1313	1.1247	1.1163	1.1106	
100	ϵ	35.00	33.10	31.20	29.55	106.12
	d	1.1320	1.1260	1.1180	1.1110	

[a] ϵ = dielectric constant, M_{xy} = mean molecular weight, and d = density.

are given in Table II. Because the vapor pressures of the solvents were very small in the temperature range under study, no vapor pressure correction was applied to the emf data.

The standard molal potentials, ${}_sE_m{}^\circ$, of the Ag–AgI electrode in different solvent mixtures were obtained from a plot of the function $E^{\circ\prime}$, defined by

$$\begin{aligned} E^{\circ\prime} &= E - k\,\mathrm{pK}\,(\mathrm{HOAC}) + k \log ({}^m\mathrm{HOAC}\ {}^m\mathrm{I}^- / {}^m\mathrm{OAC}^-) \\ &= ({}_sE_m{}^\circ)\mathrm{Ag{-}AgI} - k \log (\gamma_{\mathrm{HOAC}}\,\gamma_{\mathrm{I}^-}/\gamma_{\mathrm{OAC}^-}) \\ &= ({}_sE_m{}^\circ) + f(\mu) \qquad (2) \end{aligned}$$

against μ, the ionic strength, by extrapolating to $\mu = 0$. Such a plot in 20 wt % EG + 80 wt % DEG, which is typical of the results in these mixtures, is shown in Figure 1. In Equation 2, $k = (RT/F) \ln 10$, m and γ values represent the molalities and activity coefficients of the various species. $E^{\circ\prime}$ values were computed using the p*K* (HOAC) values and the emf data from Table II. Values of pK (HOAC) required in Equation 2 were obtained from measurements on cell 1 ($X = Cl^-$) using the ${}_sE^\circ$ values for the Ag–AgCl electrode in these solvents obtained by a procedure similar to that described (*6*) previously. The pK data so obtained are in Table III. In calculating ${}_sE_m{}^\circ$ of the Ag–AgI electrode it was assumed that KI and NaOAC are completely dissociated in all of the

Table II. Data of Cell 1 from 5° to 35°C (in Volts) in Various Ethylene Glycol–Diethylene Glycol Mixtures[a]

m_{HOAC}	m_{I^-}	m_{OAC^-}	*5°C*	*15°C*	*25°C*	*35°C*
$X = 20$						
0.00740	0.00734	0.00730	0.3373	0.3413	0.3464	0.3518
0.00926	0.00956	0.00938	0.3304	0.3339	0.3387	0.3439
0.01116	0.01120	0.01119	0.3270	0.3307	0.3358	0.3413
0.01603	0.01522	0.01567	0.3172	0.3202	0.3263	0.3308
0.01695	0.01580	0.01636	0.3178	0.3207	0.3242	0.3285
0.01916	0.01921	0.01911	0.3141	0.3172	0.3214	0.3250
0.02048	0.02042	0.02038	0.3127	0.3152	0.3192	0.3234
0.02546	0.02618	0.02579	0.3068	0.3101	0.3138	0.3182
0.04059	0.04028	0.04039	0.2962	0.2984	0.3008	0.3038
$X = 40$						
0.00686	0.00980	0.00697	0.3482	0.3513	0.3554	0.3593
0.01082	0.01005	0.00922	0.3447	0.3484	0.3524	0.3568
0.01252	0.01286	0.01121	0.3385	0.3418	0.3454	0.3491
0.01473	0.01392	0.01266	0.3380	0.3406	0.3448	0.3460
0.01654	0.02070	0.01682	0.3288	0.3306	0.3332	0.3369
0.02008	0.02111	0.01819	0.3262	0.3288	0.3308	0.3329
0.02261	0.02235	0.01983	0.3234	0.3253	0.3276	0.3319
0.04637	0.04146	0.04216	0.3094	0.3120	0.3156	0.3178
$X = 60$						
0.00285	0.00266	0.00248	0.3836	0.3875	0.3917	0.3961
0.00342	0.00329	0.00371	0.3818	0.3856	0.3914	0.3959
0.00438	0.00422	0.00475	0.3760	0.3802	0.3852	0.3893
0.00454	0.00391	0.00451	0.3773	0.3818	0.3855	0.3908
0.00483	0.00502	0.00545	0.3726	0.3768	0.3815	0.3863
0.00642	0.00635	0.00603	0.3652	0.3682	0.3716	0.3755
0.00923	0.00960	0.01041	0.35785	0.3605	0.3637	0.3676
0.01185	0.01232	0.01336	0.3518	0.3536	0.3567	0.3588
$X = 80$						
0.00153	0.00148	0.00154	0.4134	0.4194	0.4249	0.4304
0.00262	0.00253	0.00263	0.4027	0.4066	0.4106	0.4153
0.00328	0.00333	0.00346	0.3965	0.3993	0.4036	0.4076
0.00408	0.00422	0.00423	0.3895	0.3932	0.3978	0.4021
0.00449	0.00447	0.00465	0.3889	0.3919	0.3957	0.3994
0.00555	0.00537	0.00576	0.3828	0.3866	0.3909	0.3939
0.01016	0.00982	0.01020	0.3704	0.3713	0.3741	0.3784

[a] X = wt % diethylene glycol.

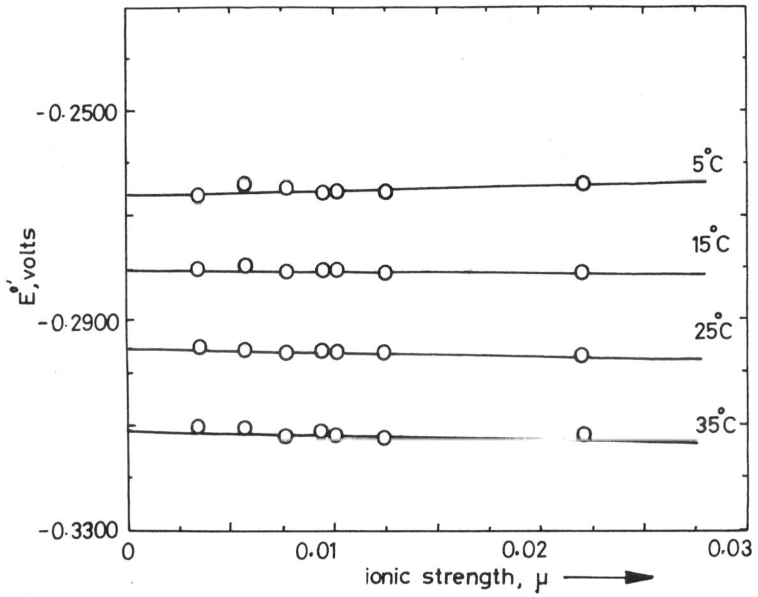

Figure 1. Extrapolation of E°′ *in 80 wt % diethylene glycol at various temperatures*

solvents and that the dissociation of HOAC is so small that it does not affect the effective concentration of HOAC and OAC^-. Table IV gives the ${}_sE_m{}^\circ$ values for the Ag–AgI electrode at different temperatures and ${}_sE_c{}^\circ$ and ${}_sE_N{}^\circ$—i.e., the $E_s{}^\circ$ values on the molar and mole fraction scales at 25°C calculated from

$$ {}_sE_c{}^\circ = {}_sE_m{}^\circ + 2\,k \log d_o \quad (3) $$

and

$$ {}_sE_N{}^\circ = {}_sE_m{}^\circ - 2\,k \log (1000/M_{xy}) \quad (4) $$

Table III. pK_a Values of Acetic Acid in Ethylene Glycol–Diethylene Glycol Mixtures at Various Temperatures

Temperature	*5°C*	*15°C*	*25°C*	*35°C*
Diethylene glycol, wt %				
20	8.62	8.53	8.44	8.42
40	8.64	8.55	8.48	8.45
60	9.00	8.92	8.87	8.82
80	9.48	9.40	9.34	9.28

Table IV. ${}_{s}E_m°$ at Different Temperatures and ${}_{s}E_N°$ and ${}_{s}E_C°$ at 25°C in Ethylene Glycol–Diethylene Glycol Mixtures

Diethylene Glycol, wt %	*0*[a]	*20*	*40*	*60*	*80*	*100*[b]
Temperature						
5°C	−0.2732	−0.2568	−0.2390	−0.2527	−0.2657	−0.3222
15°C	−0.2820	−0.2698	−0.2517	−0.2663	−0.2801	−0.3500
25°C	−0.2932	−0.2788	−0.2649	−0.2810	−0.2954	−0.3711
$E_N°$	−0.4360	−0.4172	−0.3984	−0.4091	−0.4175	−0.4864
$E_C°$	−0.2879	−0.2732	−0.2593	−0.2754	−0.2898	−0.3654
35°C	−0.3006	−0.2932	−0.2800	−0.2968	−0.3111	−0.3804

[a] Data from Ref. *9*.
[b] Data from our earlier unpublished work.

In Equations 3 and 4, d_o is the density of the solvent, and M_{xy} is the mean molecular weight of the solvent as defined (*4*) earlier. The standard error in $E_m°$ is ± 0.2 mV in all the mixtures at different temperatures. $E_m°$ can be expressed as a function of temperature according to

$$ {}_{s}E_m° = E_m° \ (25°C) + b(t - 25) + c(t - 25)^2 \quad (5) $$

where t is the temperature in degrees Celsius, and b and c are empirical coefficients. These data are summarized in Table V. The standard deviations in ${}_{s}E_m°$ calculated from Equation 5 were within ± 0.2–± 0.3 mV. The standard free energy change, $\Delta G_t°$, associated with the transfer of 1 mol of HI from ethylene glycol to the given solvent at infinite dilution according to Equation 6

$$ \text{HI (ethylene glycol)} = \text{HI (ethylene glycol} - \text{diethylene glycol mixture)} \quad (6) $$

was calculated on the mole fraction scale for the various solvent mixtures at 25°C from Equation 7:

$$ \Delta G_t° = F({}_{EG}E_N° - {}_{s}E_N°) \quad (7) $$

Here ${}_{EG}E_N°$ and ${}_{s}E_N°$ are the standard potentials of the Ag–AgI electrode in ethylene glycol and the solvent on the mole-fraction scale. The standard entropy of transfer, $\Delta S_t°$, was calculated from

$$ \Delta S_t° = (-d/dT)F({}_{EG}E_N° - {}_{s}E_N°) = F(b_s - b_{EG}) + 2\,(C_s - C_{EG})\,(t - 25°) + (K_{EG} - K_s) \quad (8) $$

Table V. Coefficients of the Empirical Equation: $_{s}E_{m}^{\circ} = {_{s}E_{m}^{\circ}} (25°C + b(t - 25) + c(t - 25)^2)$

Diethylene Glycol, wt %	$_{s}E_{m}^{\circ}$ *(25°C)*	$b \times 10^3$	$c \times 10^6$
0	−0.2923	−8.990	−3.500
20	−0.2802	−1.217	−3.500
40	0.2651	−1.422	−6.000
60	−0.2810	−1.526	−5.625
80	−0.2953	−1.5475	−3.250
100	−0.3703	−1.494	46.200

where b_s, C_s and b_{EG}, C_{EG} are the coefficients from Table V in the solvent and ethylene glycol, respectively, and $K_{EG} = 2R/F \ln (1000/62.07)$ and $K_s = (2R/F) \ln (1000/M_{xy})$. The last term arises in the conversion of $_{s}E_{m}^{\circ}$ to $_{s}E_{N}^{\circ}$. The standard enthalpy of transfer, ΔH_t°, was then calculated from Equation 9:

$$\Delta H_t^{\circ} = \Delta G_t^{\circ} + T\Delta S_t^{\circ} \tag{9}$$

All these thermodynamic quantities are recorded in Table VI. The ΔG_t° values are accurate to $\pm$ 40 J mol^{-1}, and the expected errors in ΔS_t° and ΔH_t° are 0.5 J K^{-1} mol^{-1} and 150 J mol^{-1}, respectively. Figure 2 shows how these quantities vary with solvent composition.

This figure also shows that ΔG_t° is negative and decreases up to about 40 wt % diethylene glycol, passes through a minimum at about this composition, and then increases, reaching a large positive value in pure diethylene glycol. Considering that the transfer process (Equation 6) involves the transfer of charged species—i.e., H^+ and I^- ions from ethylene glycol to the mixed solvents—ΔG_t° can be expressed as:

$$\Delta G_t^{\circ} = \Delta G_t^{\circ}{}_{(el)} + \Delta G_t^{\circ}{}_{(non\text{-}el)} \tag{10}$$

Table VI. Thermodynamic Quantities of Transfer of HI from Ethylene Glycol to Various Ethylene Glycol–Diethylene Glycol Mixtures at 25°C on the Mole-Fraction Scale

Diethylene Glycol, wt %	*0*	*20*	*40*	*60*	*80*	*100*
ΔG_t° [a]	0	−1810	−3630	−2600	−1790	+4860
$T\,\Delta S_t^{\circ}$ [a]	0	−8,640	−14,130	−16,600	−16,690	−14,450
ΔH_t° [a]	0	−10,450	−17,760	−19,200	−18,480	−9590

[a] Values are in J mol^{-1}.

where $\Delta G^\circ_{(el)}$ and $\Delta G_t^\circ{}_{(non\text{-}el)}$ represent the electrostatic and non-electrostatic transfer free energy contributions, respectively. Although $\Delta G_t^\circ{}_{(el)}$ arises mainly from the differences in the dielectric constants of the solvents, $\Delta G_t^\circ{}_{(non\text{-}el)}$ reflects the contributions of solvation and other specific ion–solvent interactions that depend on the basicity of the solvent. It is expected that $\Delta G_t^\circ{}_{(el)}$ will be small in the present case, and thus the variation of ΔG_t° with solvent composition is largely controlled by the changes in the $\Delta G_t^\circ{}_{(non\text{-}el)}$. Table VII gives the transfer free energy of the proton, $\Delta G_t^\circ{}_{(H^+)}$ (mole fraction scale), determined on the basis of the ferrocene reference method in these mixtures at 25°C, as described earlier (*11*). The transfer free energy data of the iodide ion, $\Delta G_t^\circ{}_{(I^-)}$ obtained from Equation 11:

$$\Delta G_t^\circ{}_{(HI)} = \Delta G_t^\circ{}_{(H^+)} + \Delta G_t^\circ{}_{(I^-)} \tag{11}$$

are also recorded in Table VII. It is seen that although $\Delta G_t^\circ{}_{(H^+)}$ generally decreases with the addition of diethylene glycol, $\Delta G_t^\circ{}_{(I^-)}$ initially decreases, passes through a minimum at about 40 wt % diethylene glycol, and subsequently increases. The overall variation of $\Delta G_t^\circ{}_{(HI)}$ with solvent composition is largely controlled by the changes in $\Delta G_t^\circ{}_{(I^-)}$.

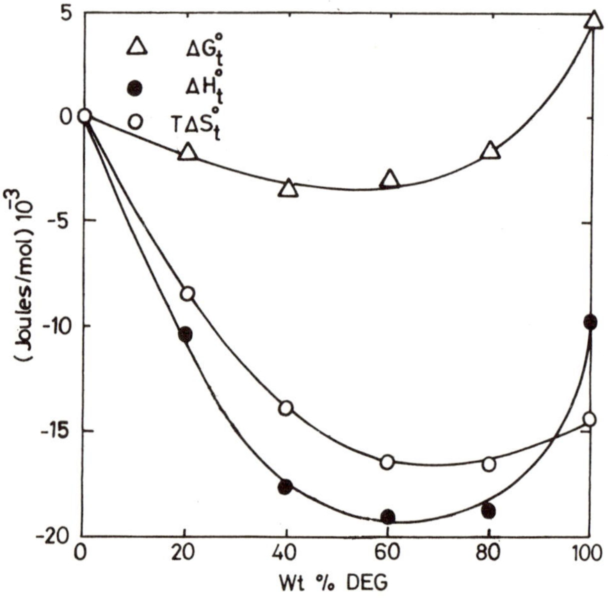

Figure 2. Variation of ΔG_t°, $T\Delta S_t^\circ$, *and* ΔH_t° *of HI with solvent composition at 25°C*

Table VII. Transfer Free Energies of H^+ and I^- Ions from Ethylene Glycol to Ethylene Glycol–Diethylene Glycol Mixtures at 25°C on the Mole Fraction Scale

Diethylene Glycol, wt %	*0*	*20*	*40*	*60*	*80*	*100*
$\Delta G_t^\circ(H^+)$ [a]	0	25	−39	−144	−332	−1267
$\Delta G_t^\circ(I^-)$ [a]	0	−1835	−3591	−2456	−1458	6127

[a] Values are in $J \cdot g\ ion^{-1}$.

The variations of ΔH_t° and ΔS_t° are generally similar to the variation of ΔG_t°. They decrease with the addition of diethylene glycol, pass through a minimum at about 60 wt % diethylene glycol, and then increase. According to Franks and Ives (*12*) and Feakins and Voice (*13*), the effect of the ionic fields on the structure of the solvent will appear as compensating contributions in ΔH_t° and $T\ \Delta S_t^\circ$, and the structural effects of the solvent on the transfer process are revealed through these quantities. The continuous decrease of ΔH_t° with the addition of diethylene glycol is associated with structure-making ion–solvent interactions although this effect decreases significantly beyond 60 wt % diethylene glycol. An examination of ΔS_t° in the presence of increasing amounts of diethylene glycol supports the above conclusions. The decrease of ΔS_t° up to 60 wt % diethylene glycol is indicative of the net structure-making effect of the ions, the effect being somewhat smaller at higher compositions of diethylene glycol. A comparison of the thermodynamics of transfer of hydrogen iodide in this mixed-solvent system with that of water + diethylene glycol (reference solvent in water) (*14*) shows that although the variations of ΔG_t° and ΔH_t° are similar in both cases, ΔS_t° passes through a maximum at compositions less than 40 wt % diethylene glycol in the latter case, which supports the view (*12*) that the addition of small amounts of alcohols promote the hydrogen-bonded structure of water. This structural promotion by the addition of organic component to water is clearly absent in this mixed-solvent system.

Literature Cited

1. Bennetto, H. P., Spitzer, J. J., *J. Chem. Soc., Faraday Trans 1* (1973) **69,** 1491.
2. Kundu, K. K., Rakshit, A. K., Das, M. N., *J. Chem. Soc., Dalton Trans.* (1972) 381.
3. Kalidas, C., Siva Prasad, P., Venkatram, U. V., *Z. Naturforsch.* (1977) **32a,** 791.
4. Kalidas, C., Srinivas Rao, V., *J. Chem. Eng. Data.* (1974) **19,** 201.
5. Kalidas, C., Srinivas Rao, V., *J. Chem. Eng. Data* (1976) **21,** 314.
6. Siva Prasad, P., Kalidas, C., *Bull. Chem. Soc. Jpn.,* (1978) **51,** 2710.
7. Kalidas, C., Palit, S. R., *J. Chem. Soc.* (1961) 3998.

8. Janz, G. J., "Reference Electrodes," D. J. G. Ives and G. J. Janz, Eds., p. 209, Academic Press, New York, 1961.
9. Kundu, K. K., Jana, D., Das, M. N., *J. Phys. Chem.* (1970) **74**, 2625.
10. Owen, B. B., *J. Am. Chem. Soc.* (1935) **57**, 1526.
11. Srinivas Rao, V., Kalidas, C., *Indian J. Chem.* (1975) **13**, 1303.
12. Franks, F., Ives, D. J. G., *Q. Rev.* (1966) **20**, 1.
13. Feakins, D., Voice, P. J., *J. Chem. Soc., Faraday Trans 1* (1972) **68**, 1390.
14. Kalidas, C., Srinivas Rao, V., *J. Chem. Eng. Data,* in press.

RECEIVED February 27, 1978.

Electromotive Forces and Thermodynamic Functions of the Cell Pt, H_2 | HBr(*m*), *X*% Alcohol, *Y*% Water | AgBr–Ag in Pure and Mixed Solvents

ALLEN F. ROBINETTE[1] and EDWARD S. AMIS

Dunhall Pharmaceuticals, Inc., Gravette, Arkansas 72736

*Electromotive force measurements of the cell Pt, H_2 | HBr(m), X% alcohol, Y% water | AgBr–Ag were made at 25°, 35°, and 45°C in the following solvent systems: (1) water, (2) water–ethanol (30%, 60%, 90%, 99% ethanol), (3) anhydrous ethanol, (4) water–*tert*-butanol (30%, 60%, 91% and 99%* tert*-butanol), and (5) anhydrous* tert*-butanol. Calculations of standard cell potential were made using the Debye–Huckel theory as extended by Gronwall, LaMer, and Sandved. Gibbs free energy, enthalpy, entropy changes, and mean ionic activity coefficients were calculated for each solvent mixture and temperature. Relationships of the standard potentials and thermodynamic functons with respect to solvent compositions in the two mixed-solvent systems and the pure solvents were discussed.*

Many studies of electromotive forces of the cells

$$H_2\ (1\ \text{atm})\ |\ H + X - (m)\ |\ AgX\text{–}Ag, \text{ where } X \text{ is } Cl^-, Br^-, \text{ or } I^- \quad (1)$$

recently have been made in pure and mixed solvents (*1–10*). Amis and associates (*8, 9, 10*) have investigated the properties of the above cells involving pure and mixed hydroxylic solvents. Unusual results were

[1] To whom inquiries should be addressed.

0-8412-0428-4/79/33-177-355$05.75/1

observed in these systems, and it was deemed important to extend the work in other hydroxylic solvent systems to ascertain whether the effects noted are of a general nature. In addition, a careful study was made of experimental techniques necessary for preparing a stable silver–silver bromide electrode.

The cell where X^- is Br^- has been investigated in anhydrous ethanol but not in mixed ethanol–water systems. This study involves the cell using water, 30%, 60%, 90%, and 99% ethanol–water, and anhydrous ethanol at 25°, 35°, and 45°C, and similar compositions and temperatures for the water–*tert*-butanol system and anhydrous *tert*-butanol. This cell had not been studied previously in this latter solvent system.

Experimental

Purification of Nitrogen and Hydrogen Gases. Hydrogen and nitrogen were obtained from the Air Reduction Company. Both gases were purified by being passed through a purification train of copper turnings heated to 600°C, through a drying tower filled with concentrated sulfuric acid, then through an empty drying tower to trap droplets of sulfuric acid, and finally through tubes containing Drierite, Ascarite, and again Drierite. Tests for the presence of carbon dioxide in the purified gas were made by bubbling a rapid stream of the gas through a saturated solution of barium hydroxide. If no turbidity occurred after 1 hr, the gas was assumed to be free of carbon dioxide. The passing of a rapid stream of the gas through an alkaline pyrogallol solution for 1 hr, with no change in color of the pyrogallol, together with stability in the electromotive force of the galvanic cells being studied, indicated the absence of oxygen in the gas. The effectiveness of Drierite and concentrated sulfuric acid in removing moisture from a stream of gas is well known, so no special test was made for moisture in the purified gas.

Purification of Electrode Materials. Chemically pure silver bromate from A. D. Mackay, Inc. was recrystallized eight times from conductivity water in near total darkness. The small, snowy white crystals of purified material were stored in a vacuum desiccator over Drierite in the dark until used. The method of Ives and Janz (*11*) was used in preparing the high-purity silver oxide used in making the electrodes. Silver oxide was precipitated from silver nitrate by dilute sodium hydroxide. It was purified by decanting several times, then extracting in a Soxhlet extractor with conductivity water for two days, with the conductivity water being changed twice each day. This procedure yielded a wash water with a specific conductance of 10^{-5} ohm^{-1} cm^{-1}. The purified material was stored in the dark over Ascarite in a vacuum desiccator until used.

Electrode-Preparation and Aging. Electrode preparation procedures are given by Keston (*14*) and by Ives and Janz (*11*). Electrodes were prepared from a paste containing 90% silver oxide and 10% silver bromate placed on a platinum spiral and heated in a furnace for 7 min at 650°C. Janz and Taniguchi (*12*) have reviewed the preparation, reproducibility, and stability of this electrode. Taylor and Smith (*13*) found the equilibrium potential to be stable within 0.02 mV. Electrodes so prepared are very stable and reproducible (*14*). Electrode preparation was carried out in near darkness since using fluorescent lighting gave electrodes a grayish color and a potential normally several millivolts higher than those prepared in near darkness. Small amounts of light could be tolerated without a noticeable effect. The freshly prepared electrodes were aged (*16*) by immersion in dilute HBr solution in the aging cell. The electrodes were interconnected with copper wire and the cell and its contents slowly were heated to 75°C, slowly cooled to room temperature, and left to stand in the dark for 12 to 18 hr with purified nitrogen flowing through the cell. The measured bias potential between pairs of electrodes was normally ±0.01 mV. Any electrodes having bias potentials greater than ±0.02 mV were discarded. Rule and La Mer (*17*) reported that silver–silver chloride showed an increase of bias potentials of no more than ±0.04 mV after standing for 6 weeks.

The hydrogen electrodes used were of the classical Hildebrand electrode type (*18*), as modified by Hills and Ives (*19*) and Popoff et al. (*20*). Bates procedure (*15*) for cleaning the electrode surface prior to platinizing was chosen. The plating solution and the procedure for plating the electrodes were described by Hills and Ives (*19*). The resulting lightly platinized electrodes used in water and *tert*-butanol–water systems had a dark gray appearance with the original metallic sheen still clearly visible. In this work greater electrode stability in ethanol–water and in anhydrous ethanol solutions was obtained by electrodes having heavier platinum black coatings. Before use, the electrode potential was measured against an older hydrogen electrode (*21*). After aging in the desired solvent for several hours at room temperature, the bias potentials measured in this way were routinely less than ±0.01 mV.

Preparation of Reagents. Pure dry hydrogen bromide was prepared by the method of Booth (*22*) from tetrahydronaphthalene and bromine according to the reaction:

$$C_{10}H_{12} + 4Br_2 \rightarrow C_{10}H_8Br_4 + 4HBr \quad (2)$$

The HBr purified as described (*22*) was bubbled through a bottle containing the solvent under investigation by permitting the HBr to evaporate from the collecting cold trap and dissolving in a bottle containing the selected solvent until the desired concentration of hydrobromic acid was reached. The outlet of the solution bottle was protected by means of a capillary tube ending in a guard tube containing Drierite and Ascarite.

The solution was standardized, either gravimetrically as silver bromide, or by titration with sodium hydroxide to a phenolphthalein end point. Such hydrobromic acid has been sufficiently pure for accurate electrochemical measurements (*9*).

Ethanol was dried by the method of Riddick and Bunger (*23*). The dry ethanol was forced by dry pure nitrogen into a still previously flushed with dry nitrogen. The ethanol was distilled under a constant stream of pure dry nitrogen, and the middle fraction collected in a dilution bottle equipped with a ground glass joint having a Teflon stopper and Teflon stop cocks. A guard tube having a layer of Ascarite sandwiched between layers of Drierite, and terminating in several inches of capillary tubing, protected the solution from the atmosphere. The potentials of cells are not influenced by traces of benzene present in absolute ethanol (*12*), hence it was not considered necessary to remove the traces of benzene in the ethanol used. Water (0.05%) was found in the alcohol using a Precision Karl Fischer Titrator and a Perkin–Elmer Model 900 gas chromotograph equipped with a flame ionization detector and a Porapak-qs column. Using the gas chromatograph and a 4% SE-30 column, a trace of benzene was detected in the ethanol. The small amount of water and the trace of benzene were the only impurities detected.

tert-Butanol was purified by the same procedures used for ethanol. It contained no measurable water by Karl Fisher titration and no impurities were detected by gas chromatographic analysis.

Water was purified by passing distilled water through two Barnstead demineralizer cartridges connected in series. The water was purified further by a second distillation using an all-Pyrex still equipped with ground glass and Teflon connections. A small amount of potassium permanganate was added to the water in the still. A stream of dry nitrogen was passed continuously through the still during the distillation process. The water passed from the still condenser into a solution bottle; the outlet was protected with a guard tube to prevent entry of atmospheric gases into the bottle. The conductance of the collected water was 2.4×10^{-7} ohm^{-1} at room temperature.

Solution Preparation. All solution bottles were fitted with ground glass joints equipped with Teflon cuffs and Teflon stop cocks. When not in use all outlet tubes were plugged with ground glass stoppers. In addition a positive pressure of pure dry nitrogen was maintained in the solution bottles at all times to prevent entry of atmospheric gases of water vapor. Ethanol–water and *tert*-butanol–water solvents of 30%, 60%, 90%, and 99% by weight of the respective alcohols were prepared using a large solution balance. The desired pure alcohol and water solutions were forced into the solution bottle by dry purified nitrogen. Contact of the solvents with the air was prevented at all times by keeping the system closed except for Ascarite–Drierite guard tubes ending in long capillary tubes.

Instruments and Methods of Measurements. A Leeds and Northrup Type K-3 universal potentiometer, in conjunction with a General Electric Model 29 galvanometer, was used to measure electromotive force. The potentiometer was calibrated by means of a Weston Standard Cell which had been calibrated against a National Bureau of Standards (NBS) certified standard cell. Galvanic cells which were maintained at constant temperatures of 25°, 35°, and 45°C ± 0.01° by being immersed in a water bath at the desired temperature. The temperatures of the baths were set using a Fisher Scientific calibrated standard thermometer, with calibration traceable to the NBS. An adaptation of the cell sketched by Ives and Janz (*11*) was used. The modification of the cell was that described by McIntrye and Amis (*10*).

The cell containing the silver–silver bromide and hydrogen electrodes was placed in the bath at the proper temperature and was purged by passing pure dry nitrogen through it for 30 min. The HBr solution of the proper solvent composition previously prepared and stored under pure dry nitrogen, was added to the cell. Potential measurements were made at intervals until three determinations made over a 30-min period agreed within 0.05 mV. A sample of the cell solution was forced by dry nitrogen into a tared weighing bottle, the bottle and contents were weighed, and the solution was titrated with standard sodium hydroxide solution to a phenolphthalein end point. Pure solvent of the correct composition was added to the galvanic cell by means of a similar weighing bottle, a new hydrogen electrode was inserted, the cell was purged with pure dry nitrogen, and a new series of potential measurements was made at a lower HBr concentration. Using this procedure several potential measurements could be made at a given solvent composition and temperature before the HBr became too dilute to titrate accurately.

Data and Its Treatment

The measured electromotive forces of the cell

$$H_2 \text{ (1 atm)} \mid HBr(m), X\% \text{ alcohol}, Y\% \text{ water} \mid AgBr\text{–}Ag \qquad (3)$$

at 25°C, 35°C, and 45°C, and various molalities of HBr in 0%, 30%, 60%, 90%, 99%, and 100% ethanol and in 0%, 60%, 91%, 99%, and 100% *tert*-butanol were measured and are recorded in Table I.
The standard cell potential is calculated by

$$E_m{}^{\circ\prime} \equiv E + 2k \log m = E^\circ - 2k \log \gamma_\pm \qquad (4)$$

where $\gamma_\pm{}^2 = \gamma_+\gamma_-$ is the square of the mean activity coefficient. Equation 4 is used to determine the standard potential of the silver–silver bromide electrode from the intercept on the E° axis of $E_m{}^{\circ\prime}$ or $E + 2k \log m$ vs. $2k \log \gamma_\pm$. E in Equation 4 is obtained from the observed electromotive force E_{obs} using the equation

$$E = E_{obs} + \frac{RT}{2F} \ln \frac{760}{P_{bar} - P_{solvent}} \qquad (5)$$

where P_{bar} is the observed barometric pressure and $P_{solvent}$ is the vapor pressure of the solvent being studied at absolute temperature T. Since $\gamma_\pm$ is taken as 1 at $m = 1$, a determination of the standard potential E° may be made by evaluating E° at infinite dilution.

The Debye–Hückel theory (*24*) is used to evaluate the activity coefficients by adding a term linear in m (*25*). The equation is

$$\log \gamma_\pm = -\frac{\mathbf{A}(d_o m)^{\frac{1}{2}}}{1 + \mathbf{B}\mathring{a}(d_o m)^{\frac{1}{2}}} - \log(1 + 0.002\, m\overline{M}_{xy}) + bm \qquad (6)$$

Table I. Electromotive Force Measurements of the Galvanic Cell Containing Hydrobromic Acid in Ethanol–Water and *tert*-Butanol–Water Solvents, and in the Separate Solvent Components

25°C		*35°C*		*45°C*	
Concentration (m)	*emf* *(V)*	*Concentration* (m)	*emf* *(V)*	*Concentration* (m)	*emf* *(V)*
		100% Water			
0.085483	0.20840	0.089895	0.020801	0.086646	0.20671
0.059188	0.22642	0.059766	0.22645	0.060899	0.22525
0.034959	0.25254	0.033960	0.25510	0.035871	0.25280
0.022188	0.27361	0.021088	0.27818	0.023375	0.27409
0.013918	0.29669	0.012914	0.30289	0.015066	0.29690
0.009329	0.31624	0.008424	0.32449	0.010279	0.31671
0.007403	0.33356	0.005759	0.34433	0.007343	0.33452
		30% Ethanol			
0.009879	0.30278	0.010119	0.30387	0.010040	0.30572
0.005096	0.33647	0.005149	0.33924	0.005345	0.34055
0.002660	0.36889	0.002572	0.37529	0.002992	0.37229
0.001578	0.39598	0.001340	0.40976	0.001757	0.40067
0.001005	0.42349	0.000732	0.42256	0.001075	0.42834
0.000437	0.45676	0.000323	0.48759	0.000585	0.46150
		60% Ethanol			
0.011112	0.27875	0.010622	0.27950	0.010932	0.27552
0.006784	0.30326	0.004646	0.32264	0.005779	0.30982
0.002561	0.33956	0.002361	0.35637	0.003178	0.34171
0.001770	0.36392	0.001335	0.38739	0.002029	0.36740
0.001199	0.39030	0.000776	0.41255	0.001224	0.39501
0.000679	0.42160	0.000407	0.44657	0.000639	0.43141
		90% Ethanol			
0.012672	0.20951	0.014138	0.17621	0.013153	0.19435
0.006644	0.24148	0.009071	0.20755	0.007418	0.22310
0.003808	0.26788	0.004221	0.25713	0.004541	0.24837
0.002267	0.29374	0.003010	0.27538	0.003101	0.26681
0.001412	0.31861	0.001349	0.29949	0.001859	0.29449
0.000756	0.35128	0.000961	0.33441	0.000983	0.32902
		99% Ethanol			
0.014212	0.11326	0.007031	0.13925	0.005636	0.14310
0.007495	0.14364	0.003390	0.16619	0.003151	0.16405
0.003917	0.16760	0.001724	0.19929	0.002053	0.18514
0.002166	0.19347	0.000911	0.23295	0.001399	0.20487
0.001368	0.21812	0.000537	0.26829	0.000926	0.22414
0.000790	0.24363				

Table I. Continued

25°C Concentration (m)	*25°C* emf (V)	*35°C* Concentration (m)	*35°C* emf (V)	*45°C* Concentration (m)	*45°C* emf (V)
		100% Ethanol			
0.017322	0.05074	0.017747	0.03449	0.007541	0.08014
0.009832	0.07368	0.009951	0.06710	0.003697	0.10696
0.005333	0.09821	0.005515	0.09339	0.001409	0.15469
0.002993	0.12333	0.003293	0.11915	0.000525	0.21975
0.001441	0.15975	0.001269	0.16238		
0.000616	0.20283	0.000492	0.22115		
		30% tert-*Butanol*			
0.027814	0.25044	0.027275	0.25275	0.028229	0.25036
0.010590	0.26729	0.019989	0.26572	0.019798	0.26817
0.012989	0.28690	0.013850	0.28485	0.014506	0.28374
0.007930	0.31145	0.008456	0.30861	0.010419	0.30096
0.005447	0.32873	0.005553	0.33047	0.007082	0.32587
0.003772	0.34810				
		60% tert-*Butanol*			
0.018571	0.24378	0.018580	0.24124	0.018623	0.23781
0.011179	0.26572	0.009837	0.26976	0.010505	0.26428
0.006587	0.28890	0.005817	0.29402	0.006424	0.28733
0.004252	0.30946	0.003742	0.34164	0.004245	0.30726
0.002921	0.32729	0.002528	0.33352	0.002937	0.32506
				0.002126	0.34245
		91% tert-*Butanol*			
0.014874	0.15629	0.013665	0.14601	0.014951	0.13716
0.007811	0.17395	0.007881	0.16384	0.005945	0.16180
0.004531	0.18993	0.004075	0.18052	0.003491	0.17831
0.002971	0.20374	0.002971	0.19532	0.002122	0.19371
0.001915	0.21714	0.001801	0.21024	0.001340	0.21025
0.001073	0.23631	0.000963	0.23182	0.000645	0.23666
0.000555	0.26125	0.000439	0.26223	0.000283	0.27864
		99% tert-*Butanol*			
0.024563	0.05703	0.024387	0.04412	0.024582	0.02683
0.014660	0.06532	0.014389	0.05314	0.014614	0.03442
0.009112	0.07456	0.008112	0.16512	0.008322	0.05213
0.005836	0.08481	0.004273	0.07961	0.004803	0.06711
0.003898	0.09384	0.002279	0.09703	0.002600	0.08287
0.002766	0.10511	0.001231	0.11362	0.001520	0.00931
		100% tert-*Butanol*			
0.005092	0.05692	0.005642	0.06496	0.005133	0.03806
0.002882	0.06387	0.002455	0.11527	0.002291	0.06078
0.001324	0.10454	0.000856	0.094:0	0.000613	0.09603
		0.000233	0.14587	0.000163	0.14720

The term log $(1 + 0.002\ m\ \mathrm{M}_{xy})$ arises from the conversion of the rational activity coefficient, $f_{\pm}$, measured on the mole-fraction scale, to the molal activity coefficient $\gamma_{\pm}$ according to the relationship

$$\log \gamma_{\pm} = \log f_{\pm} - \mathit{log}\left[1 + \frac{\nu m\,\overline{\mathrm{M}}_{xy}}{1000}\right] \quad (7)$$

where ν is 2 for a 1–1 electrolyte, m is the molal concentration, and $\overline{\mathrm{M}}_{xy}$ is the average molecular weight of the solvent system given by

$$\overline{\mathrm{M}}_{xy} = \frac{100}{(X/\overline{\mathrm{M}}_x + Y/\overline{\mathrm{M}}_y)} \quad (8)$$

where X and Y are the respective weight percentages of the solvent components x and y of molecular weights $\overline{\mathrm{M}}_x$ and $\overline{\mathrm{M}}_y$, respectively.

In Equation 6, **A** and **B** are constants for a given solvent and temperature and are given by the equations

$$\mathbf{A} = \frac{1}{2\mathrm{k} \ln 10}\left[\frac{2\pi \mathrm{N}\mathrm{e}^6}{1000(\mathrm{D}kT)^3}\right]^{\frac{1}{2}} \quad (9)$$

and

$$\mathbf{B} = \left[\frac{8\ \mathrm{N}\mathrm{e}^2}{1000\mathrm{D}kT}\right]^{\frac{1}{2}} \quad (10)$$

In Equations 6–10 d_0 is the density of the solvent which was measured at each solvent composition at each temperature, $\mathring{a}$ is the ion-size parameter, z is an empirical constant, N is Avagadro's number, D is the dielectric constant of the solvent, k is the Boltzman constant, and e is the electronic charge. Other terms have been defined already.

The values of E, log $\gamma_{\pm}$, M_{xy}, **A**, and **B** from Equations 5–10 substituted into Equation 4, make it possible to calculate $E_m{}^{\circ\prime}$ at known molalities of hydrobromic acid, solvent compositions, and temperatures. By plotting values of $E_m{}^{\circ\prime}$ at a given solvent composition and temperature vs. molality, one can find the standard electrode potential E° of the Ag–AgBr electrode at that solvent composition and temperature from the value of $E_m{}^{\circ\prime}$ extrapolated to infinite dilution. This method has been used successfully in water and in organic solvent–water mixtures of higher dielectric constants, but if the mixed solvents have low dielectric constants, ca. 50 or below, the curvatures of the $E_m{}^{\circ\prime}$ vs. m plots are sufficient to prevent accurate determinations of $E_m{}^{\circ\prime}$ and hence of E°.

To overcome this difficulty the Debye–Hückel theory was expanded for symmetrical valence-type electrolytes, and the complex functions in the expansion $(1/2\, X_3 - 2Y_3)$ and $(1/2\, X_5 - 4Y_5)$ calculated and published (*26*). The result of these expansions is to add a term E_{ext} to the equation for the activity coefficient given in Equation 2. For symmetric valence-type electrolytes such as hydrobromic acid this term is

$$E_{ext} = \frac{2k}{\ln 10}\left\{\left(\frac{z^2e^2}{DkTa}\right)^3 \frac{1}{2}X_3 - 2Y_3 + \left(\frac{z^2e^2}{DkTa}\right)^5 \frac{1}{2}X_5 - 4Y_5\right\}. \tag{11}$$

The terms $1/2\, X_3 - 2Y_3$ and $1/2\, X_5 - 4Y_5$ in Equation 11 are complicated functions of $\mathcal{H}$ where $\mathcal{H}$ is defined as

$$\mathcal{H} = \left[\frac{8\pi e^2 N d_o m}{1000 DkT}\right]^{\frac{1}{2}}. \tag{12}$$

The standard potential of the galvanic cell (Equation 3) which is also the standard potential of the silver–silver bromide electrode is given by

$$E^\circ = E_m{}^\circ = E_m{}^{\circ\prime} + E_{ext}. \tag{13}$$

When the proper choice of the constants å and b are made, the function $(E_m{}^{\circ\prime} + E_{ext})$ should be constant within the limits of the extended Debye–Huckel theory. In calculating $E_m{}^{\circ\prime}$ the value of the equation for $\log \gamma_\pm$ (Equation 6) which must be substituted into Equation 4 becomes

$$\log \gamma_\pm = \frac{\mathbf{A}(d_o m)^{\frac{1}{2}}}{1 + \mathbf{B}å(d_o m)^{\frac{1}{2}}} - \log(1 + 0.002\, m\, \overline{M}_{xy}) + bm + E_{ext}. \tag{14}$$

For the thermodynamic functions for the process occurring in the cell at standard state, the standard free energy, ΔG°, standard enthalpy, ΔH°, and standard entropy, ΔS°, are given by the respective equations:

$$\Delta G^\circ = -nFE^\circ \qquad (\text{cal/mol}); \tag{15}$$

$$\Delta H^\circ = \frac{\partial(\Delta F^\circ/T)}{\partial(1/T)} \ (\text{cal/mol}); \tag{16}$$

and

$$\Delta S^\circ = \frac{\Delta H^\circ - \Delta F^\circ}{T} \ (\text{eu}). \tag{17}$$

The symbol n is the number of Faradays per mole of cell electrolyte reacting: $n = 1$ for the cell involving HBr. Table II contains the calculated values for $E°$ (V), $\Delta G°$ (cal/mol), $\Delta H°$ (cal/mol), and $\Delta S°$ (eu) for the cell defined by Equation 3 when it contains the indicated solvents.

Table III contains the experimental quantities (except the potential, E) and the constants used to determine the standard potentials of the cell (Equation 3). The ion-size parameter å for water and *tert*-butanol–water solvents is 5.50 Å, and for ethanol and ethanol–water it is 5.00 Å.

Discussion

The value for $E°$ at 25°C in water solvent for the cell represented in Equation 6 is 0.07104 V, which is in good agreement with the values recorded in the literature: 0.07105 V (*1*); 0.07106 V (*5*); 0.07118 V (*9*); and 0.07111 V (*14*). At 35°C the standard potential of the cell in water was 0.06584 V compared with the literature values of 0.06577 V (*1*) and 0,06585 V (*5*). At 45°C in water $E°$ was 0.06104 V while values from the literature are 0.05999 (1) and 0.06102 (*5*).

In 100% ethanol at 25°C, $E°$ for the cell represented by Equation 6 was −0.19301 V while Nunez and Day (*27*) reported the value of $E°$ to be −0.1816 V. The two values differ by 11.4 mV. The work of Woolcock and Hartley (*28*) agree with our value of $E°$. Scatchard (*29*) found that 0.1% water contamination in absolute ethanol caused a change of 12 mV in $E°$ of the cell defined by Equation 3.
These results were substantiated further by Goldenberg and Amis (*30*), who reported sharp drops in the equivalent conductance of perchloric acid when 0.3 wt % water was added to either anhydrous ethanol or anhydrous methanol. These authors postulated that the conductance drop could be caused by a decreased Grothous conductance along the hydrogen-bonded chains in the pure alcohols, which resulted from the rupture of the chains by preferential solvation of the alcohol molecules by water rather than by other alcohol molecules. It, therefore, seems reasonable to assume that a small amount of water contamination ($\sim$ 1 mL of water per 25.5 L of ethanol) was possibly present in the alcohol used by Nunez and Day, and that the correct value of $E°$ for the cell at 45°C in 100% ethanol should be −0.19301 V (expressed on the molal scale) as reported here.

No comparison exists for the standard potentials in *tert*-butanol–water mixtures or in anhydrous *tert*-butanol. However, the trends in the data obtained here in *tert*-butanol are similar to the trends observed in this work in ethanol–water and in anhydrous ethanol. *See* Figures 1 and 2 in which the standard potentials $E°$ for the silver–silver bromide elec-

Table II Calculated Values of $E°$, $\Delta G°$, $\Delta H°$, and $\Delta S°$ for the Galvanic Cell Containing Hydrobromic Acid in Ethanol–Water, *tert*-Butanol–Water Solvents, and in the Separate Solvent Components

Solvent	T *(K)*	E° *(V)*	ΔG° *(cal/mol)*	ΔH° *(cal/mol)*	ΔS° *(eu)*
	298	0.07104	−1632		
100% H_2O	308	0.06584	−1524	−5356	−12.44
	318	0.06014	−1382		
	298	0.06355	−1488		
30% Ethanol	308	0.05714	−1346	−6146	−15.58
	318	0.05083	−1176		
	298	0.03457	−870		
60% Ethanol	308	0.02932	−672	−5844	−16.79
	318	0.02292	−537		
	298	−0.02690	605		
90% Ethanol	308	−0.04486	1036	−11182	−39.67
	318	−0.06107	1396		
	298	−0-13294	3081		
99% Ethanol	308	−0.14653	3334	−9020	−40.11
	318	−0.16877	3893		
	298	−0.19301	4467		
100% Ethanol	308	−0.20182	4648	−561	−16.91
	318	−0.21082	4085		
	298	0.05516	−1278		
30% *tert*-Butanol	308	0.04663	−1072	−4650	−11.62
	318	0.04441	−1051		
	298	0.11401	−324		
60% *tert*-Butanol	308	0.00252	−55	−8366	−26.98
	318	−0.00952	216		
	298	−0.14864	3502		
91% *tert*-Butanol	308	−0.17402	4110	−14256	−59.63
	318	−0.20090	4693		
	298	−0.27377	6533		
99% *tert*-Butanol	308	−0.31084	7489	−20850	−92.01
	318	−0.34358	8370		
	298	−0.30069	7164		
100% *tert*-Butanol	308	−0.33884	7992	−15367	−75.84
	318	−0.36929	8675		

Table III. Experimental Quantities Except the Potential *E* and the Constants Used in Calculating the Standard Potentials E° of the Cell in Equation 3. Radius is 5.00 Å for Water, Ethanol and Ethanol–Water and 5.50 Å for *tert*-6 Butanol and *tert*-Butanol–Water Solvent

Solvent	T *(K)*	D	d_o *(g/mL)*	P_{vapor} *(mm)*	*b*
	298.16	78.34	0.99708	23.76	0.00
100% Water	308.16	75.00	0.99406	42.18	0.00
	318.16	71.59	0.99025	71.88	0.00
	298.16	52.71	0.94600	25.48	0.90
30% *tert*-Butanol	308.16	49.79	0.93868	45.39	1.20
	318.16	46.88	0.93088	77.72	1.50
	298.16	27.94	0.87559	28.63	1.40
60% *tert*-Butanol	308.16	25.99	0.86717	51.29	1.70
	318.16	24.02	0.85802	88.42	2.10
	298.16	12.46	0.80146	36.72	−3.70
91% *tert*-Butanol	308.16	11.44	0.79200	66.43	−1.80
	318.16	10.46	0.78223	115.89	−0.70
	298.16	10.08	0.78350	41.27	−5.00
99% *tert*-Butanol	308.16	9.07	0.77348	74.93	−3.50
	318.16	8.30	0.76201	131.33	−1.50
	298.16	9.90	0.78150	42.00	1.20
100% *tert*-Butanol	308.16	8.85	0.77090	76.30	1.40
	318.16	8.30	0.76043	133.80	1.60
	298.16	60.98	0.95117	28.82	3.00
30% Ethanol	308.16	58.04	0.94427	51.01	3.30
	318.16	55.16	0.93701	86.54	3.70
	298.16	43.38	0.88404	36.79	6.50
60% Ethanol	308.16	40.96	0.87874	64.92	7.00
	318.16	38.66	0.86972	109.64	7.90
	298.16	28.12	0.81401	51.20	9.70
90% Ethanol	308.16	26.44	0.80507	90.09	10.10
	318.16	24.86	0.79560	151.40	11.00
	298.16	24.57	0.78759	58.11	0.30
99% Ethanol	308.16	23.15	0.77852	102.20	1.70
	318.16	21.79	0.76947	171.40	4.60
	298.16	24.20	0.78506	59.00	2.70
100% Ethanol	308.16	22.79	0.77641	103.70	5.80
	318.16	21.53	0.76761	174.00	14.30

trode are plotted against weight-percent ethanol and weight-percent *tert*-butanol, respectively. The plots in the two figures are quite similar. Similar trends were observed for the silver–silver bromide electrodes in the methanol–water system (*9*) and for the silver–silver iodide electrode in the methanol–water system (*10*). MacInnes (*31*) also points out similar trends.

All cell potentials reached equilibrium in 1 or 2 hr, except when the solvent was anhydrous *tert*-butanol, in which the electrodes reached equilibrium only in dilute soltuions of HBr and even then only in a sluggish manner. This sluggish behavior has been reported (*27*) for the silver–silver bromide electrode in anhydrous ethanol when the acid was concentrated. In the dilute hydrobromic acid solutions used here, this phenomena was not observed in anhydrous ethanol. It is estimated that the standard electrode potential of the silver–silver bromide electrode in anhydrous *tert*-butanol is accurate to only ±1 mV. However, these data are reported to the same degree of precision found in the other *tert*-butanol–water solvents in order to facilitate comparisons of the emf's in the various dilutions of *tert*-butanol used.

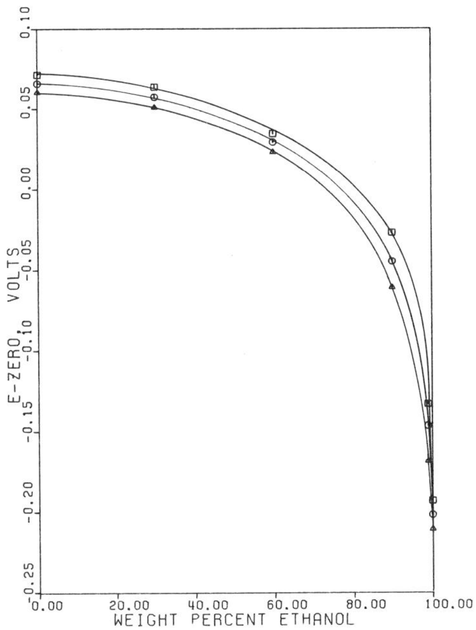

Figure 1. Standard potentials of the cell Pt, | H_2 HBr(m), X% alcohol, Y% water | AgBr–Ag in ethanol–water (radius = 5.00): (□), 25°C; (○), 35°C; (△), 45°C.

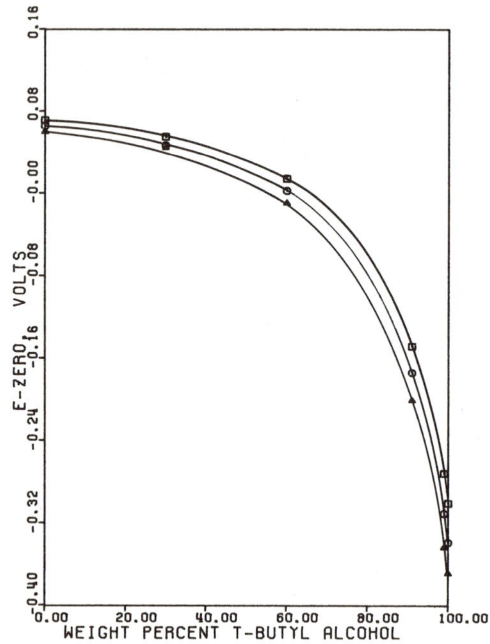

*Figure 2. Standard potentials of the cell Pt, H_2 | HBr(*m*), X% alcohol, Y% water | AgBr–Ag in* tert-*butanol–water (radius = 5.50): (□), 25°C; (○), 35°C; (△), 45°C.*

Figures 3 and 4 contain graphs of the functions $\Delta G°/T$ vs. $1/T$ for the ethanol–water and the *tert*-butanol–water systems, respectively. These plots were used in calculating the standard enthalpies, $\Delta H°$, (Equation 16) for the respective solvent systems. The curves show slopes of the same general signs and orders as found for the silver–silver bromide electrode (9) and for the silver–silver iodide electrode (*10*) in methanol–water solvents. Figure 5 is a plot of $\Delta H°$ vs. weight-percent alcohol for the respective ethanol–water and *tert*-butanol–water solvent systems. The *tert*-butanol data show the same trends as the corresponding data for the silver–silver chloride electrode in the *tert*-butanol–water solvent system (*32*). They are also of the same type as those by Amis and co-workers (*8, 10*). These data can be explained possibly as follows (*10, 33*). After an initial decrease in the structure of highly hydrogen-bonded water caused by the addition of small amounts of *tert*-butanol, further addition of alcohol increases the structure of the solvent system, probably by the selective solvation of the acid ions by alcohol. Alcohol molecules will fit into the lattice structure of water until the composition of alcohol becomes greater than that of water (*34*). This should occur at 80–90 wt % *tert*-butanol (49.3–68.6 mol % *tert*-butanol). A change in slope of the enthalpy curves occurs at 90 wt % *tert*-butanol indicating a possible change in ion solvation at this solvent composition.

The rapid rise in enthalpy between 99 and 100 wt % *tert*-butanol may arise from the breaking down of long chain associations found in pure alcohol by adding a very small amount of water (*30, 35*). The effect of water on the structure of alcohols must depend largely on competitive hydrogen bonding similar to that found in acid–base equilibrium (*36*). The order of increasing basic strength of alcohols is (*37, 38*): CH_3OH < ethanol < *n*-propanol < *tert*-butanol, and it also is, from IR studies (*39*), the order of an increasing tendency to act as a proton acceptor in hydrogen bond formation. Thus *tert*-butanol should have a strong structure-making effect when added to *tert*-butanol–water mixtures. It has been noted (*40*) that additions of very small amounts of water to *tert*-butanol causes a decrease in the dielectric constant. The present data indicate a decrease in hydrogen-bonded structure in *tert*-butanol when small amounts of water are added to the pure alcohol. This could be a factor in the rapid increase in enthalpy and entropy from 99% to 100% *tert*-butanol. It has been stated (*41*) that the region between 90% and 100% alcohol is a critical one structurally, and that for acid solutions in this region some "uncompensated" structural effects clearly become im-

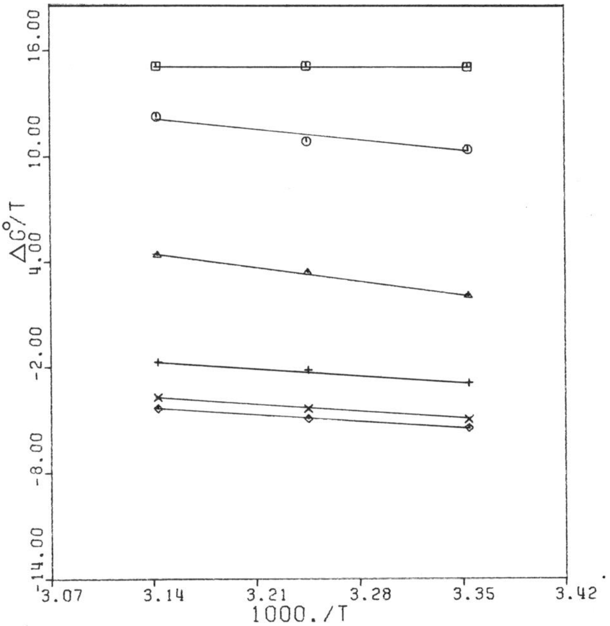

Figure 3. ΔG°/T as a function of 1/T for the ethanol–water system. Ethanol concentrations: (□), 100%; (○), 99%; (+), 60%; (×), 30%. (◇), 100% water.

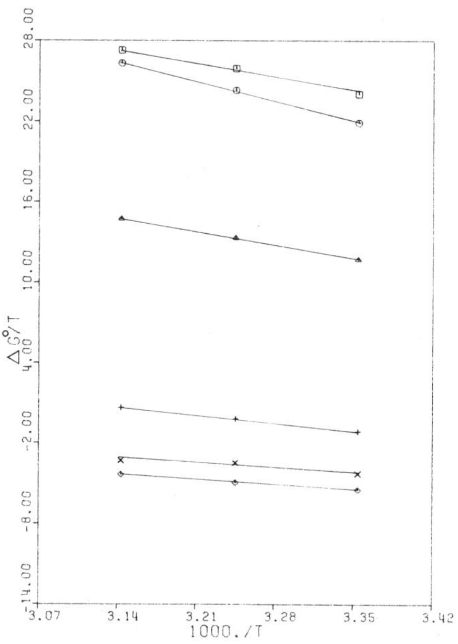

Figure 4. *ΔG°/T as a function of 1/T for the* tert-*butanol–water system.* Tert-*Butanol concentrations:* (□), *100%;* (○), *99%;* (△), *91%;* (+), *60%;* (×), *30%.* (◇) *100% water.*

portant. The situation has been summed up by stating (*42*) that alcohol–water mixtures are subject to the mixed control of competing ordering influences peculiar to the pure compounds as well as to others arising from specific cooperating influences between them. This consequently presents a wealth of structural problems.

$\Delta H°$ and $\Delta S°$ vs. solvent composition for the ethanol–water solvent systems (*see* Figures 5 and 6) show a more complex behavior at about 60% ethanol than the corresponding plots for the *tert*-butanol–water solvent systems (*see* Figures 6 and 8 at about 60% *tert*-butanol. The ethanol–water system shows a slight minimum and then a maximum in the 0 to 60% ethanol range, while *tert*-butanol–water system shows only a maximum in the 0 to 60% *tert*-butanol range.

Table IV contains the calculated values of $-\log \gamma_{\pm}$, where γ is the activity coefficient of hydrobromic acid for the solvent systems investigated here.

The activity coefficients show the generally expected decreasing trends ($-\log \gamma_{\pm}$ show the increasing trends) as the acid concentration increases, and as the dielectric constant of the solvent medium decreases. These trends are in close agreement with those obtained in other alcohol–

water systems and with other halogen acids (*8–12*). The trends may be explained as the results of ion pairing favored by high concentrations of electrolytes and low dielectric constants of the solvents.

As was observed by Melton and Amis (*9*) in 90% methanol, we find that b has a large positive value in ethanol. However, in 90% *tert*-butanol, we find b to be negative. In 99% ethanol and 99% *tert*-butanol b approaches an apparent minimum and has a negative sign at all three temperatures investigated. These minima at 99% alcohol perhaps can arise from the decrease of hydrogen-bonded structure as mentioned in the observed minimums in plots of enthalpy and entropy. It might be noted in the case of plots of $\Delta H°$, $\Delta S°$, and b (*see* Figures 5, 6, and 7) that there are similarities in the curves in that they all show maxima and minima and there is a marked increase in these functions in going from 99% alcohol to pure alcohol. In both alcohol systems similar trends of b vs. alcohol composition are shown with respect to temperature—the plot at 25°C falling below the plot at 35°C which is below the plot at 45°C. The increase of b with temperature in alcohol–water systems would support the observation that there is decreased hydrogen-bonded structure involved in the phenomena. It is reasonable to assume that the higher the temperature, the less hydrogen-bonded structure can exist.

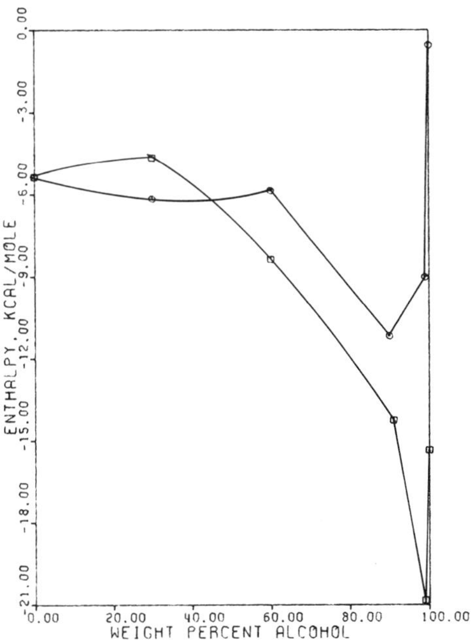

Figure 5. Enthalpy as a function of solvent composition in alcohol–water mixtures: (□), tert-*butanol; (○), ethanol.*

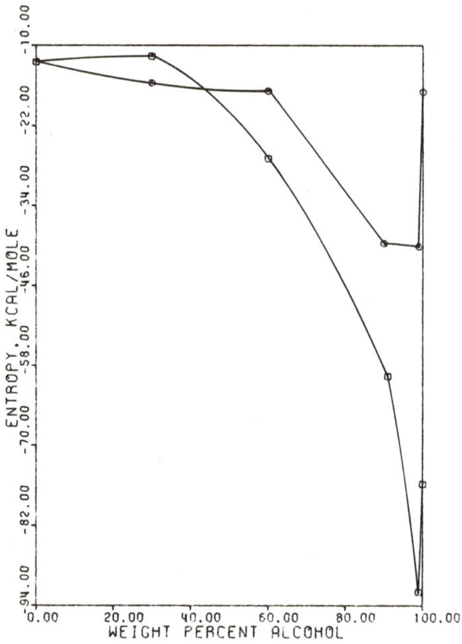

Figure 6. Entropy as a function of solvent composition in alcohol–water mixtures: (□), tert-butanol; (○), ethanol.

Table IV. Activity Coefficients of Hydrobromic Acid

25°C		*35°C*		*45°C*	
Molality	*—Log* $\gamma_\pm$	*Molality*	*—Log* $\gamma_\pm$	*Molality*	*—Log* $\gamma_\pm$
		100% Water			
0.085483	0.2282	0.089895	0.2359	0.086646	0.2374
0.059188	0.2014	0.059766	0.2055	0.060899	0.2107
0.034959	0.1662	0.033960	0.1672	0.035871	0.1739
0.022188	0.1393	0.021088	0.1389	0.023375	0.1474
0.013918	0.1152	0.012914	0.1136	0.015066	0.1235
0.009329	0.0972	0.008424	0.0947	0.010279	0.1052
		0.005759	0.0802	0.007343	0.0910
		30% Ethanol			
0.009879	0.1412	0.010119	0.1456	0.10040	0.1482
0.005096	0.1058	0.005149	0.1085	0.005345	0.1128
0.002660	0.0789	0.002572	0.0793	0.002992	0.0870
0.001578	0.0620	0.001340	0.0586	0.001757	0.0681
0.001005	0.0501	0.000732	0.0440	0.001075	0.0541
0.000487	0.0355	0.000323	0.0297	0.000585	0.0405

Table IV. Continued

25°C		35°C		45°C	
Molality	*—Log* $\gamma_{\pm}$	*Molality*	*—Log* $\gamma_{\pm}$	*Molality*	*—Log* $\gamma_{\pm}$
		60% Ethanol			
0.011112	0.2323	0.010622	0.2348	0.010932	0.2462
0.006770	0.1887	0.004598	0.1646	0.005779	0.1882
0.001871	0.1065	0.002361	0.1223	0.003178	0.1446
0.002116	0.1127	0.001303	0.0930	0.002029	0.1181
0.001199	0.0867	0.000776	0.0730	0.001224	0.0935
0.000679	0.0664	0.000407	0.0537	0.000639	0.0689
		90% Ethanol			
0.012672	0.4340	0.021360	0.5471	0.013153	0.4726
0.006644	0.3341	0.012015	0.4395	0.007418	0.3757
0.003808	0.2636	0.004349	0.2981	0.004541	0.3057
0.002267	0.2097	0.003010	0.2464	0.003101	0.2592
0.001412	0.1693	0.001849	0.1983	0.001859	0.2064
0.000756	0.1268	0.000918	0.1439	0.000983	0.1543
		99% Ethanol			
0.014212	0.5376	0.007031	0.4214	0.004829	0.3726
0.007495	0.4175	0.003390	0.3095	0.003151	0.3102
0.003917	0.3182	0.001724	0.2294	0.002053	0.2568
0.002166	0.2455	0.000911	0.1714	0.001368	0.2139
0.001368	0.1996	0.000537	0.1340	0.000926	0.1788
0.000790	0.1551				
		100% Ethanol			
0.047675	0.8364	0.017747	0.6172	0.007541	0.4565
0.028807	0.7083	0.009951	0.4942	0.003697	0.3383
0.018557	0.6056	0.005515	0.3885	0.001409	0.2206
0.013849	0.5427	0.003190	0.3075	0.000525	0.1396
0.009227	0.4631	0.001285	0.2050		
0.006747	0.4080	0.000388	0.1173		
0.004371	0.3402				
		30% tert-*Butanol*			
0.027814	0.2599	0.027275	0.2655	0.028229	0.2798
0.019590	0.2272	0.019989	0.2358	0.019798	0.2443
0.012989	0.1928	0.013850	0.2039	0.014506	0.2159
0.007930	0.1570	0.008456	0.1662	0.010419	0.1885
0.005447	0.1335	0.005553	0.1388	0.007082	0.1602
0.003772	0.1135				

Table IV. Continued

25°C		35°C		45°C	
Molality	*−Log* $\gamma_\pm$	*Molality*	*−Log* $\gamma_\pm$	*Molality*	*−Log* $\gamma_\pm$
		60% tert-*Butanol*			
0.018571	0.5193	0.018580	0.5461	0.018623	0.5806
0.011179	0.4279	0.009837	0.4280	0.010505	0.4669
0.006587	0.3456	0.005817	0.3455	0.006424	0.3830
0.004252	0.2875	0.003742	0.2866	0.004245	0.3219
0.002921	0.2443	0.002528	0.2416	0.002937	0.2746
				0.002126	0.2382
		91% tert-*Butanol*			
0.014874	1.3802	0.013665	1.4309	0.014951	1.5872
0.007811	1.0898	0.007881	1.1682	0.005945	1.1254
0.004531	0.8781	0.004075	0.8980	0.003493	0.9054
0.002971	0.7371	0.002971	0.7870	0.002122	0.7321
0.001915	0.6108	0.001801	0.6349	0.001340	0.5990
0.001073	0.4734	0.000963	0.4811	0.000645	0.4314
0.000555	0.3510	0.000439	0.3370	0.000283	0.2971
		99% tert-*Butanol*			
0.024563	2.1313	0.024387	2.3250	0.024582	2.4950
0.014660	1.8038	0.014389	1.9627	0.014614	2.1140
0.009112	1.5249	0.008112	1.6004	0.008322	1.7334
0.005836	1.2833	0.004273	1.2375	0.004803	1.3913
0.003898	1.0886	0.002279	0.9467	0.002600	1.0699
0.002766	0.9423	0.001231	0.7209	0.001520	0.8431
		100% tert-*Butanol*			
0.005092	1.2447	0.005642	1.4290	0.005133	1.4312
0.002882	0.9818	0.002455	1.0067	0.002291	1.0121
0.001324	0.6994	0.000856	0.6318	0.000613	0.5585
		0.000233	0.3519	0.000163	0.3091

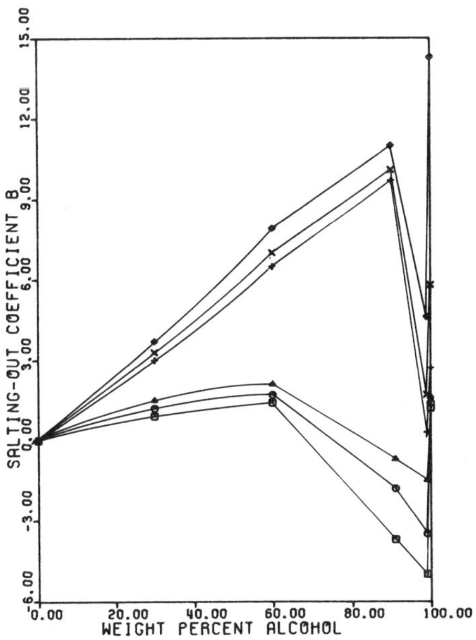

*Figure 7. Salting-out coefficient **B** as a function of weight-percent alcohol.* tert-*Butanol temperatures: (□), 25°C; (○), 35°C; (▲), 45°C. Ethanol temperatures: (+), 25°C; (×), 35°C; (◇), 45°C.*

Until more data is collected in the various alcohol–water systems, and better models for these systems are developed, it will be impossible to definitely delineate the processes occurring in these solvent systems.

Acknowledgment

This chapter has been abstracted in part from a thesis submitted by A. F. Robinette in partial fulfillment of the requirements for the degree Doctor of Philosophy, University of Arkansas. Appreciation is expressed for financial support in the form of a research fellowship provided by Dunhall Pharmaceuticals, Inc. during a portion of this investigation.

Literature Cited

1. Harned, H. S.; Keston, A. S.; Donelson, J. G. *J. Am. Chem. Soc.* **1936,** *58,* 989.
2. Owen, B. B.; Foering, L. *J. Am. Chem. Soc.* **1936,** *58,* 1575.
3. Oiwa, I. T. *J. Phys. Chem.* **1956,** *60,* 754.
4. Towns, M. B.; Greely, R. S.; Leitzke, M. H. *J. Phys. Chem.* **1960,** *64,* 1861.
5. Hetzer, H. B.; Robinson, R. A.; Bates, R. G. *J. Phys. Chem.* **1962,** *66,* 1423.
6. Feakins, D.; Watson. *J. Chem. Soc.* **1963,** 4686.
7. Harned, H. S.; Owen, B. B. In "Physical Chemistry of Electrolytic Solutions," 2nd ed.; Reinhold: New York, 1958.
8. Hefly, J. D.; Amis, E. S. *J. Electrochem. Soc.* **1965,** *112,* 336.
9. Melton, S. L.; Amis, E. S. *J. Chem. Eng. Data* **1968,** *13,* 429.
10. McIntyre, J. M.; Amis, E. S. *J. Chem. Eng. Data* **1968,** *13,* 371.
11. Ives, D. J. G.; Janz, G. J. "Reference Electrodes," Academic: New York, 1961, p. 198.
12. Janz, G. J.; Taniguchi, H. *Chem. Rev.* **1953,** *53,* 397.
13. Taylor, J. K.; Smith, E. R. *J. Res. Natl. Bur. Stand.* **1939,** *22,* 307.
14. Keston, A. S. *J. Am. Chem. Soc.* **1935,** *57,* 1671.
15. Bates, R. G. "Electromotive pH Determinations," John Wiley and Sons: New York, 1954; p. 206.
16. Greely, R. S.; Smith, W. T.; Leitzke, M. H.; Stronghton, R. W. *J. Phys. Chem.* **1960,** *64,* 652.
17. Rule, C. K.; LaMer, V. K. *J. Am. Chem. Soc.* **1936,** *58,* 2339.
18. Hildebrand, J. H. *J. Am. Chem. Soc.* **1913,** *35,* 847.
19. Hills, G. J.; Ives, D. J. G. *J. Chem. Soc.* **1951,** 305.
20. Popoff, Stephen; Kunz, A. H.; Snow, R. D. *J. Phys. Chem.* **1928,** *32,* 1056.
21. Beans, H. T.; Hammett, L. P. *J. Am. Chem. Soc.* **1925,** *47,* 1215.
22. "Inorganic Synthesis," Booth, H. S., Ed.; McGraw Hill: New York, 1939; Vol. 1, pp. 149–157.
23. Riddick, J. A.; Bunger, W. B. "Techniques of Chemistry," Vol. 2, 3rd ed.; Organic Solvents," Wiley Interscience: New York, 1970, p. 645.
24. Debye, P.; Huckel E. *Phys. Z.* **1923,** *24,* 185.
25. Hitchcock, D. I. *J. Am. Chem. Soc.* **1928,** *50,* 2067.
26. Gronwall,(T. H.; LaMer, V. K.; Sandved, K. *Phys. Z.* **1928,** *29,* 358.
27. Nunez, L. J.; Day, M. C. *J. Phys. Chem.* **1961,** *65,* 164.
28. Woolcock, J. W.; Hartely, H. *Philos. Mag.* **1928,** *7,* 5, 1133.
29. Scatchard, G. *J. Am. Chem. Soc.* **1926,** *48,* 2026.
30. Goldenberg, N.; Amis, E. S. *Z. Phys. Chem.* **1962,** *31,* 145.

31. MacInnes, D. A. "The Principles of Electrochemistry," Reinhold; New York, 1939.
32. Roy, R. N.; Vernon, W.; Bothwell, A. L. M. *J. Chem. Soc.* **1971,** *A,* 1242.
33. McIntyre, J. M. Ph.D. Dissertation, University of Arkansas, Fayetteville, AR, 1968.
34. Franks, D.; Ives, D. J. G. *Quart. Rev.* **1966,** *20.*
35. Franks, F. In "Physico Chemical Processes in Mixed Solvents," Franks, F., Ed.; American Elseviere Publishing Company: New York, 1939.
36. Gordy, W.; Stanford, S. C. *J. Chem. Phys.* **1940,** *8,* 170.
37. Gordy, W.; Stanford, S. C. *J. Chem. Phys.* **1941,** *9,* 204.
38. Gerrard, W.; Maclin, E. D. *Chem. Rev.* **1959,** *59,* 1105.
39. Kuhn, L. P. *J. Am. Chem. Soc.* **1952,** *74,* 2492.
40. Brown, A. C.; Ives, D. J. G. *J. Chem. Soc.* **1962,** 1608.
41. Feakins, D.; Watson, P. *J. Chem. Soc.* **1963,** 4734.
42. Brown, A. C.; Ives, D. J. G. *J. Chem. Soc.* **1962,** 1608.

RECEIVED February 21, 1978.

INDEX

INDEX

A

A coefficients 174
Acetic acid 7–10
  in EG–DEG mixtures at various temperatures, pK_a values of 349*t*
  from wood-pulping liquors, recovery of 1–10
Acetone 6–10
Acetone–water
  aggregation numbers of ionic micelles in 122*t*
  *n*DTMABr and SDS in 115*t*, 123
  pK_a* of acids in 238*t*–239*t*
  single-ion standard Gibbs free energy of transfer from water to 117*f*
  tetrabutylammonium particle in 118*t*
  transference number of NaBr in . 114*t*
Acetonitrile (ACN) 100
  HPi in 89
  –water
    $\Delta H°$ of Bu_4NBr in 101*t*
    $\Delta H°$ of $LiClO_4$ in 105
    mixed-solvent systems 230
Acid(s)
  in acetone–water at 25°C, pK_a* of 238*t*–239*t*
  –anion complexes 87
  –base interactions 82
  –base relationship, Brönsted–Lewis equation of 53
  vs. dielectric constant function, ionization constants of 235*f*
  in ethanol–water, ionization constants of 241*f*
  in ethanol–water at 25°C, pK_a* (molal scale) of various 236*t*–237*t*
  heats of mixing of ether with sulfolane solutions of 155*t*
  in methanol–water at 25°C, pK_a* 238*t*–239*t*
  minimum distance approach of . 244*t*
  in nonaqueous solvents, behavior of weak 86
  in organic–water solvents at 25°C, pK_a* of 238*t*–240*t*
  in solvent, free energy change of ionization of 232
  in solvents, ionization constants of 234*f*
  vapor pressure of sulfolane solutions of 150*f*
Acidity Function, Hammett's 53
ACN (*see* Acetonitrile)
Activity behavior of
  metallic ion species in aqueous solution 299–321
  $NdNO_3$–HNO_3–H_2O–HDEHP–AMSCO system 323–343
  single-component lanthanide salt in aqueous solution 323
Activity coefficient(s)
  of an electrolyte in binary mixtures 263, 268
  of HBr in HBr + $(Et)_4NBr$ + H_2O 269
  of the HDEHP dimer
    in AMSCO 335
    calculations 341*t*
    in *n*-octane 335
  on HNO_3 and NO_3^- as a function of nitric acid concentration . 332*f*
  of hydrobromic acid 372*t*–374*t*
  in mixed solvents 259
  molal 362
  of the $Nd(NO_3)_3$–HNO_3–H_2O system at 25°C 299–321
  of the neodymium ion in aqueous nitrate solution 306*f*, 307*t*
  of the neodymium ion in neodymium nitrate aqueous solutions 307*t*
  rational 362
  in water 259
Ag–AgBr electrode, standard electrode potential ($E°$) of 362
Ag–AgCl electrode
  bulk dielectric constants of solvent, standard potentials of 228*t*
  computation of standard potential in 216
  curve-fitting polynomial for computation of $E°$ of 222
  curve-fitting technique for evaluation of $E°$ of 222
  E_{obs} of 218
  standard potential of 225*t*
    variation of 227*f*
Ag–AgI electrode in glycol + diethylene glycol, standard potentials of 346
Aggregation numbers of ionic micelles in water–acetone mixtures 122*t*
Aggregation numbers for ionic and non-ionic micelles in aqueous binary mixtures, variations of . 126*f*

Alcohol
increasing basic strength of 369
salting-out coefficient as a function of 375*f*
–TMS mixtures 93
–water 370
enthalpy as a function of solvent composition in 371*f*
entropy as a function of solvent composition in 372*f*
–inorganic salt systems 12
trends of K_n(corr) values in .. 140
Alkali bromides 100
Alkali halides in methanol–water mixtures, viscosity of dilute solutions of 167–176
Aminosugar research, solvent systems in 178–188
Ammonium bromide–ethanol–water system 14
at fixed-salt concentrations 18*f*
isobaric vapor–liquid equilibrium data for 16*t*
salt effect parameter for 21*t*
salt effect parameter as a function of the liquid composition ratio for 19*f*
at saturation 18*f*
Ammonium bromide on the ethanol–water system, salt effect of 15*f*
AMSCO, HDEHP dimer activity coefficient in 335
AMSCO Odorless Mineral Spirits . 335
Anilinium ion 68
Anion(s)
–acid complexes 87
in aqueous binary mixtures, behavior for inorganic 118
carboxylate 69
–TMS interactions 84
Anionic radii, hydrated 241
Apparent charge
of ionic micelles in aqueous binary solvents 109–128
of micelles in mixed solvents ... 121
of ionic micelles with solvent composition, variation of .. 125*f*
Aprotic solvents
dipolar 77
enthalpies of solution of electrolytes in mixtures of water with 99–107
ion–ion interactions in 78
ion–solvent interactions in 78
Aqueous
binary mixtures, behavior for inorganic anions in 118
binary mixtures, variations of aggregation numbers for ionic and non-ionic micelles in 126*f*
binary solvents, apparent charge of ionic micelles in 109–128
Aqueous (*continued*)
media, thermodynamic behavior of tetraalkylammonium bromide and hydrobromic acid in 263–274
micelles 210
mixed-salt solutions, solubility of carbon dioxide in 189–205
mixtures, $\Delta H°$ of Bu_4NBr in ... 106*f*
$Nd(NO_3)_3$–HNO_3, static vapor pressure measurements of . 310
neodymium concentration, equilibrium 334*t*
nitrate solution, activity of the neodymium ion in 305*f*, 306*f*
phase, equilibrium metal concentrations in 332
phase, pH electrode data for equilibrated 333*t*
salt solution, liquid–liquid extraction of the lanthanides from 323–343
single-salt solutions, applicability of the Setschenow equation on the salt concentration in 190
solution(s)
activities of neodymium ion in neodymium nitrate 307*t*
activity behavior of a metallic ion species in 299–321
activity coefficients of neodymium ion in neodymium nitrate 307*t*
lanthanide complexes in 301
lanthanum nitrate complexes in . 301
of nitric acid, nitrate ion activity coefficient in 331*t*
Association constant(s) 80*t*, 85
on the dielectric constant(s)
for HCl in water–TMS, dependence of 90*f*
for LiCl in methanol–TMS, dependence of 94*f*
in water–TMS, dependence of 91*f*
of picric acid in ethanol–TMS .. 95*f*
of picric acid in *tert*-butyl alcohol–TMS 95*f*
Association phenomena 85
importance of solute–solvent interactions to 86
in water–TMS, ionic 92–93

B

B coefficients 175
in the water-rich region 175
Bases in nonaqueous solvents, behavior of weak 86
Basicity in mixtures of water with organic solvents 53
Bending energy 211

Binary
electrolyte model, Q coefficients for Harned's ... 316f
electrolyte solutions ... 310
vapor pressure of $Nd(NO_3)_3$–HNO_3–H_2O ... 311f
water activity in the $Nd(NO_3)_3$–HNO_3–H_2O ... 313f
mixed-salt solutions, carbon dioxide solubility in ... 191–202
mixed-salt systems ... 198t
mixture(s)
activity coefficient of an electrolyte in ... 263
behavior for inorganic anions in aqueous ... 118
of electrolytes, activity coefficient of an electrolyte in . 268
of electrolytes, osmotic coefficient of an electrolyte in 268
of EG–DEG, transfer of hydrogen iodide in ... 345–353
excess Gibbs free energy of .. 317
involving sulfolane (TMS) and protic solvents, conductance and ionic association of electrolytes in ... 77–99
ion–solvent interactions in ... 81
$Nd(NO_3)_3$–HNO_3–H_2O electrolyte mixture, excess Gibbs free energy of ... 318f
$Nd(NO_3)_3$–HNO_3–H_2O electrolyte solution vapor pressure data, Harned's plot of ... 315f
solvents, apparent charge of ionic micelles in aqueous ... 109–128
Biochemical oxygen demand (BOD) ... 8
Black liquors from kraft pulping .. 10
BOD (biochemical oxygen demand) ... 8
Born electrostatic model ... 232
Born equation ... 57, 118
Bromine, Harned interaction coefficient vs. total molality of ... 267f
Brönsted–Guggenheim Equation . 267
Brönsted–Guggenheim parameters, relationship of the Harned coefficient to ... 268
Brönsted–Lewis equation of an acid–base relationship ... 53
Bu_4NBr (tetrabutylammonium bromide)
$\Delta H°$ of ... 100
in aqueous mixtures ... 106f
in water–ACN at 25°C, $\Delta H°$ of . 101t
in water–EC at 45°C, $\Delta H°$ of .. 101t
Bulk component ... 64
Bulk dielectric constants
in mixed solvents ... 223
of solvents at 25°C ... 236t
of solvents, standard potentials of Ag–AgCl electrode ... 228t
tert-BuOH–TMS mixtures ... 93
1-Butanol ... 225t
tert-Butanol ... 64, 225t
standard electrode potential of the silver–silver bromide electrode in anhydrous ... 367
–TMS ... 82t
pK values of picric acid in ... 82t
–water ... 368
$E°$ for hydrobromic acid in .. 365t
electromotive force measurements of the galvanic cell containing hydrobromic acid in ... 360t
$\Delta G°/T$ as a function of $1/T$ for ... 370f
standard potentials of Pt, $H_2|HBr(m)$, $X\%$ alcohol, $Y\%$ water|AgBr–Ag in .. 368f
Butyl acetate–butanol
–calcium chloride, preferential solvation number in ... 34f
–calcium chloride system, prediction for ... 37f
solubility of calcium chloride in 30f
tert-Butyl alcohol–TMS, association constants of picric acid in ... 95f
tert-Butyl alcohol–TMS, ionization of picric acid in ... 96f

C

CH_3OH–H_2O system ... 130
elution profile of ... 133f
$(CH_3)_2SO$–H_2O system ... 130
eluting profile of ... 134f
C_2H_5OH–H_2O system ... 130
elution profile of ... 133f
Calcium chloride
in butyl acetate–butanol, solubility of ... 30f
in methanol–ethyl acetate, solubility of ... 28f
in methyl acetate–methanol, solubility of ... 29f
Cane sugar refining ... 3–4
Carbon dioxide solubility
in aqueous mixed-salt solutions 189–205
in binary mixed-salt solutions 191–202
in potassium chloride–calcium chloride solutions ... 194f, 196f
in potassium chloride–sodium chloride–calcium chloride solution ... 193t
in sodium chloride–sodium sulfate–ammonium chloride solutions ... 193t, 195f, 197f
in sodium chloride–magnesium sulfate–potassium nitrate solution ... 193t
in ternary mixed-salt solutions 193–202

Carboxylate anions 69
$\Delta G_t°$(carboxylate)$_e$ against mole fraction of *tert*-butanol with 70*f*
Cation–cation interaction, doublet 269
Cation–TMS interactions 84
Cationic micelle 119
Cavity formation 116
Cell A, Nernst Equation for 252
Cell A from 5°–45°C, standard 256*t*–257*t*
Chelated ligand 132
Chloride single-ligand complexes of lanthanides 301
Chromium(III)
coordination shell of 130
discrimination of 131
equilibrated in 1,3-propanediol(G)–H_2O system, elution profile for species of 134*f*
ion in mixed solvents, solvation of 129–144
kinetic inertness of 130
random solvation of 131
solvation in mixed-solvent systems 141
species in the water–1,3- propanediol system 140
cmc (*see* Critical micelle concentrations)
Coefficients of $_sE_m° = {}_sE_m°$ (25°C $+ b(t - 25) + c(t - 25)^2$) 351*t*
Coefficients, medium activity 136
Complex formation stability constants 302
Complex fraction data 308*t*
Concentrated solutions, extraction of 1–10
Conductance(s)
of electrolytes in binary mixtures involving sulfolane (TMS) and protic solvents 77–99
of HCl in pure TMS 87*f*
limiting equivalent 78*t*–80*t*
of sulfolane solutions of acids as a function of diethyl ether concentration 153*f*
of sulfolane solutions of acids as a function of water concentration 152*f*
Conductivity equations, Evans' ... 113
Conductivity for the Et_2O–$HSbCl_6$ system 151
Conductometric behavior of ions in water 79
Confidence region(s) 46
for methanol–water–sodium chloride 48*f*, 49*f*
Conjugation phenomena 86
Contact ion pair, formation of ... 92
Coordinate bonds 129
Coordination shell of chromium (III) 130
Co-solvent(s)
$\Delta G_t°$(anilinium$^+$)$_e$ against 69*f*
$\Delta G_t°$(BPh_4^-) against 63*f*
Co-solvent(s) (*continued*)
$\Delta G_t°$($CH_3CH_2COO^-$)$_e$ against mole fraction of 66*f*
$\Delta G_t°$(CH_3COO^-)$_e$ against mole fraction of 65*f*
$\Delta G_t°$(CH_3COO^-)$_e$ against reciprocal of the dielectric constant with 64*f*
$\Delta G_t°$($C_6H_5COO^-$)$_e$ against mole fraction of 67*f*
$\Delta G_t°$(Cl^-) against 60*f*
$\Delta G_t°$($HCCO^-$)$_e$ against mole fraction of 66*f*
$\Delta G_t°$(K^+) against 59*f*
$\Delta G_t°$(OH^-) against 62*f*
$\Delta G_t°$(*n*-propylammonium$^+$)$_e$ against 68*f*
influence of 105
in mixture, structural effect of .. 65
Coulomb potential for the micelle 113
Coulombic stretching force 211
Counterions 120
Critical micelle concentrations (cmc) 110, 114
of micelles in mixed solvents ... 121
of POE_{23} 125
with solvent composition, variation of 110, 121*f*
values of *n*DTMABr in water .. 115
values of SOS in water 115
$Cr(OH_2)_3(OS(CH_3)_2)_3^{3+}$ from $Cr(OH_2)_4(OS(CH_3)_2)_2^{3+}$, yields of 136t
$Cr(OH_2)_{6-n}(OS(CH_3)_2)_n^{3+}$, relative stabilities of isomeric species 135*t*
$Cr(OH_2)_{6-n}(OS(CH_3)_2)_n^{3+}$ system, identification of isomers in ... 135
Crystalline phase, lamellar liquid . 211
stability of 213
Curve-fitting polynomial for the computation of $E°$ of Ag–AgCl electrode 222
Curve-fitting technique for evaluation of $E°$ of Ag–AgCl electrode 222

D

D-values 45
Debye–Hückel
Equation 122, 212
extend terms of 220
reduced 148
extended equation, extrapolation of emf in dioxane–water mixtures computed by 217*f*
potential 112
theory 300, 359
for symmetrical valence-type electrolytes 363
DEG (*see* Diethylene glycol)

Density–concentration coefficients . 170
Density parameters for salt solutions in methanol–water 174*t*
Dielectric constant(s)
with co-solvents, $\Delta G_t°(CH_3COO^-)_e$ against reciprocal of 64*f*
function 234*f*
ionization constants of acids vs. 235*f*
for HCl in water–TMS, dependence of association constants on 90*f*
for LiCl in methanol–TMS, dependence of association constants on 94*f*
in mixed solvents, bulk 223
of nonaqueous solvents 250
of solvents at 25°C, bulk 236*t*
of solvents, standard potentials of Ag–AgCl electrode bulk . 228*t*
of TMS 86
in water–TMS, dependence of association constants on ... 91*f*
Dielectric continuum, spheres in .. 85
Dielectric saturation 86
Diethyl ether concentration, conductance of sulfolane solutions of acids as a function of 153*f*
Diethylene glycol (DEG) 346
at various temperatures, $E°'$ in . 349*f*
–ethylene glycol mixtures, physical constants of 347*t*
Di-2-ethylhexyl phosphoric acid (HDEHP) 324
dimer activity 342*f*
coefficient in AMSCO 335
coefficient calculations 341*t*
dimer concentration in the equilibrated organic phase . 340*t*
dimer in *n*-octane, activity coefficients of 335
extraction reactions of lanthanum by 325
ion exchange mechanism for the extraction of lanthanide ions by 325
trimers, formation of 325
2,6-Dihydroxylbenzoic acid in TMS 88
Dilution, enthalpies of 100
Dimer
activity, HDEHP 342*f*
coefficients calculations 341*t*
concentration in the equilibrated organic phase, HDEHP ... 340*t*
molarity 335
1,2-Dimethoxyethane 225*t*
N,N-Dimethylacetamide (DMA) . 99
N,N-Dimethylformamide (DMF) 99, 226*t*
–water 102
$\Delta H°$ of salts in 104*f*
$\Delta H°$ of solutes in 102*f*
$\Delta H°$ of tetraalkylammonium bromides in 103*f*
Pr_4NBr in 104
Dimethylsulfoxide (DMSO) 99, 145, 226*t*
–water 102, 138, 235
$\Delta H°$ of solutes in 102*f*
Dioxane 186
–water 82
addition of an electrolyte to .. 186
computed by the Debye–Hückel extended equation, extrapolation of emf in 217*f*
Dipolar aprotic (DPA) solvents .77, 145
solvation enthalpy of gaseous ions in 162*t*–163*t*
sulfolane as 146
thermodynamic properties of electrolytes in 146
Discrimination factor 132*f*
Dissociation
of ionic micelles in mixed solvents, degree of 119–121
of ionic micelles with solvent composition, degree of 120*f*
of a micelle at 298.15 K, degree of 115*t*
of a micelle at 308.15 K, degree of 116*t*
of micelles, determination of the degree of 113
steps of glycine 278
Distance approach of acids, minimum 244*t*
DMA (*N,N*-dimethylacetamide) . 99
DMF (*see N,N*-Dimethylformamide)
DMSO (*see* Dimethylsulfoxide)
Doublet cation–cation interaction . 269
DPA (*see* Dipolar aprotic solvents)
*n*DTMABR (*see* Trimethyldecylammonium bromide)
Durbin–Watson test 42

E

$E°$ (standard electrode potential)
of the Ag–AgBr electrode 362
of the Ag–AgCl electrode, curve-fitting polynomial for the computation of 222
of the Ag–AgCl electrode, curve-fitting technique for the evaluation of 222
for hydrobromic acid in *tert*-butanol–water 365*t*
for hydrobromic acid in ethanol–water 365*t*
in mixed solvents 57
of the silver–silver bromide electrode in anhydrous *tert*-butanol 367
in water 57
$E°'$ (standard emf) 252, 347
of Cell A from 5°–45°C ...256*t*–257*t*
data of Pt; H_2(g, 1 atm)/HBr(m_1), $(C_2H_5)_4NBr(m_2)$/AgBr; Ag 265*t*

$E^{\circ\prime}$ (standard emf) (*continued*)
in DEG at various temperatures 349*f*
as a function of the molality of HBr mixed solvents at 25°C 256*f*
with molality, variation of 224*f*
with temperature, variation of .. 224*f*
${}_sE_c{}^{\circ}$ in EG–DEG mixtures 350*t*
$E_m{}^{\circ\prime}$ (standard cell potential) ... 359
${}_sE_m{}^{\circ}$ in EG–DEG mixtures 350*t*
${}_sE_m{}^{\circ} = {}_sE_m{}^{\circ}$ (25°C + $b(t - 25)$ + $c(t - 25)^2$), coefficients of 351*t*
${}_sE_N{}^{\circ}$ in EG–DEG mixtures 350*t*
E_{obs} (observed electromotive force) 359
of Ag–AgCl electrode 218
EC (ethylene carbonate) 100
–water, ΔH° of Bu_4NBr in 101*t*
EG (*see* Ethylene glycol)
Electric
double layer formation 212
double layer as an inverse micelle 212
potential of micelle, calculation of 112
Electrode
materials, purification of 356
potentials, standard (*see* E°)
preparation and aging 356
Electrolyte(s)
in binary mixture(s)
activity coefficient of 263, 268
involving sulfolane (TMS) and protic solvents, conductance and ionic association of 77–99
osmotic coefficient of 268
binary, solutions 310
Debye–Hückel theory for symmetrical valence-type 363
in DPA solvents, thermodynamic properties of 146
in a mixed-aqueous solvent, ionization constant of 233
in mixed solvents, behavior of .. 145
in mixtures of water with aprotic solvents, enthalpies of solution of 99–107
preferential solvation of 145–165
in solution, equilibrium behavior of 278
in sulfolane (TMS) 78
of symmetrical valence type 219
vapor pressure of sulfolane solutions of 149*f*
in water, ionization constant of . 233
Electromotive force
measurements of the galvanic cell containing hydrobromic acid in ethanol–water and *tert*-butanol–water solvents . 360*t*
of Pt H_2|HBr(m), X% alcohol, Y% water|AgBr–Ag in pure and mixed solvents 355–376
of Pt; H_2(g, 1 atm)|HBr(m) in NMA/H_2O|AgBr; Ag ..254*t*–255*t*
Electromotive force (*continued*)
of Pt/H_2(g)/$G^{\pm}(m_1)$, NaG(m_2), KBr(m_3) in water + 10, 30, and 50 mass % THF/AgBr/Ag 282*t*–283*t*
of Pt/H_2(g)GHBr(m_1), $G^{\pm}(m_2)$ in water + 10, 30, and 50 mass % THF/AgBr/Ag 280*t*–281*t*
Electrostatic model, Born 232
Elution profile
of H_2O–CH_3OH system 133*f*
of H_2O–$(CH_3)_2SO$ system 134*f*
of H_2O–C_2H_5OH system 133*f*
for species of chromium(III) equilibrated in 1,3-propanediol(G)–H_2O system 134*f*
Emf, standard (*see* $E^{\circ\prime}$)
Energy
constants, Wilson 42
effect of errors on standard deviation of 44*f*
parameters, Wilson 44
Enthalpic effect of hydrophobic hydration 102
Enthalpy(ies)
of dilution 100
as a function of solvent composition in alcohol–water 371*f*
of mixing, excess 67
of solution (*see* ΔH°)
of transfer (*see* $\Delta H_t{}^{\circ}$)
Entropy
as a function of solvent composition in alcohol–water 372*f*
changes (*see* ΔS°)
of mixing, excess 67
of transfer (*see* $\Delta S_t{}^{\circ}$)
Equilibrated
aqueous phase, pH electrode data for 333*t*
organic phase, HDEHP dimer concentration in 340*t*
organic phase, nitrate concentration in 336*f*
Equilibrium
aqueous neodymium concentration 334*t*
behavior of electrolytes in solution 278
constants
for formation of iomeric species 136
for reactions 156*t*–157*t*
for solvent species at a single solvation site 168
thermodynamic (K_n) 136
metal concentrations in the aqueous phase 332
metal concentrations in the organic phase 332
organic neodymium concentration 334*t*
Error analysis, application of 42

Error analysis of isobaric liquid–vapor equilibrium data for mixed solvents 39–51
π-Error 42
Et_2O–HCF_3SO_3 160
Et_2O–$HSbCl_6$ system, conductivity for 151
Ethanol, solubility of inorganic salts in boiling solutions of .. 25*t*
Ethanol–TMS 82*t*
association constants of picric acid in 95*f*
ionization of picric acid in 96*f*
pK values of picric acid in 82*t*
Ethanol–water
$E°$ for hydrobromic acid in 365*t*
electromotive force measurements of the galvanic cell containing hydrobromic acid in .. 360*t*
$\Delta G°/T$ as a function of $1/T$ for . 369*f*
ionization constants of acids in 241*f*
pK_a* (molal scale) of various acids in 236*t*–237*t*
salt effect
of ammonium bromide on ... 15*f*
parameters for inorganic salts in 22*t*
of potassium iodide on 15*f*
of sodium acetate on 20*f*
solubility of inorganic salts in boiling solutions of 25*t*
standard potentials of Pt,|H_2 HBr(m), X% alcohol, Y% water|AgBr–Ag in 367*f*
Ether, protonation of water and .. 161
Ether, reaction of solvated proton with 161
Ethylene carbonate (EC) 100
Ethylene glycol (EG) 346
–DEG mixtures
data of Cell 1 from 5° to 35°C in 348*t*
${}_sE_m°$, ${}_sE_N°$, and ${}_sE_C°$ in 350*t*
transfer of HI in binary mixtures of 345–353
at 25°C, transfer of HI from EG to 351*t*
at various temperatures, pK_a values of acetic acid in . 349*t*
to EG–DEG, transfer energies of H^+ and I^- ions from 353*t*
EtOH–TMS mixtures 93
Evans' conductivity equation 113
Excess
enthalpy of mixing 67
entropy of mixing 67
Gibbs free energy 318*t*
of the binary electrolyte mixture 317
of binary $Nd(NO_3)_3$–HNO_3–H_2O electrolyte mixture . 318*f*
of the $Nd(NO_3)_3$–HNO_3–H_2O system at 25°C 299–321
Extended terms contribution 217

F

Fixed-liquid compositions 12
salt effect in vapor–liquid equilibrium at 11–25
Fixed-salt concentrations, ammonium bromide–ethanol–water at 18*f*
Formalism according to Pitzer ... 268
Formamide 99
Free energy change of ionization of acid in solvent 232
Free energy of transfer between water and the mixed solvent . 56
Fuoss equation 85

G

$\Delta G°$ 363
for glycine in THF–H_2O 286
(ROH_2) 55
T^{-1} as a function of $1/T$ for ethanol–water 369*f*
T^{-1} as a function of $1/T$ for *tert*-butanol–water 370*f*
ΔG^E in tetraethyl ammonium bromide, specific effects on 272
$\Delta G_t°$ (Gibbs free energy of transfer) 110, 233, 293, 350
$(anilinium^+)_e$ against co-solvents 69*f*
$(anion)_e$ 65
(BPh_4^-) against co-solvents 63*f*
$(carboxylate)_e$ against mole fraction of *tert*-butanol with carboxylate anions 70*f*
$(CH_3CH_2COO^-)_e$ against mole fraction of co-solvent 66*f*
$(CH_3COO^-)_e$ against mole fraction of co-solvent 65*f*
$(CH_3COO^-)_e$ against reciprocal of the dielectric constant with co-solvents 64*f*
$(C_6H_5COO^-)_e$ against mole fraction of co-solvent 67*f*
(Cl^-) against co-solvents 60*f*
as a function of NMA in mixed solvent 260*f*
(H^+) against solvents 58*f*
for $HBr(H_2O) \rightarrow HBr(H_2O/NMA)$ 258*t*
$(HCCO^-)_c$ against mole fraction of co-solvent 66*f*
(HOH) 57
(K^+) against co-solvents 59*f*
(OH^-) against co-solvents 62*f*
$(n\text{-Propylammonium}^+)_e$ against co-solvents 68*f*
for transfer of HBr from water to mixed solvent 257
from water to water–acetone mixtures, single-ion standard .. 117*f*
$\Delta G_t°(+)$ 118
$\Delta G_t°(-)$ 118

$\Delta G_t°$ (diss), transfer Gibbs energy of dissociation 290
$\Delta G_t°_{(I^-)}$ (transfer free energy data of the iodide ion) 352
Galvanic cell, standard potential of 363
Gaseous ions in DPA solvents, solvations enthalpy 162*t*–163*t*
Gaseous ions in water, solvations enthalpy of 162*t*–163*t*
Gases in single-salt solutions, two-parameter equation for the solubility of 197
Gaussian elimination technique, modified 222
Gibbs–Duhem equation 43
of lanthanide nitrate 302
for $M(NO_3)_3$–HNO_3–H_2O 303
to the $Nd(NO_3)_3$–HNO_3–H_2O system, application of 313
Gibbs free energy of the binary electrolyte mixture, excess ... 317
Gibbs free energy of transfer (*see* $G_t°$)
Glycerol 226*t*
Glycine ($\overset{+}{N}H_3$–CH_2–COO) 284
dissociation steps of 278
in 10 mass % THF–water, thermodynamic functions for the dissociations of 288*t*–290*t*
in 30 mass % THF–water, thermodynamic functions for the dissociations of ..288*t*–289*t*, 291*t*
in 50 mass % THF–water, thermodynamic functions for the dissociations of 290*t*–292*t*
in THF–H_2O
$\Delta G°$ for 286
$\Delta H°$ for 286
heat capacity for 286
$\Delta S°$ for 286
at varying temperatures, thermodynamic study of 277
transfer dissociation energies of 292*t*–293*t*
transfer free energies for 294*f*
Guggenheim–Scatchard equations . 300

H

H^+ diffusion effect 329
$\Delta H°$ (enthalpies of solution) ...100, 363
–composition profile 105
of electrolytes in mixtures of water with aprotic solvents 99–107
for glycine in THF–H_2O 286
of $LiClO_4$ in W–ACN 105
(M) (standard enthalpy of solution in a mixture of water) 101
of salts in DMF–water 104*f*
of a solute in various solvents .. 99
$\Delta H°$ (*continued*)
of solutes in DMF–water 102*f*
of solutes in DMSO–water 102*f*
of tetraalkylammonium bromides in DMF–water 103*f*
of tetrabutylammonium bromide (Bu_4NBr) 100
in aqueous mixtures 101*t*, 106*f*
$\Delta H°_t$ (enthalpy of transfer) 99, 258*t*, 351
as a function of NMA in mixed solvent 260*f*
for $HBr(H_2O) \rightarrow HBr(H_2O/NMA)$ 258*t*
for reactions 156*t*
HBr
+ $(Et)_4NBr$ + H_2O mixtures, activity coefficient of HBr in 269
$(H_2O) \rightarrow HBr(H_2O/NMA)$, enthalpy and entropy changes for 258*t*
$(H_2O) \rightarrow HBr(H_2O/NMA)$, $\Delta G_t°$ for 258*t*
from H_2O to mixed solvent, $\Delta G_t°$ for transfer of 257
from H_2O to mixed solvent, $\Delta S_t°$ for transfer of 257
mean activity coefficient of 252
for mixed solvents, $E°'$ as a function of molality of 256*f*
+ NH_4Br, Pitzer interaction parameters for 273*t*
in NMA/H_2O, ionic activity coefficients of 258*t*
+ tetraalkylammonium bromides, Pitzer parameters for 273*t*
HCF_3SO_3
–Et_2O 160
–H_2O 160
proton affinity of 164
in sulfolane 147
vapor pressure of 148
HCl
mean activity coefficient of 229
+ NH_4Cl, Pitzer interaction parameters for 273*t*
in pure TMS, conductance of .. 87*f*
in water–TMS, dependence of association constants on the dielectric constant for 90*f*
HDEHP (*see* Di-2-ethylhexyl phosphoric acid)
HF, hydrated 160
HF, undissociated 160
HI (*see* Hydrogen iodide)
HNO_3 as a function of nitric acid concentration, activity coefficient of 332*f*
HNO_3–H_2O data 314*t*
HPi (*see* Picric acid)
$HSbCl_6$ in sulfolane 147
nature of 148
H_3O^+ 55
heat of solvation of 162
H_2S, protonation of 163

H_2SO_4, nitrate electrode data for organic phase samples back-extracted with 336t–337t
Hammett's Acidity Function 53
Harned's
binary electrolyte model, Q coefficients for 316f
coefficient to the Brönsted–Guggenheim parameters, relationship of 268
interaction coefficient vs. total molality of Br 267f
plot of binary $Nd(NO_3)_3$–HNO_3–H_2O electrolyte solution vapor pressure data 315f
Rule 265
Heat capacity for glycine in THF–H_2O 286
Heats of mixing of ether with sulfolane solutions of acids .. 155t
Heats of mixing of water with sulfolane solutions of electrolytes 154t
Hexaaquachromium(III) ion, stability of 141
n-Hexane, trimer formation in ... 325
Hydrated anionic radii 241
Hydrated HF 160
Hydrated proton in water, structures for 56f
Hydrates, proton 161
Hydration constant of $CF_3SO_3^-$ in sulfolane 159
Hydration, hydrophobic 100, 102
Hydrobromic acid 363
activity coefficients of 372t–374t
in aqueous media, thermodynamic behavior of tetraalkylammonium bromide and .263–274
in *tert*-butanol–water, $E°$ for ... 365t
in *tert*-butanol–water solvents, electromotive force measurements of the galvanic cell containing 360t
in ethanol–water, $E°$ for 365t
in ethanol–water, electromotive force measurements of the galvanic cell containing ... 360t
in NMA–water mixtures, thermodynamic behavior of .249–260
transfer free energies for 294f
Hydrogen
bonding 68
bromide, preparation of 357
gas, purification of 356
iodide (HI) 350
in binary mixtures of EG–DEG, transfer of 345–353
from EG to EG–DEG, transfer of 351t
ion activity in the $Nd(NO_3)_3$–HNO_3–H_2O electrolyte solution 319f
ion molality, apparent 284
Hydrophobic
bonding 68
character of a tetraalkylammonium salt, influence of substitution groups on 104
hydration behavior of tetraalkylammonium salts in water .. 100
hydration, enthalpic effect of .. 102
solute, size of 103
Hydroxyl ion concentration 285

I

Inertness of chromium(III), kinetic 130
Inorganic anions in aqueous binary mixtures, behavior for 118
Inorganic ionic behavior 83
Inorganic salts
in boiling solutions of ethanol, water, and ethanol–water, solubility of 25t
in ethanol–water mixtures, salt effect parameters for 22t
physical properties of saturated solutions of 23t
Invariant point confirmation 185
Inverse micellar phase 211
Inverse micellar solution of pentanol, solubility of sodium chloride in 205
Inverse micelle(s) 206f, 209f
electric double layer of 212
stability of 210
Iodide ion ($G_t°_{(I^-)}$), transfer free energy data of 352
Ion(s)
exchange mechanism for the extraction of lanthanide ions by HDEHP 325
interaction parameters 263
–ion interactions in aprotic and protic–aprotic solvents 78
pair, formation of 92
reassociation 24
-size parameter 218
–solvent interactions in aprotic and protic–aprotic solvents 78
–solvent interactions in binary mixtures 81
structure-breaking capacities of . 65
structure-forming capacities of . 65
transfer between solvents 231
in water, conductometric behavior of 79
Ionic
activity coefficients of HBr in NMA/H_2O 258t
association of electrolytes in binary mixtures involving sulfolane (TMS) and protic solvents 77–99
association phenomenon in water–TMS 92–93

Ionic (*continued*)
behavior, inorganic 83
constituents of micelles in mixed solvents, thermodynamic behavior of 116–119
equilibrium processes in mixed-organic–aqueous solvent systems 216
micelles
in aqueous binary mixtures, variations of aggregation numbers for 126*f*
in aqueous binary solvents, apparent charge of ...109–128
in mixed solvents, degree of dissociation of 119–121
with solvent composition, degree of dissociation of . 120*f*
with solvent composition, variation of apparent charge of 125*f*
in water–acetone, aggregation numbers of 122*t*
in water–*n*-propanol 126
processes in water–organic mixed solvents, computational techniques of 215–246
Walden products 79
normalized to mol % organic solvents 85*f*
normalized to mol % TMS .. 83*f*
Ionization of acid in solvent, free energy change of 232
Ionization constant(s)
of acids vs. dielectric constant function 235*f*
of acids in ethanol–water mixtures 241*f*
of acids in solvents 234*f*
of electrolyte in water 233
of electrolyte in a mixed-aqueous solvent 233
of weak monoprotic acids in organic–water mixed solvents 231
Ionization of picric acid in ethanol–TMS and *tert*-butyl alcohol–TMS 96*f*
Isobaric vapor–liquid equilibrium data for
ammonium bromide–ethanol–water system 16*t*
mixed solvents, error analysis of 39–51
potassium acetate–ethanol–water system 18*t*, 19*t*, 21*t*
sodium acetate–ethanol–water system 17*t*
potassium iodide–ethanol–water system 14*t*
Isomeric species $Cr(OH_2)_{6-n}(OS(CH_3)_2)_n^{3+}$, relative stabilities of 135*t*
Isomeric species, equilibrium constants for the formation of ... 136
Isomers in the $Cr(OH_2)_{6-n}(OS(CH_3)_2)_n^{3+}$ system, identification of 135
Isopropanol 225*t*
Isotherms of
potassium acetate–water–dioxane 180*f*
potassium acetate–water–THF . 184*f*
potassium chloride–water–THF . 185*f*

J

Jones–Dole equation 170

K

K′ to mixed-solvent composition, insensitivity of 24
K′ salt effect parameter values 17
K_c 55
K_n (thermodynamic equilibrium constants) 136, 138
for solvent exchange reactions .. 139*t*
values in water–alcohol systems, trends of 140
$K_n/K_n°$ 142
Kinetic inertness of chromium(III) 130
Kraft pulping, black liquors from . 10
Kraft pulping, liquors from 9–10

L

Lamellar liquid crystalline phase .. 211
stability of 213
Lanthanide(s)
from an aqueous salt solution, liquid–liquid extraction of 323–343
chloride complexes 301
chloride single-ligand complexes of 301
complexes in aqueous solution .. 301
extraction by organophosphorous acids 324
ions by HDEHP, ion exchange mechanism for the extraction of 325
nitrate, Gibbs–Duhem equation of 302
nitrate single-ligand complexes of 301
salt in the aqueous solution, activity behavior of a single-component 323
Lanthanum by HDEHP, extraction reactions of 325
Lanthanum nitrate complexes in aqueous solution 301
LiCl in methanol–TMS, dependence of association constants on the dielectric constant for . 94*f*
$LiClO_4$ in W–ACN, $\Delta H°$ of 105
Ligand, chelated 132

Ligand, monodentate 132, 136
Limiting equivalent conductances 78t–80t
Linear regression analysis of
stability constants 308
Liquid
composition ratio for the ammonium bromide–ethanol–water system, salt effect parameter as a function of . 19f
composition, salt effect in vapor–liquid equilibrium at 11–25
–gas mass transfer 189
junction potential 328
layer formation of potassium acetate–water–THF 186
–liquid extraction 2
of impurities from concentrated sugar solutions ... 5
of the lanthanides from an aqueous salt solution ..323–343
–liquid tie-lines 181
–vapor equilibrium data for mixed solvents, error analysis of isobaric 39–51
Liquors from neutral semichemical pulping 9–10

M

$M(NO_3)_3$–HNO_3–H_2O, Gibbs–Duhem equation for 303
Mass transfer, liquid–gas 189
MeOH/H_2O/NaCl data 42
MeOH/H_2O/NaCl system
calculated data for 43t
effect of random error on simulated data for 46t–47t
experimental data for 43t
Mean activity coefficient of
free ions 122
HBr 252
HCl 229
polynomial for monoglyme–water mixture 230t
polynomial for THF–water mixtures 229t
Medium activity coefficients 136
Medium effect, primary 231
Metal concentrations in the aqueous phase, equilibrium 332
Metal concentrations in the organic phase, equilibrium 332
Metallic ion species in aqueous solution, activity behavior of 299–321
Methanol–ethyl acetate
–calcium chloride system, prediction for 35f
preferential solvation number in 32f
solubility of calcium chloride in 28f
Methanol–TMS 81t
dependence of association constants on the dielectric constant for LiCl in 94f
Methanol–water
density parameters for salt solutions in 174t
pK_a* of acids in 238t–239t
salts in 175t
solvents, silver–silver bromide electrode in 368
solvents, silver–silver iodide electrode in 368
viscosities of salt solutions in 172t–173t
viscosity B coefficients for salt solutions in 171f
viscosity of dilute solutions of alkali halides in 167–176
viscosity parameters for salt solutions in 174t
–sodium chloride system
confidence region for 48f, 49f
pressure difference for 48f
vapor composition difference for 49f
N-Methylacetamide (NMA) ..249, 250
–H_2O 250
|AgBr; Ag, electromotive force of Pt; H_2(g, 1 atm)|-HBr(m) in 254t–255t
mixed solvents from 5°–45°C, properties of 252t–253t
thermodynamic behavior of hydrobromic acid in ..249–260
in mixed solvent
$\Delta G_t°$ as a function of 260f
$\Delta H_t°$ as a function of 260f
$T\Delta S_t°$ as a function of 260f
Methyl acetate–methanol–calcium chloride system, prediction for 36f
Methyl acetate–methanol–calcium chloride system, preferential solvation number in 33f
Methyl acetate–methanol, solubility of calcium chloride in 29f
N-Methylformamide (NMF) 99
N-Methylpropionamide (NMP) 250, 259
Micellar phase, inverse 211
Micellar solution of pentanol, solubility of sodium chloride in an inverse 205
Micelle(s)
aqueous 210
calculation of the electric potential of 112
cationic 119
contraction, critical 114
coulomb potential for 113
degree of dissociation of ...115t, 116t
determination of 113
inverse 206f, 209f
electric double layer of 212
stability of 210
ionic, in aqueous binary solvents, apparent charge of 109–128
ionic, with solvent composition, degree of dissociation of .. 120f

Micelle(s) (*continued*)
in mixed solvents
apparent charge of 121
critical micelle concentration of 121
degree of dissociation of ionic 119–121
thermodynamic behavior of the ionic constituents of ..116–119
surface 113
Microemulsions 205
Minimum distance approach of acids 244*t*
Mixed-aqueous solvent, ionication constant of electrolyte in 233
Mixed-aqueous solvent, transfer of proton between water and .. 55
Mixed-micelle formation 122, 126
Mixed-salt solutions, carbon dioxide solubility in binary 191–202
Mixed-salt solutions, solubility of carbon dioxide in aqueous .189–205
Mixed-salt systems, binary 198*t*
Mixed-salt systems, Ostwald absorption coefficient values for 201*t*
Mixed solvent(s)
acetonitrile–water 230
activity coefficients in 259
apparent charge of micelles in . 121
behavior of electrolytes in 145
behavior to proton in 69
bulk dielectric constants in 223
chromium(III) solvation in 141
composition, insensitivity of k′ to 24
computational techniques of ionic processes in water–organic 215–246
critical micelle concentration of micelles in 121
degree of dissociation of ionic micelles in 119–121
$E^{\circ\prime}$ as a function of the molality of HBr for 256*f*
electromotive forces of Pt, H_2|-HBr(*m*), *X*% alcohol, *Y*% water|AgBr–Ag in 355–376
error analysis of isobaric liquid–vapor equilibrium data for .39–51
free energy of transfer between water and 56
ΔG_t° as a function of NMA in . 260*f*
ΔG_t° for the transfer of HBr from water to 257
ΔH_t° as a function of NMA in . 260*f*
ionization constant of weak monoprotic acids in organic–water 231
properties of NMA/H_2O ...252*t*–253*t*
ΔS_t° for transfer of HBr from H_2O to 257
solvation of chromium(III) ion in 129–144
standard electrode potentials in . 57
$T\Delta S_t^\circ$ as a function of NMA in . 260*f*
Mixed solvent(s) (*continued*)
thermodynamic behavior of the ionic constituents of micelles in 116–119
thermodynamic functions of Pt, H_2|HBr(*m*), *X*% alcohol, *Y*% water|AgBr–Ag in ..355–376
Mixing, excess enthalpy of 67
Mixing, excess entropy of 67
Molal activity coefficient 362
Molasses, sugar and 2–7
Mole-fraction scale 293
Monodentate ligands 132, 136
Monoglyme–water mixture, mean activity coefficients of the polynomial for 230*t*
Monoprotic acids in organic–water mixed solvents, ionization constant of weak 231

N

NaBr in water–acetone, transference number of 14*t*
$NdNO_3$–HNO_3–H_2O–HDEHP–AMSCO, activity behavior of 323–343
$Nd(NO_3)^{+2}$ complex, stability constant of 309*f*
$Nd(NO_3)^{+2}$ (K_1), stability constant data for 308
$Nd(NO_3)_2^{+1}$ complex, stability constant of 307*t*
$Nd(NO_3)_3$–HNO_3, stability constants for neodymium–nitrate complexes in 310
$Nd(NO_3)_3$–HNO_3, static vapor pressure measurements of aqueous 310
$Nd(NO_3)_3$–HNO_3–H_2O
activity coefficients of 299–321
application of the Gibbs–Duhem equation to 313
binary electrolyte solution, vapor pressure of 311*f*
binary electrolyte solution, water activity in 313*f*
data 310*t*
electrolyte mixture, excess Gibbs free energy of binary 318*f*
electrolyte solution
hydrogen ion activity in 319*f*
nitrate ion activity in 320*f*
vapor pressure data, Harned's plot of binary 315*f*
excess Gibbs free energy of ..299–321
–HDEHP–AMSCO extraction data 334*t*
Neodymium concentration
equilibrium aqueous 334*t*
equilibrium organic 334*t*
fraction of the total 309*f*

Neodymium ion 305
in aqueous nitrate solution, activity coefficient of . . 305*f*, 306*f*
in neodymium nitrate aqueous solutions, activity coefficients of 307*t*
Neodymium molar concentration in the organic phase 338*f*
Neodymium nitrate
aqueous solutions, activity coefficients of neodymium ion in 307*t*
complexes in $Nd(NO_3)_3$–HNO_3, stability constants for 310
–nitric acid solutions, nitrate electrode measurements on . 319*t*
–nitric acid solutions, pH electrode measurements on . . . 320*t*
water activity of 312*f*
Nernst Equation for Cell A 252
Neutral semichemical pulping, liquors from 9–10
Nitrate
concentration in the equilibrated organic phase 336*f*
concentration in the organic phase 339*f*
electrode data for equilibrated aqueous phase 330*t*
electrode data for organic phase samples back-extracted with H_2SO_4 . . . . 336*t*–337*t*
electrode measurements on neodymium nitrate–nitric acid solutions 319*t*
ion activity coefficient in aqueous solutions of nitric acid 331*t*
ion activity in the $Nd(NO_3)_3$–HNO_3–H_2O electrolyte solution 320*f*
single-ligand complexes of lanthanides 301
solution, activity coefficient of the neodymium ion in aqueous 305*f*, 306*f*
-to-neodymium concentration ratio in the organic phase . 337*f*
Nitric acid
concentration, activity coefficient of HNO_3 and NO_3^- as a function of 332*f*
concentration, reference voltage as a function of 329*f*
nitrate ion activity coefficient in aqueous solutions of 331*t*
p-Nitroaniline B 54
Nitrogen gas, purification of 356
NMA (*see N*-Methylacetamide)
NMF (*N*-methylformamide) 99
NMP (*N*-methylpropionamide) 250, 259
NO_3^- as a function of nitric acid concentration, activity coefficient of 332*f*
Nonaqueous solvents, behavior of weak acids and bases in 86
Nonaqueous solvents, dielectric constants of 250
Noncoulombic stretching force . . . 211
Non-ionic micelles in aqueous binary mixtures, variations of aggregation numbers for 126*f*

O

Observed electromotive force (E_{obs}) 359
n-Octane, activity coefficients of the HDEHP dimer in 335
Organic neodymium concentration, equilibrium 334*t*
Organic phase
equilibrated, HDEHP dimer concentration in 340*t*
equilibrated, nitrate concentration in 336*f*
equilibrium metal concentrations in 332
neodymium molar concentration in 338*f*
nitrate concentration in 339*f*
nitrate-to-neodymium concentration ratio in 337*f*
samples back-extracted with H_2SO_4, nitrate electrode data for 336*t*–337*t*
Organic solvent(s)
basicity in mixtures of water with 53
component 132
ionic Walden products normalized to mole percent . . . 85*f*
Organic-water mixed solvents, ionization constant of weak monoprotic acids in 231
Organic–water solvents, pK_a* of acids in 238*t*–240*t*
Organophosphorous acids, lanthanide extraction by 324
Osmotic coefficient of an electrolyte in a binary mixture of electrolytes 268
Ostwald absorption coefficient values for mixed-salt systems . 201*t*

P

Pentanol, solubility of sodium chloride in inverse micellar solution of 205
Pentanol solution of potassium oleate and water 208*f*
solubility area for 207*f*
Pentanol: surfactant mixtures, solubility region in 210
pH electrode data for equilibrated aqueous phase 333*t*

pH electrode measurements on neodymium nitrate–nitric acid solutions 320*t*
Phase diagram determinations concerning potassium electrolyte influence, ternary system 177–188
Physical constants of DEG–EG ... 347*t*
Pi(HPi)$_2$, formation of 89
Picric acid (HPi) 79
in acetonitrile 89
in *tert*-butanol–TMS, pK values of 82*t*
in *tert*-butyl alcohol–TMS ionization of 96*f*
in ethanol–TMS, ionization of .. 96*f*
in ethanol–TMS, pK values of .. 82*t*
solute–solvent interactions in ... 93
in TMS 89
conductometric measurements on 89
spectrophotometric data of .. 88*f*
Pitzer
Equations 269*t*, 300
formalism according to 268
interaction parameters for
$HBr + NH_4Br$ 273*t*
HBr + tetraalkylammonium bromides 273*t*
$HCl + NH_4Cl$ 273*t*
parameter 272
pK_a values of acetic acid in EG–DEG mixtures at various temperatures 349*t*
pK_a* 234–240
of acids in acetone–water ..238*t*–239*t*
of acids in methanol–water .238*t*–239*t*
of acids in organic–water solvents 238*t*–240*t*
(molal scale) of various acids in ethanol–water 236*t*–237*t*
pK_1, actual 285
pK_2 285
POE_{23} (*see* Polyoxyethylene lauryl ether)
Poisson–Boltzmann Equation 110, 112, 212
Polynomial(s)
for monoglyme–water mixture, mean activity coefficients of 230*t*
second-degree 228
for THF–water mixtures, mean activity coefficients of 229*t*
third-degree 228
Polyoxyethylene lauryl ether (POE_{23}) 124
cmc of 125
Potassium acetate
–ethanol–water system, isobaric vapor–liquid equilibrium data for 18*t*, 19*t*, 21*t*
–ethanol–water system, salt effects on 20*f*
salting-out effects owing to 186
Potassium acetate (*continued*)
–water–THF
isotherm of 184*f*
liquid layer formation of 186
plot 187*f*
solubility curve for 182*t*
tie-line data for 182*t*
–water–dioxane 186
isotherms of 180*f*
plot 187*f*
solubility curve for 179*t*
tie-line data for 179*t*
Potassium chloride
–calcium chloride solutions, solubility of carbon dioxide in 194*f*, 196*f*
–calcium chloride system, variation of constants with salt composition for 199*f*
salting-out effect owing to 187
–sodium chloride–calcium chloride solution, carbon dioxide solubility in 193*t*
–sodium chloride–calcium chloride system, values of constants for 201*t*
–water–THF 180
isotherm of 185*f*
solubility curve for 183*t*
tie-line data for 183*t*
Potassium electrolyte influence, ternary system phase diagram determinations concerning .177–188
Potassium hydroxide–water–dioxane system 177
Potassium iodide–ethanol–water system, isobaric vapor–liquid equilibrium data for 14*t*
Potassium iodide on the ethanol–water system, salt effect of .. 15*f*
Potassium oleate and water, pentanol solution of 208*f*
Potassium oleate and water, solubility area for the pentanol solution of 207*f*
Pr_4NBr in W–DMF 104
Preferential solvate system 28
Preferential solvation of electrolytes 145–165
Preferential solvation number 31
in butyl acetate–butanol–calcium chloride 34*f*
in methanol–ethyl acetate 32*f*
in methyl acetate–methanol–calcium chloride 33*f*
Pressure difference 46*f*
region for methanol–water–sodium chloride 48*f*
Primary medium effect 231
1,3 Propanediol–water system, chromium(III) species in ... 140
1,3 Propanediol (*G*)–H_2O system, elution profile for species of chromium(III) equilibrated in 134*f*

Propanediol–water system 137
1-Propanol 226t
n-Propanol–water mixtures 114
*n*DTMABr in 116t
ionic micelles in 126
n-Propylammonium ion 68
Propylene carbonate 226t
Protic–aprotic solvents, ion–ion interactions in 78
Protic–aprotic solvents, ion–solvent interactions in 78
Protic solvents, conductance of electrolytes in binary mixtures involving 77–99
Protic solvents, ionic association of electrolytes in binary mixtures involving 77–99
Proton
affinity (P_a) 61
of HCF_3SO_3 164
hydrates 161
in mixed solvent, behavior of .. 69
solvated aqueous 55
solvated, reaction with ether ... 161
between water and a mixed-aqueous solvent, transfer of 55
in water, structures for hydrated 56f
Protonation
constants 151
of H_2S 163
of water and ether 161
Pt, H_2|HBr(m), X% alcohol, Y% water|AgBr–Ag in
tert-butanol–water, standard potentials of 368f
ethanol–water, standard potentials of 367f
mixed solvents, electromotive forces and thermodynamic functions of 355–376
pure solvents, electromotive forces and thermodynamic functions of 355–376
Pt/H_2(g)/$G^{\pm}$(m_1), NaG(m_2), KBr(m_3) in water +
10 mass % THF AgBr/Ag, electromotive force of 282t–283t
30 mass % THF AgBr/Ag, electromotive force of 282t–283t
50 mass % THF AgBr/Ag, electromotive force of 282t–283t
Pt/H_2(g)GHBr(m_1), $G^{\pm}$(m_2) in water +
10 mass % THF/AgBr/Ag, electromotive force of 280t–281t
30 mass % THF/AgBr/Ag, electromotive force of 280t–281t
50 mass % THF/AgBr/Ag, electromotive force of 280t–281t
Pt; H_2(g, 1 atm)|HBr(m_1), $(C_2H_5)_4NBr$(m_2)|AgBr, Ag .. 264
Pt; H_2(g, 1 atm)|HBr(m) in NMA/H_2O|AgBr; Ag, electromotive force of 254t–255t
Pt–H_2(g, 1 atm)/HOAC (m_1), NaOAC (m_2), KX (m_3)/ AgX–Ag 346
Pulping, liquors from kraft 9–10
Pulping, liquors from neutral semi-chemical 9–10
Pure phase 110
Pure solvents, electromotive forces of Pt, H_2|HBr(m), X% alcohol, Y% water|AgBr–Ag in 355–376
Pure solvents, thermodynamic functions of Pt, H_2|HBr(m), X% alcohol, Y% water|AgBr–Ag in 355–376
Purification of
electrode materials 356
hydrogen gas 356
nitrogen gas 356

Q

Q coefficients 316t
for Harned's binary electrolyte model 316f

R

R values 84
R.A.A.D. (relative average absolute deviation) 22
Radii, hydrated anionic 241
Random error 40
comparison of experimental data and simulated data containing 48t
on simulated data of MeOH/ H_2O/NaCl, effect of 46t–47t
Random solvation of chromium(III) 131
Raoult's Law 24, 252
Rational activity coefficient 362
Raw sugar
by solvent extraction, refining of 3
solvent extraction of sugar solution from 5–6
by washing, refining of 4–5
Reference voltage as a function of nitric acid concentration 329f
Refining, cane sugar 3–4
Refining of raw sugar by washing . 4–5
Regression analysis
results of 306t
for the solution vapor pressure . 304
of stability constants, linear 308
Regression coefficients 304
Relative average absolute deviation (R.A.A.D.) 22
Repulsion force 213
ROH 54
chemical identity of 55
Runge–Kutta Equation 113

S

$\Delta S°$ (entropy changes) 258*t*, 363
for glycine in THF–H_2O 286
for $HBr(H_2O) \rightarrow HBr(H_2O/NMA)$ 258*t*
$\Delta S_t°$ (entropy of transfer) 350
for transfer of HBr from H_2O to mixed solvent 257
Salt(s)
composition for potassium chloride calcium chloride system, variation of constants with 199*f*
composition for sodium chloride–potassium chloride system, variation of constants with . 200*f*
composition for sodium chloride–sodium sulfate–ammonium chloride system, variation of constants with 200*f*
concentration on solubility of carbon dioxide in potassium chloride–calcium chloride solutions, effect of 196*f*
concentration on solubility of carbon dioxide in sodium chloride–sodium sulfate–ammonium chloride solutions, effect of 197*f*
in DMF–water, $\Delta H°$of 104*f*
effect
of ammonium bromide on the ethanol–water system ... 15*f*
parameter
for ammonium bromide–ethanol–water 21*t*
for inorganic salts in ethanol–water 22*t*
as a function of the liquid composition ratio for ammonium bromide–ethanol–water 19*f*
values, k′ 17
on potassium acetate–ethanol–water 20*f*
of potassium iodide on ethanol–water 15*f*
of sodium acetate on ethanol–water 20*f*
on vapor–liquid equilibrium . . 27–38
at fixed liquid composition . 11–25
prediction of 34
in methanol–water 175*t*
at saturation 39–51
solubility and salting-out effect . 187
solution in methanol–water
density parameters for 174*t*
at 25°C, viscosities of 172*t*–173*t*
viscosity parameters for 174*t*
viscosity B coefficients for ... 171*f*
into solvent mixture, solubility of 28–31
structure-breaking 175
Salting-out coefficient as a function of alcohol 375*f*
Salting-out effect 178
owing to potassium acetate 186
owing to potassium chloride 187
salt solubility and 187
Saturated solutions of inorganic salts, physical properties of .. 23*t*
Saturation
ammonium bromide–ethanol–water system at 18*f*
dielectric 86
salts at 39–51
Scaled-particle theory 116
Schreinemakers wet residue method, modification of 184
SDOS (sodium dodecylsulfate) .. 114
SDS (*see* Sodium decysulfate)
Separation techniques 2
Setschenow Equation, modified . 189–190
Setschenow Equation on the salt concentration in aqueous single-salt solutions, applicability of 190
Silver–silver bromide electrode
in anhydrous *tert*-butanol, standard electrode potential of . 367
in methanol–water solvents 368
standard potential of 253, 363
Silver–silver iodide electrode in methanol–water solvents 368
Single-component lanthanide salt in aqueous solution 323
Single-ion standard Gibbs free energy of transfer from water to water–acetone mixtures ... 117*f*
Single-ligand complexes of lanthanides, chloride 301
Single-ligand complexes of lanthanides, nitrate 301
Single-salt solutions, applicability of the Setschenow equation on the salt concentration in aqueous 190
Single-salt solutions, two-parameter equation for the solubility of gases in 197
Single solvation site, equilibrium constant for solvent species at 168
Sodium acetate–ethanol–water system, isobaric vapor–liquid equilibrium data for 17*t*
Sodium acetate on the ethanol–water system, salt effect of .. 20*f*
Sodium chloride
in an inverse micellar solution solution of pentanol, solubility of 205
–magnesium sulfate–potassium nitrate solution, carbon dioxide solubility in 193*t*
–magnesium sulfate–potassium nitrate system, values of constants for 201*t*

Sodium chloride (*continued*)
 –potassium chloride system, variation of constants with salt composition for 200*f*
 –sodium sulfate–ammonium chloride
  carbon dioxide solubility in 193*t*, 195*f*
   effect of salt concentration on 197*f*
  values of constants for 201*t*
  variation of constants with salt composition for 200*f*
Sodium decylsulfate (SDS) 110
 in water–acetone mixtures ..115*t*, 123
Sodium dodecylsulfate (SDOS) .. 114
Solid–liquid tie-lines 182
Solubility
 area for the pentanol solution of potassium oleate and water 207*f*
 of calcium chloride in butyl acetate–butanol 30*f*
 of calcium chloride in methanol–ethyl acetate 28*f*
 of calcium chloride in methyl acetate–methanol 29*f*
 of carbon dioxide in aqueous mixed-salt solutions 189–205
 curve for
  potassium acetate–water–dioxane 179*t*
  potassium acetate–water–THF 182*t*
  potassium chloride–water–THF 183*t*
 limit 210
 region in pentanol:surfactant mixtures 210
 of salt into solvent mixture 28–31
 of sodium chloride in an inverse micellar solution of pentanol 205
 weight percentages 181
Solute(s)
 in DMF–water, $\Delta H°$ of 102*f*
 in DMSO–water, $\Delta H°$ of 102*f*
 influence of different 102
 –solvent interaction to the association phenomena, importance of 86
 –solvent interactions in picric acid 93
 in various solvents, enthalpy of solution of 99
Solution
 enthalpies of (*see* $\Delta H°$)
 extraction of concentrated 1–10
 parameters 170
 vapor pressure, regression analysis for 304
Solvate system, preferential 28
Solvated aqueous proton 55
Solvated proton with ether, reaction of 161
Solvated species, concentrations of individual differently 132
Solvated species, correlation of stabilities of differently 138
Solvation
 of chromium(III) ion in mixed solvents 129–144
 of electrolytes, preferential ..145–165
 enthalpy of gaseous ions in DPA solvents 162*t*–163*t*
 enthalpy of gaseous ions in water 162*t*–163*t*
 number, preferential 31
  in butyl acetate–butanol–calcium chloride 34*f*
  in methanol–ethyl acetate ... 32*f*
  in methyl acetate–methanol–calcium chloride 33*f*
Solvation II 27–38
Solvent(s)
 apparent charge of ionic micelles in aqueous binary 109–128
 bulk dielectric constants of 236*t*
 component, organic 132
 composition 137
  degree of dissociation of ionic micelles with 120*f*
  dependence 139*t*
  transport numbers and 169
  variation of apparent charge of ionic micelles with ... 125*f*
  variation of cmc with 110, 121*f*
 enthalpy of solution of a solute in 99
 exchange reactions, K_n for 139*t*
 extraction, refining of raw sugar by 3
 extraction of sugar solutions ... 6–7
  from raw sugar 5–6
 free energy change of ionization of acid in 232
 $\Delta G_t°(H^+)$ against 58*f*
 ion transfer between 231
 ionization constants of acids in . 234*f*
 mixture, solubility of salt into ..28–31
 properties 170
 -separated ion pair, formation of 92
 species at a single solvation site, equilibrium constant for .. 168
 standard potentials of Ag–AgCl electrode bulk dielectric constants of 228*t*
 systems in aminosugar research 178–188
 systems, ionic equilibrium processes in mixed-organic–aqueous 216
 transfer of proton between water and mixed-aqueous 55
Sorting effect 82
SOS in water, cmc values of 115
Spheres in a dielectric continuum 85
Stability constant(s) 308*t*
 data for $NdNO_3^{+2}$ (K_1) 308
 linear regression analysis of 308
 of the $Nd(NO_3)_2^{+1}$ complex 307*f*

Stability constant(s) (*continued*)
of the $Nd(NO_3)^{+2}$ complex 309*f*
neodymium–nitrate complexes in $Nd(NO_3)_3$–HNO_3 310
Standard
cell potential $E_m{}^{\circ\prime}$ 359
deviation of energy, effect of errors on 44*f*
deviation of the *y*-error, effect of errors on 45*f*
electrode potential (*see* E°)
emf (*see* $E^{\circ\prime}$) ....
enthalpy of solution in a mixture of water ($\Delta H^\circ(M)$) 101
enthalpies of transfer ($\Delta H_{tr}{}^\circ$) .. 99
free energy 220
Gibbs free energy of transfer from water to water–acetone mixtures, single-ion 117*f*
potential(s) 366*t*
of Ag–AgCl electrode 225*t*
bulk dielectric constants of solvents 228*t*
computation of 216
variation of 227*f*
of the Ag–AgI electrode in EG–DEG 346
of the galvanic cell 363
of Pt,|H_2 HBr(*m*), *X*% alcohol, *Y*% water|AgBr–Ag in ethanol–water 367
of Pt, H_2|HBr(*m*), *X*% alcohol, *Y*% water|AgBr–Ag in *tert*-butanol–water 368*f*
of the silver–silver bromide electrode 253, 363
Static vapor pressure measurements of aqueous $Nd(NO_3)_3$–HNO_3 . 310
Stretching energy 211
Stretching force, coulombic 211
Stretching force, noncoulombic ... 211
Structure-breaking
capacities of ions 65
effects 82
salts 175
Structure-forming capacities of ions 65
Structure-forming effect, types of 68
Sugar
and molasses 2–7
refining costs 3
solutions, liquid–liquid extraction of impurities from concentrated 5
solutions, solvent extraction of .. 6–7
from raw sugar 5–6
from wood-pulping liquors 1–10
Sulfolane (TMS) 78
–alcohol 93
–anion interactions 84
–*tert*-BuOH 93
–*tert*-butanol 82*t*
–cation interactions 84
conductance of electrolytes in binary mixtures involving .77–99
Sulfolane (TMS) (*continued*)
dielectric constant of 86
2,6-dihydroxylbenzoic acid in .. 88
as DPA solvent 146
electrolytes in 78
–ethanol 82*t*
–EtOH 83
HCF_3SO_3 in 147
$HSbCl_6$ in 147
hydration constant of $CF_3SO_3^-$ in 159
ionic association of electrolytes in binary mixtures involving 77–99
ionic Walden products normalized to mole % 83*f*
–methanol 81*t*
nature of $HSbCl_6$ solutions in .. 148
picric acid in 89
conductometric measurements on 89
spectrophotometric data of .. 88*f*
pure 86–89
conductance of HCl in 87*f*
reversal of basicity order in 162
solutions of acids
as a function of diethyl ether concentration, conductance of 153*f*
as a function of water concentration, conductance of .. 155*t*
vapor pressure of 150*f*
solutions of electrolytes, heats of mixing of water with 154*t*
solutions of electrolytes, vapor pressure of 149*f*
vapor pressure of HCF_3SO_3 solutions in 148
–water 80*t*, 89–93
dependence of association constants on the dielectric constant in 90*f*, 91*f*
ionic association phenomenon in 92–93
Systematic error 40

T

T-error 40, 42
$T\Delta S_t{}^\circ$ as a function of NMA in mixed solvent 260*f*
Tailing-off 12
TBP (tri-*n*-butyl phosphate) 324
Ternary mixed-salt solutions, carbon dioxide solubility in 193–202
Ternary system phase diagram determinations concerning potassium electrolyte influence 177–188
Tetraalkylammonium
bromide series 12
bromide and hydrobromic acid in aqueous media, thermodynamic behavior of 263–274
bromides in DMF–water, ΔH° of 103*f*

Tetraalkylammonium (*continued*)
salt, influence of substitution groups on the hydrophobic character of 104
salts in water, hydrophobic hydration behavior of 100
Tetrabutylammonium bromide (Bu_4NBr), $\Delta H°$ of 100
Tetrabutylammonium particle in water–acetone 118*t*
Tetraethylammonium bromide ... 12
specific effects on ΔG^E in 272
Tetrahydrofuran 225*t*
Tetrahydrofuran–water
addition of an electrolyte to ... 186
$\Delta G°$ for glycine in 286
$\Delta H°$ for glycine in 286
heat capacity for glycine in 286
mean activity coefficients of the polynomial for 229*t*
$\Delta S°$ for glycine in 286
thermodynamic functions for the dissociations of glycine in
10 mass % 288*t*–290*t*
30 mass % 288*t*–289*t*, 291*t*
50 mass % 290*t*–292*t*
at varying temperatures, thermodynamic study of glycine in 277
Thermodynamic behavior of hydrobromic acid in NMA–water mixtures 249–260
Thermodynamic behavior of the ionic constituents of micelles in mixed solvents 116–119
Thermodynamic equilibrium constants (k_n) 136
Thermodynamic properties of electrolytes in DPA solvents 146
Tie-line data for
potassium acetate–water–dioxane 179*t*
potassium acetate–water–THF . 182*t*
potassium chloride–water–THF . 183*t*
Tie-lines, liquid–liquid 181
Tie-lines, solid–solid 182
TMS (*see* Sulfolane)
Transfer dissociation energies for glycine 292*t*–293*t*
Transfer free energy(ies)
data of the iodide ion ($\Delta G_t°_{(I^-)}$) 352
for glycine 294*f*
of H^+ and I^- ions from EG to EG–DEG 353*t*
for hydrobromic acid 294*f*
for tricine 294*f*
Transfer Gibbs energy of dissociation ($\Delta G_t°$(diss)) 290
Transference number 114
of NaBr in water–acetone 114*t*
Transport numbers and solvent composition 169
Tri-*n*-butyl phosphate (TBP) 324
Tricine, transfer free energies for . 294*f*
Trimer formation in *n*-hexane 325
Trimers, formation of HDEHP ... 325
Trimethyldecylammonium bromide (*n*DTMABr) 110
in water–acetone mixtures ...115, 123
in water, cmc values of 115
in water–*n*-propanol mixtures .. 116*t*
Triplet-ion interactions 273
Two-parameter equation for the solubility of gases in single-salt solutions 197

U

Undissociated HF 160

V

Vapor composition difference 47*f*
for methanol–water–sodium chloride 49*f*
Vapor composition, sources of error in 40
Vapor–liquid equilibrium
data for the potassium iodide–ethanol–water system, isobaric 14*t*
data for the ammonium bromide–ethanol–water system, isobaric 16*t*
data for the potassium acetate–ethanol–water system, isobaric 18*t*, 19*t*, 21*t*
data for the sodium acetate–ethanol–water system, isobaric 17*t*
at fixed liquid composition, salt effect in 11–25
prediction of salt effect on 34
salt effect on 27–38
Vapor pressures 156
of $Nd(NO_3)_3$–HNO_3–H_2O binary electrolyte solution 311*f*
of sulfolane solutions of acids .. 150*f*
of sulfolane solutions of electrolytes 149*f*
Vaporization 2
Viscosity
A coefficient 169
B coefficients for salt solutions in water–methanol mixtures .. 171*f*
of dilute solutions of alkali halides in methanol–water 167–176
parameters for salt solutions in methanol–water 174*t*
of salt solutions in methanol–water 172*t*–173*t*

W

W–DMF (*see* DMF–water)
Walden products, ionic 79
normalized to mole % organic solvents 85*f*
normalized to mole % TMS . . . 83*f*
Walden's rule 169
Water
activity coefficients in 259
activity in the $Nd(NO_3)_3$–HNO_3–H_2O binary electrolyte solution 313*f*
activity of neodymium nitrate . . 312*f*
conductometric behavior of ions in 79
and ether, protonation of 161
–organic mixed solvents, computational techniques of ionic processes in 215–246
with organic solvents, basicity in mixtures of 53
-rich region, B coefficients in . . . 175
solubility of inorganic salts in boiling solutions of 25*t*
solvation enthalpy of gaseous ions in 162*t*–163*t*
standard electrode potentials in 57
structures for the hydrated proton in 56*f*
Water (*continued*)
to water–acetone mixtures, single-ion standard Gibbs free energy of transfer from 117*f*
Wet residue method, modification of Schreinemakers 184
Wilson energy constants 42
Wilson energy parameters 44
Wood-pulping liquors, recovery of acetic acid from 1–10
Wood-pulping liquors, refining of sugar from 1–10

X

x-error 42

Y

y-error 40
effect of errors on the standard deviation of 45*f*

Z

Z_1-error 42
Z_2-error 42
Zone refiner 250

The text of this book is set in 10 point Caledonia with two points of leading. The chapter numerals are set in 30 point Garamond; the chapter titles are set in 18 point Garamond Bold.

The book is printed offset on Text White Opaque 50-pound. The cover is Joanna Book Binding blue linen.

Jacket design by John Sinnett.
Editing and production by Candace A. Deren.

The book was composed by Service Composition Co., Baltimore, MD, printed and bound by The Maple Press Co., York, PA.